Lecture Notes in Physics

Edited by H. Araki, Kyoto, J. Ehlers, München, K. Hepp, Zürich
R. Kippenhahn, München, H. A. Weidenmüller, Heidelberg
J. Wess, Karlsruhe and J. Zittartz, Köln

Managing Editor: W. Beiglböck

310

W. C. Seitter H. W. Duerbeck
M. Tacke (Eds.)

Large-Scale Structures in the Universe Observational and Analytical Methods

Proceedings of a Workshop, Held at the Physikzentrum
Bad Honnef, Fed. Rep. of Germany, December 9–12, 1987

Springer-Verlag
Berlin Heidelberg GmbH

Editors

W. C. Seitter
H. W. Duerbeck
M. Tacke
Astronomisches Institut, Westfälische Wilhelms-Universität
Wilhelm-Klemm-Str. 10, D-4400 Münster, FRG

ISBN 978-3-662-13709-3 ISBN 978-3-540-45938-5 (eBook)
DOI 10.1007/978-3-540-45938-5

Originally published by Springer-Verlag Berlin Heidelberg New York in 1988
Softcover reprint of the hardcover 1st edition 1988

2158/3140-543210 – Printed on acid-free paper

Preface

A workshop on **Observational and Analytical Methods Relating to Large-Scale Structures in the Universe** was held in the Physikzentrum, Bad Honnef, F.R. Germany, from December 9 to 12, 1987. The suggestion to bring together an international group for the discussion of current work and general topics on observational cosmology came from the Deutsche Forschungsgemeinschaft, with the intent to further the exchange of ideas and experiences in a field which requires the handling of large amounts of data in both the observational and the theoretical approach.

In the spirit of a workshop (and of lecture notes), the authors have written their contributions with an emphasis on methods and have illustrated the procedures and techniques with their results. It is hoped that in this way parts of the proceedings can be used in lieu of a yet unwritten introductory textbook on "cosmology with large numbers of data".

The general introduction, intended to be the obligatory historical chapter, has grown unexpectedly long, as has the list of "co-authors". As more material was assembled, showing the deep engagement of the early cosmologists and their unsurpassed clarity of thought, it seemed that we could still benefit from reading parts of their original contributions.

Authors and editors hope that the reader will enjoy the mixture of textbook, review article, and "latest news" from our fascinating topic: the Universe.

We thank all contributors for their timely submission of "soft" or typed manuscripts; also Professor Dr. W.M. Lippe, the Institut für Numerische und Instrumentelle Mathematik–Informatik, Dr. W. Held and the Rechenzentrum of the WWU for use of their installations, and Dipl.-Phys. R. Budell, who was responsible for the photographic work. Our appreciation also extends to the Deutsche Forschungsgemeinschaft and to the staff of the Physikzentrum Bad Honnef, Dr. J. Debrus, Ms. Kluth, Ms. Offerzier and their staff, for creating the pleasant surroundings of the meeting.

Last but not least, we are indebted to Springer-Verlag, represented by Ms. C. Pendl, for continuing support during the preparation of these proceedings.

Münster, June 1988

Waltraut C. Seitter
Hilmar W. Duerbeck
Markus Tacke

Contents

Introduction

Large Scales – Large Numbers – Large Efforts: Historical Annotations

W.C. Seitter
Astronomisches Institut
Westfälische Wilhelms-Universität
Münster, F.R. Germany

With the raptur'd Poet may we not justly say

> O, what a Root! O what a Branch is here!
> O what a Father! what a Family!
> Worlds! Systems! and Creations!

and Consequence of this

> In an Eternity, what Scenes shall strike?
> Adventures thicken? Novelties surprize?
> What Webs of wonder shall unravel there?

Night Thoughts

Edward Young (1745)
as quoted by Thomas Wright of Durham
in the 9th letter of his book
"An Original Theory of the Universe" 1750

Abstract

The roots and branches, systems and creations, not so much of the Universe itself – as adressed in the quotation – but of our *picture* of the Universe, are briefly traced. It is intended to show small portions of the structure on which our present work is built, to provide a background onto which the data and discussions of the workshop can be projected.

Section 1 lists some major historical and contemporary large-scale surveys, in two and three dimensions, of galaxies and of clusters of galaxies. Section 2 introduces historical and modern definitions of various large-scale structures and illustrates the connections sought between the observations on the one side and mathematical and physical theory (statistics and evolution) on the other. Sections 3 and 4 are devoted to cosmological theory and, in particular, to the growth of concepts in relativistic cosmology, to the introduction of parameters and the difficult process of finding relations between theoretical and observational quantities. In Sections 5 and 6, we introduce the role of the data, the necessary corrections to be applied to direct measurements, and observational diagrams employed to make use of the data. Section 7, the conclusion, again voices the intent of these annotations.

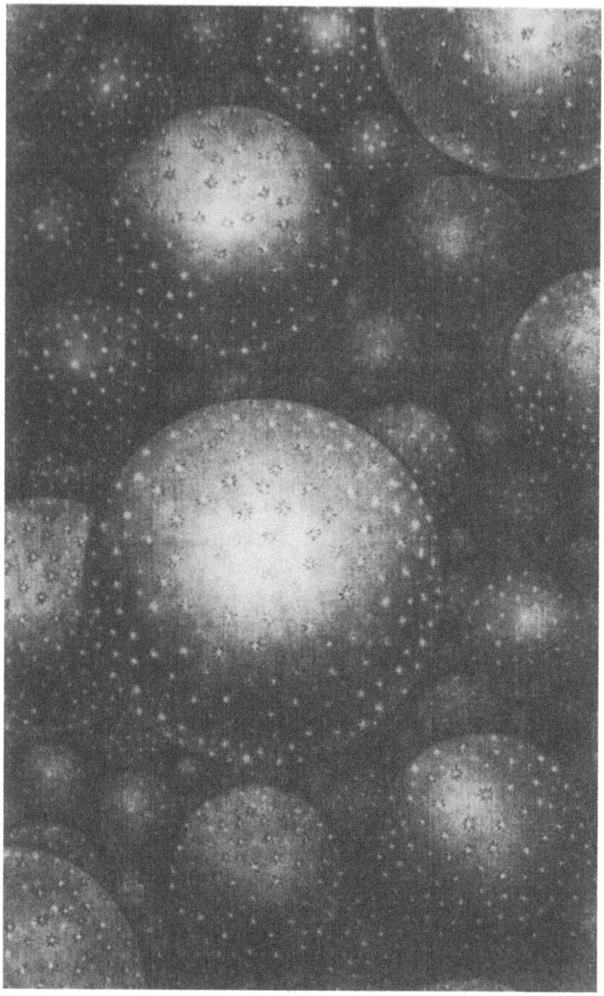

Fig. 1. From the 9th letter of Thomas Wright's "An Original Theory of the Universe" (1750) Plate XXXI, about which he writes:

> "...that as the visible Creation is supposed to be full of sidereal Systems and planetary Worlds, so on, in like similar Manner, the endless Immensity is an unlimited Plenum of Creations not unlike the known Universe. See Plate XXXI. which you may if you please, call a partial View of Immensity, or without much Impropriety perhaps, a finite View of Infinity ...
>
> That this in all Probability may be the real Case, is in some Degree made evident by the many cloudy Spots, just perceivable by us, as far without our starry Regions, in which tho' visibly luminous Spaces no one Star or particular constitutent Body can possibly be distinguished; those in all likelyhood may be external Creation, bordering upon the known one, too remote for even our Telescopes to reach."

1 Structures in the universe

1.1 Large-scale structures (1750 - 1967)

Antiquity and Middle Ages have seen attempts of visualizing worlds beyond our world. The concept of the plurality of solar systems marks the beginning of modern times. Yet, the first to actually picture – in the literal sense – a cosmos of organized *stellar systems* appears to be Thomas Wright of Durham. *Fig. 1* is taken from his book "An Original Theory of the Universe" (1750).

During the same century first attempts were made to catalogue nebulous objects, especially those which are not resolved into stars (the five original ἀστέρες νεφελοειδεῖς – nebulous stars – of Ptolemaios were clusters or loose groups of stars). Within little more than a century the "Catalogue of Nebulae and Clusters" (J. Herschel 1864) had been assembled: the work of a single family – William, Caroline and John Herschel. The first complete picture of the distribution of nebulae, which are not obviously associated with the Milky Way, based on the New General Catalogue and the two Index Catalogues (Dreyer 1888, 1895, 1908), was published by Charlier (1922). *Fig. 2* shows his presentation of 11 475 nebulae. The inhomogeneity in the distribution, by then long recognized (W. Herschel 1811), is clearly apparent. Among the counts of nebulae made in the early 20th century Fath's list (1914) obtained from photographs of 139 selected areas is mentioned here, because of the extensive interpretation of the data by Seares (1925) and his comment[1]:

> "Further, the Selected Areas are too widely spaced for a satisfactory determination of the effect of local irregularities in distribution; but, in spite of the limitations, the data merit special attention because of the freedom from any selection favoring regions in which nebulae were known to exist."

Both considerations are important because they are still disputed in connection with modern surveys. More detail on the early history of mapping nebulae is given *e.g.* by Lundmark (1927) and Flin (1988).

By the mid-twenties about 10 000 mostly faint galaxies had been accumulated in the Heidelberg nebular lists (No. 1-15, Wolf 1901 - 1916; continued by Reinmuth, 1916 - 1940).

A sample of 44 000 galaxies was available by the mid-thirties. Excesses and deficiencies of galaxies in certain areas were discussed (Hubble 1934); *Fig. 3* is taken from his paper.

The largest total sample – before the advent of the Lick Survey – was accumulated at the Harvard College Observatory under the leadership of Harlow Shapley. A plot of 78 000 from a total of 392 780 is shown in *Fig. 4*. Based on this material Shapley (1938) first claimed that structures on such large scales suggest "gradients" rather than clustering.

The last one of the catalogues assembled without the use of automatic procedures is

---

[1]Explanations in quotes are given in square brackets; the quotes from Einstein, Heckmann, Weizsäcker, Weyl and Wirtz are translations from the German originals.

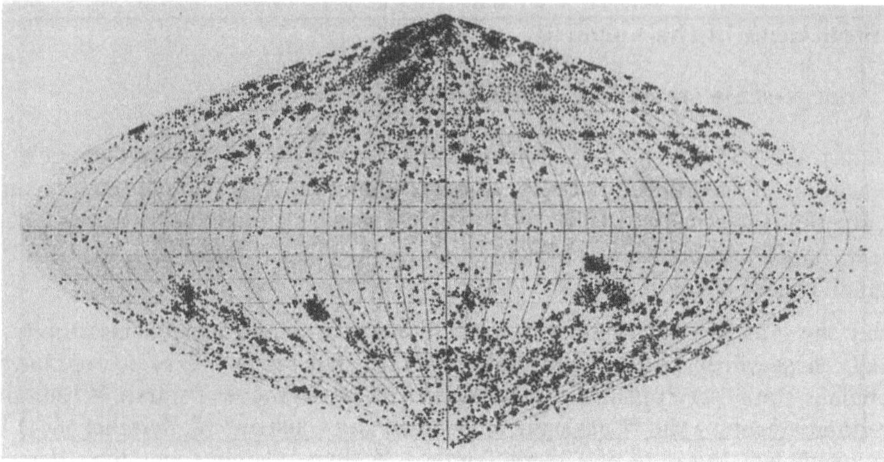

Fig. 2. Charlier's map of the nebulae (Charlier 1922).

"A glance at this plate suffices for stating how the Milky Way, which is designed by the great axis of the chart is systematically avoided by the nebulae.

A remarkable property of the image is that the nebulae seem to be piled up in clouds (as also the stars in the Milky Way). Such a clouding of the nebulae may be a real phenomenon, but it may also be an accidental effect ..."

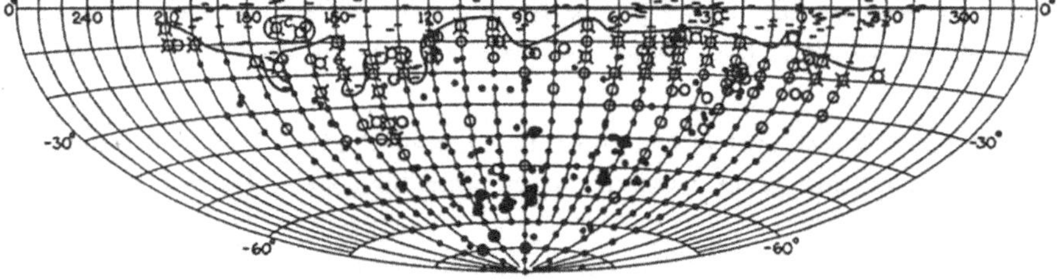

Fig. 3. Hubble's distribution of 44 000 galaxies (Hubble 1934).

"Distribution of extra-galactic nebulae when observed data are corrected for the latitude effect. Small crosses represent normal distribution (log $N = 1.82 - 2.11$); small disks and circles, moderate excesses (log $N = 2.12 - 2.26$) and deficiencies (log $N = 1.67 - 1.81$); large disks and circles, considerable excesses (log $N = 2.27 - 2.56$) and deficiencies (log $N = 1.37 - 1.66$), with crosses added for log $N > 2.56$ and log $N < 1.37$. Fields with no nebulae are omitted."

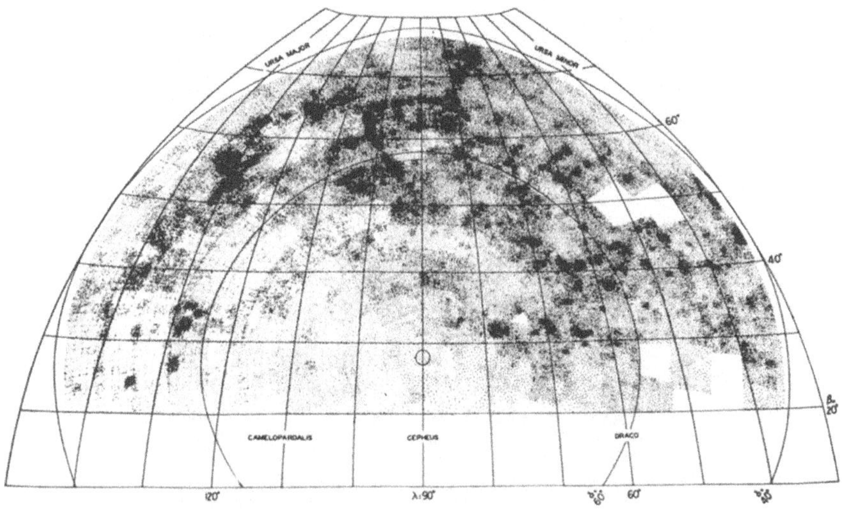

Fig. 4. Part of the Harvard Survey (Shapley 1957).

"Plot of 78,000 individual galaxies in the Canopy area."

"...the far extension of the 'Cepheus flare' or cloud of absorbing material...comes out of the Milky Way. This flare of absorption covers the north celestial pole. Supported by this survey is the evidence that from galactic latitude +40° to the north *galactic* pole there is no appreciable net increase of population density with latitude."

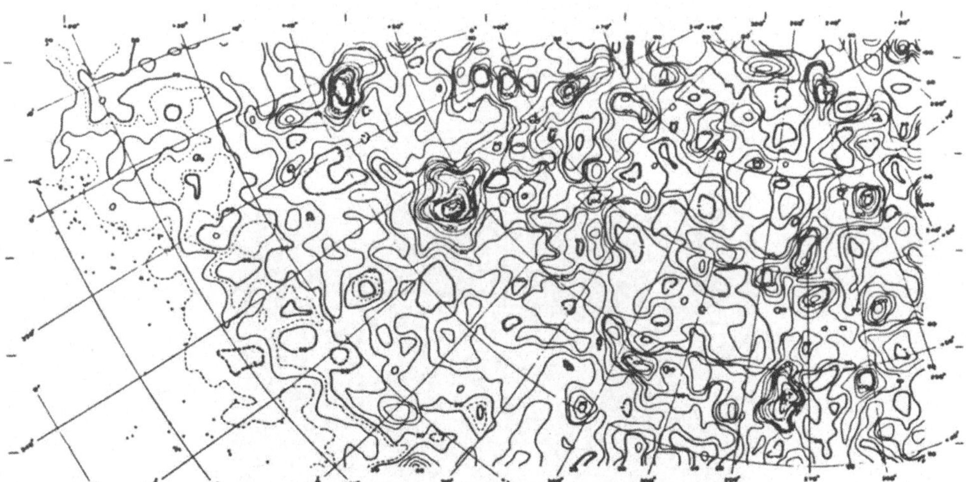

Fig. 5. First section of the Lick Survey (Shane and Wirtanen 1954).

"Total area available. Contour map, showing smoothed numbers of galaxies per square degree."

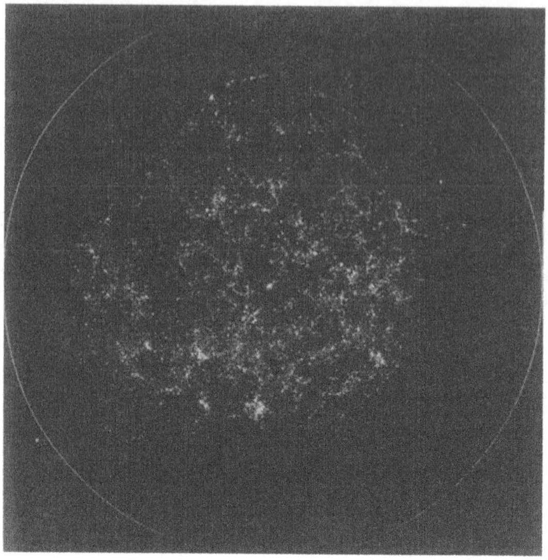

Fig. 6a. Princeton presentation of Lick Catalogue (Seldner *et al.* 1977).

"Map of galaxy counts in the northern galactic hemisphere. The north galactic pole is at the center, the galactic equator is at the edge, and galactic latitude is a linear function of radius. Galactic longitude increases in the clockwise direction with $l^{II} = 0°$ at the bottom of the map."

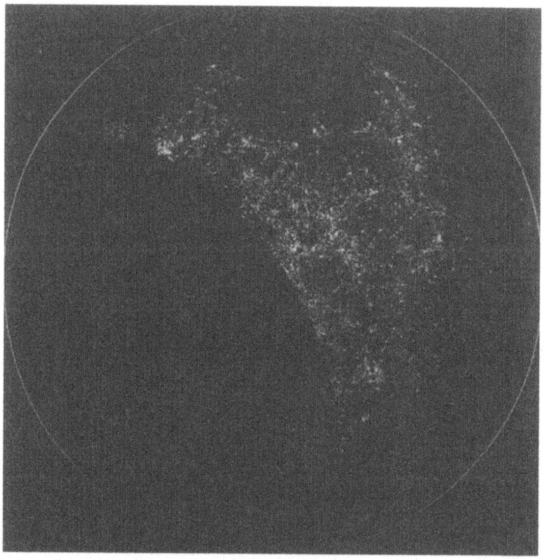

Fig. 6b. Princeton presentation (cont.)

"Map of galaxy counts in the southern galactic hemisphere. Galactic longitude increases in the counterclockwise direction from $l^{II} = 0°$ at the bottom of the map."

the Lick Survey, described by Shane and Wirtanen in 1950 and completed 17 years later (Shane and Wirtanen 1967). Even the first results presented by the authors in 1954 made an immense impact, stimulating theoretical work by Limber, Neyman and Scott (Sect. 2.3.1). The first Lick survey map is shown in *Fig. 5.* Also important because of its far reaching influence was the Princeton presentation of the Lick catalogue and its analysis made 10 years later (Seldner *et al.* 1977). It is shown for the northern and southern hemispheres in *Fig. 6.*

1.2 Surveys in other wavelength regions

The first survey conducted outside the optical region of the spectrum resulted in the Cambridge Catalogue of radio sources published by Ryle *et al.* in 1950. The following sequence of Cambridge Catalogues, reaching sucessively fainter objects, is well known as a powerful tool to locate very distant extragalactic sources. The potential of deep probing by radio surveys seems to have been first realized by Mills (1952). On the basis of his own catalogue, which suggested isotropic distribution of all sources outside the Milky Way and an intensity distribution conforming to the -1.5 power law, he considered the origin of radio radiation from extragalactic objects an equally probable hypothesis to that of the then generally favoured "radio stars" in the solar vicinity. *Fig. 7* is taken from the Second Cambridge Catalogue (Shakeshaft *et al.* 1955) where the weaker sources, however, were not confirmed by the southern survey with the 'Mills cross' (Pawsey 1957) and by the Third Cambridge catalogue. They are now considered to be artefacts. The first quasars identified (3C48 and 3C273, and others) obtained their names from the Third Cambridge Catalogue (Edge *et al.* 1959).

With the advent of satellite astronomy short wavelength surveys became possible. The first results from the UHURU satellite were published in 1971. The final catalogue of this successful mission includes 339 sources (Forman *et al.* 1978) as shown in *Fig. 8.* Hard x-rays and γ-rays were observed with Cos B. While most of the individual sources are located in the Milky way, the diffuse high energy background has cosmological significance (Pinkau 1979), with successively lower density fluctuation expected at $z \simeq 10$ (soft x-rays) and $z \simeq 100$ (hard x-rays and γ-rays) (Silk 1970).

A far reaching infrared survey was carried out with the IRAS satellite in 1985/86. It has influenced observational cosmology significantly. The brightest and most massive galaxies seem to be strong infrared emitters. Quasars show infrared brightness correlated with their x-ray fluxes. The infrared background radiation has become important as the short wavelength end of the microwave background. Another infrared background with density fluctuations of the order 1 appears at $z \simeq 1$ (Partridge 1988).

1.3 Automated two-dimensional surveys since 1980

In the 1980s two centers of two-dimensional surveys became established in Cambridge and in Edinburgh in collaboration with Durham University. Both surveys use glass copies of photographic atlas plates obtained with the UK Schmidt telescope, fast flying spot-type measuring machines and fully automized reduction procedures. A third group, Muenster, has joint them recently, using the same basic material, but film

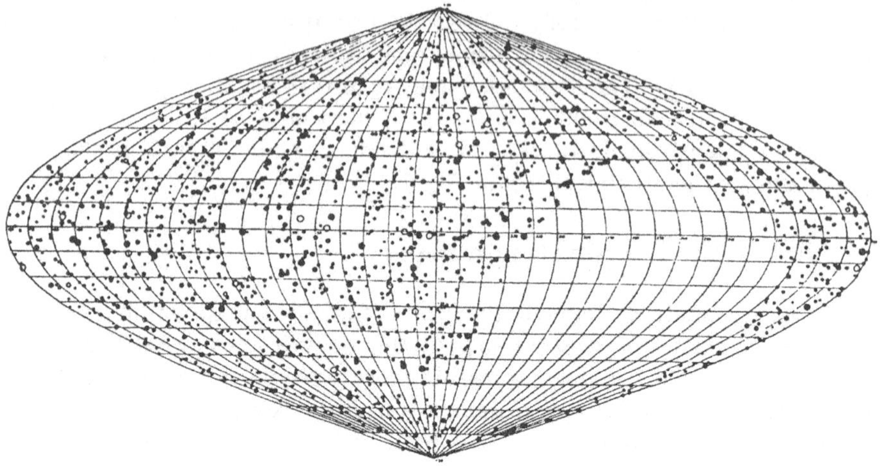

Fig. 7. 2C–Radio map of the sky (Shakeshaft *et al.* 1955).

"Map showing the distribution of radio sources in galactic coordinates. The open circles represent the sources of large angular diameter, and in both cases the sizes indicates the flux density of the sources"

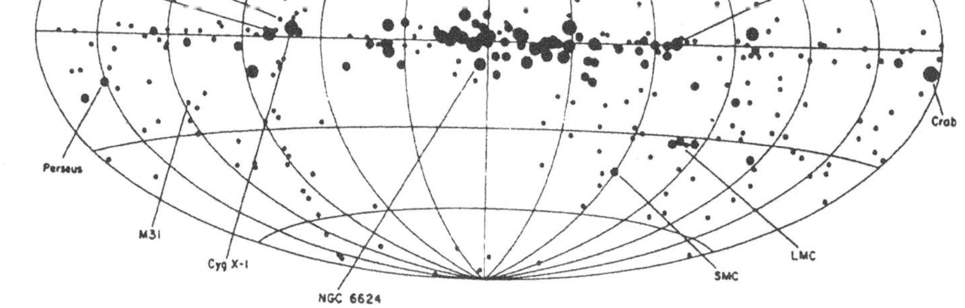

Fig. 8. 4U–X-ray map of the sky (Forman *et al.* 1978).

"The sources ... are displayed in galactic coordinates. The size of the symbols representing the sources is proportional to the logarithm of the peak source intensity." The 'extragalactic distribution' in the $\log N - \log S$ diagram follows the $S^{-1.5}$ power law.

copies, and a double-slit measuring machine. The large numbers of data have led to sophisticated procedures of interpretation which are discussed during this workshop. The increasing number of objects available through two-dimensional surveys with time is displayed in *Fig. 9.*

1.4 The third dimension

When it became evident on the basis of the first measurements of galaxy redshifts by Slipher (1913) and others, a third dimension had become available for studies of galaxy distributions. De Sitter (1917) was the first to recognize that redshifts might be used as distance indicators, revealing the structure of the universe (Sect. 3.1.2).

In 1918 Wirtz gave an interpretation of the 15 redshifts available at this date (most of them obtained by Slipher):

> "One can see from the values v of the spiral nebulae that they cannot be best represented by an apex direction and a velocity alone. The preponderance of one sign and the absolute size of the values show that a constant systematic error must be included in the least squares analysis ...

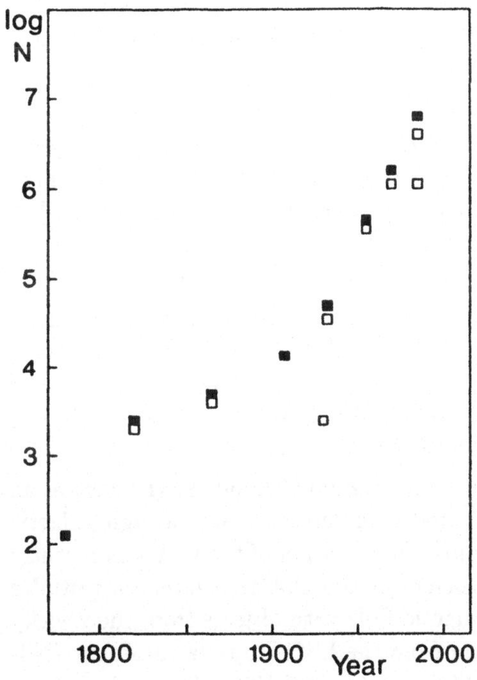

Fig. 9. The increase in number of identified galaxies from the late 18th to the late 20th century. Open symbols: individual catalogues; filled symbols: cumulative data.

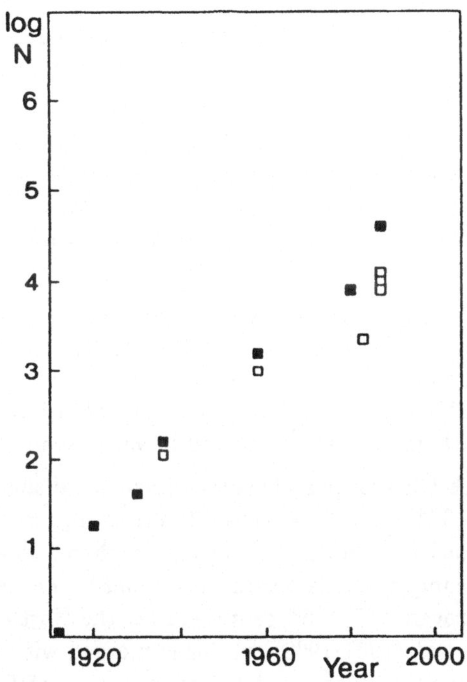

Fig. 10. The increase in number of measured redshifts from 1912 to 1987. Open symbols: individual catalogues; filled symbols: cumulative data.

Through the introduction of the constant k a considerable gain is achieved in the representation of the data ...

It is remarkable that our system of fixed stars should have such an incredibly large displacement of 820 km/sec, and equally strange is the interpretation of the systematic constant $k = +656$ km. If we give this value a literal interpretation it means that the system of spiral nebulae relative to the momentary position of the solar system as center disperses with a velocity of 656 km ...

One will criticize the attempt made here to understand the characteristic arrangement and motion of the nebulae and argue, that everything is built on material which is much too incomplete and much too uncertain. True! Against this, however, I may hold two arguments, which are actually one. On the one side, experience shows again and again, that the law of large numbers arises already at remarkably small quantities of things, and then, W. Herschel deduced his apex of 1783 (with $A = 262°, D = +26°$ for 1900) with only 13 stars ...

...in the case of the nebulae, one may expect that we hold in our hands a fabric whose pattern we cannot yet unravel. One sees, however, in which direction observations have to be pushed forward in order to obtain the simplest description of the computational results connected with the nebulae."

Application of the Doppler formula to the redshifts led to two *velocity-distance relations* originally derived by Wirtz (1922, 1924). One of them is reconstructed in *Fig. 30.* Galaxies could now be placed in velocity space and, with proper calibrations and corrections, in three-dimensional real space.

The progress in assembling redshifts z is slower than that of obtaining galaxy magnitudes and spherical positions. This is due to the loss of light in the spectrograph and the loss of light concentration through spectral dispersion. Processes which minimize this effect, *e.g.* the use of low-resolution objective prism spectra and redshift determination through colour measurements promise a much steeper increase in the number of measured redshifts for the future. *Fig. 10* shows the rise up to now.

The first major step after the initial phase of redshift measurements (165 redshifts up to $v = 42\,000\,\mathrm{km\,s^{-1}}$ by Hubble, 1936) was the presentation of more than 800 galaxy redshift measurements by Humason *et al.* (1956). The redshift catalogue assembled by Palumbo *et al.* (1983) lists a total of 8 250 galaxies with measured redshift, including 21 cm-redshifts, observed in various surveys until 1980.

A truly large-scale three-dimensional survey is the original CfA-survey (Huchra *et al.* 1983) with 2 400 redshifts up to $m_{pg} = 14.5$ and complete coverage of high galactic latitudes in the northern hemisphere. The southern extension of the CfA was recently completed (da Costa *et al.* 1988). The extension of the CfA to fainter magnitudes covers $117° \times 6°$, centered near the North Galactic Pole with objects from the Zwicky *et al.* (1961-1968) catalogue merged with data from the ESO-Uppsala catalogue (Nilson 1973) up to the limit $m_{B(0)} = 15.5$. The results for 1 100 galaxies have been published (de Lapparent *et al.* 1986), 7 500 redshifts were measured in 1987, a total of 15 000 redshifts is expected for the complete survey (Huchra 1988). A second extension was begun about 1985. It will cover $100° \times 1°$ and reach $m_{B(0)} = 17.5$ (Geller *et al.* 1987). Still fainter surveys over tens of square degrees with modern multi-slit techniques are projected by different groups in both hemispheres.

Very low dispersion spectral surveys were started in Edinburgh and Cambridge and have shown that reliable methods for the determination of redshifts can be found, that the bulk of material, however, must be subjected to fully automatic procedures in order to get results from sufficiently large volumes of space. Algorithms to achieve this have recently been implemented by Schuecker (1988).

The use of colour measurements for the determination of redshifts was suggested by Zwicky (1959):

> "It is only necessary to count the populations of the clusters [of galaxies] in various colour ranges $C_1, C_2, \ldots, C_k$. One may for instance obtain photographs of a cluster in the colour range C_1 using a number of exposure times t_k. The ratios of the numbers of galaxies $n(C_1, t_k)$ are then studied for different clusters. After certain calibrations, these ratios can give accurate information on the redshifts of extremely distant clusters whose characteristics cannot be studied by any of the methods proposed so far."

Baum (1962) used measurements in different wavelength bands to derive cluster redshifts photometrically with relatively high accuracy. Several groups have adopted the method since. One such project is the determination of redshifts from CCDs taken with a set of interference filters by Loh (1988a, b) to obtain limits on q_0 and Λ, and information on cluster evolution. A similar measuring method was discussed by Koo (1985).

1.5 Two-dimensional and three-dimensional cluster surveys

An early projection of the distribution of *clusters of anagalactic nebulae* is given by Lundmark (1927). In 1958 Abell published *The Distribution of Rich Clusters of*

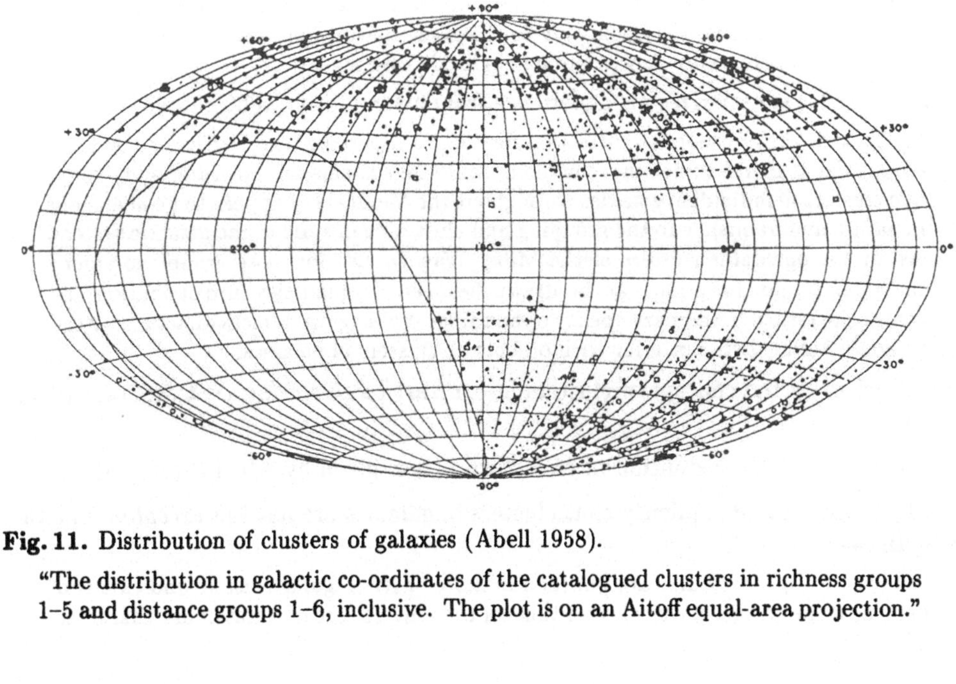

Fig. 11. Distribution of clusters of galaxies (Abell 1958).

"The distribution in galactic co-ordinates of the catalogued clusters in richness groups 1–5 and distance groups 1–6, inclusive. The plot is on an Aitoff equal-area projection."

Galaxies, the first catalogue devoted entirely to the listing of positions and characteristics of prominent clusters of galaxies. For this, Abell developed and applied a classification scheme of *richness classes* and *distance classes*. Abell's presentation of the cluster distribution is shown in *Fig. 11*. Three years later the first volume of the *Catalogue of Galaxies and Clusters of Galaxies* (Zwicky et al. 1961-1968) was published. Recently, a southern extension of the Abell Catalogue became available (Olowin 1987). For more detail on the work of Zwicky and Abell, see Sect. 2.1.

Three-dimensional cluster surveys are largely based on the Abell catalogue and its southern extension. For very distant clusters information is also drawn from radio data, since strong radio galaxies are frequently found at the center of these clusters. Redshifts have now reached $z = 1.5$ for radio galaxies and $z = 0.9$ for radio-quiet galaxies (Kron 1988).

2 The use of the surveys

2.1 Structural properties on large scales

The observed distribution of galaxies is characterized by numerous authors (from W. Herschel in 1811 onwards) as gradients, clustering, clusters, superclusters, voids, bubbles, arrangements in chains, filaments, sheets, and others. A brief history of some of these concepts is given in the following.

2.1.1 Clusters of galaxies

The term cluster is not uniquely defined. Hubble (1934) wrote:

> "The usage of the term "cluster" is quite arbitrary, ranging from the conservative practice which applies the name only to the great conspicuous examples to the other extreme in which almost any grouping is glorified by the title."

Shapley (1934) summarizes his experience from a sample of 43 000 galaxies as follows:

> "Even with the appropriate correction for 'optical doubling' the physical association of galaxies is found to be very common. Multiple systems are not rare, and groups and clusters of individual galaxies throughout the Metagalaxy appear to be analogous in nature and frequency to the grouping and clustering of stars in the galactic system. As for the open star clusters of the Milky Way, so also for loose groupings among the galaxies, no sharp line can be drawn between irregularities in distribution and coarse clustering. Lundmark speaks of hundreds of recognized 'metagalactic clusters' whereas Hubble lists but a few rich objects as clusters of galaxies."

For criticism on the concept of clusters in contrast to *clustering* see Carpenter (Sect. 2.3.1).

The first *quantitative* definition of *rich* clusters was given by Abell (Sect. 1.5).

Zwicky (1938) stated explicitly that clusters of galaxies are not the *exception* but the *rule*. He wrote:

> "a) Practically all nebulae are bunched in more or less regular clusters and clouds of nebulae if the general physical conditions of the universe are of a stationary character.

> b) In a typical cluster the number of nebulae per unit volume as a function of the distance from the center may be derived from Emden's theory of the radial distribution of mass in an isothermal gravitational gas sphere.
>
> c) The process of clustering results in a segregation of nebular types inasmuch as the most massive nebulae exhibit the greatest tendency toward clustering."

Recently, observational evidence has accumulated that members of massive cluster galaxies serve as gravitational lenses which make more distant clusters visible (Soucail *et al.* 1987). The following is a reminder to the origin of this concept.

Fifty years ago, Zwicky (1937a, b) published two brief communications:

> "Einstein recently [1936] published some calculations concerning a suggestion made by R.W. Mandl, namely, that a star B may act as a 'gravitational lens' for light coming from another star A which lies closely enough on the line of sight behind B. As Einstein remarks the chance to observe this effect for stars is extremely small.
>
> Last summer Dr. V.K. Zworykin (to whom the same ideas had been suggested by Mr. Mandl) mentioned to me the possibility of an image formation through the action of gravitational fields. As a consequence I made some calculations which show that extragalactic *nebulae* offer a much better chance than *stars* for the observation of gravitational lens effects."

In his second communication Zwicky includes an interesting footnote:

> "Dr. G. Strömberg of the Mt. Wilson Observatory kindly informs me that the idea of stars as gravitational lenses is really an old one. Among others, E.B. Frost, late director of the Yerkes Observatory, as early as 1923 outlined a program for the search of such lens effects among *stars*."

Zwicky pointed out again:

> "The problem in question, however, takes on a radically different aspect, if, instead of stars we think in terms of *extragalactic nebulae.* Provided that our present estimates of the masses of *cluster nebulae* are correct, the probability that nebulae which act as gravitational lenses will be found becomes practically a *certainty.*"

2.1.2 Superclusters

Charlier's (1922) hierarchical universe assumes clustering on increasingly larger scales:

> "N1 stars together form a galaxy G1,
> N2 galaxies form together a Galaxy of the second order G2
> N3 Galaxies of the second order galaxies form together a Galaxy of the third order G3, a.s.f."

Lundmark (1925) wrote:

> "*Our Stellar system and the system of spiral nebulae are constructed according to the conceptions expressed in the Lambert-Charlier cosmogony.*"

So far observational evidence has been found for G2 (clusters of galaxies), G3 (superclusters of galaxies), and possibly G4 (super-superclusters).

Shapley (1934) used the terms *cluster of galaxies* and *supergalaxy* interchangeably ("Undoubtedly the most important of the clusters of galaxies now known is the supergalaxy in Virgo"). With his reference to the *double supergalaxy* in Hercules he also introduced higher order clustering.

The term 'supergalaxy' assumes a new quality in the paper by de Vaucouleurs (1953). Here the expression clearly refers to the *local supercluster*, as is apparent from his figure [here *Fig. 12*] in which he also includes the *southern supergalaxy.* The local supercluster has recently been extended to dimensions of the order of hundreds of Megaparsec (Tully 1987). Dimensions of other superclusters, approaching the 10^3 Mpc regime, are reported (Ford *et al.* 1981).

It should be noted that the basic data used by de Vaucouleurs are *redshifts* obtained by Rubin (1951). Velocities, especially those indicating systematic deviations from the Hubble flow, have significantly contributed to the development of the concept of mass concentrations.

As late as 1959, Zwicky rejected the idea of second order clustering for galaxies. A chapter heading in his article "Clusters of Galaxies" reads "Superclustering non-existent".

Contrary to this view Abell (1958) had concluded from a detailed statistical analysis of his cluster catalogue that:

> "An analysis of the distribution [of rich clusters of galaxies] yields evidence that suggests the existence of second-order clusters, that is, clusters of clusters of galaxies. A statistical test reveals no incompatibilities between the observed distribution and one of complete second-order clustering of galaxies."

A much stronger statement was made in 1982 (Einasto *et al.*):

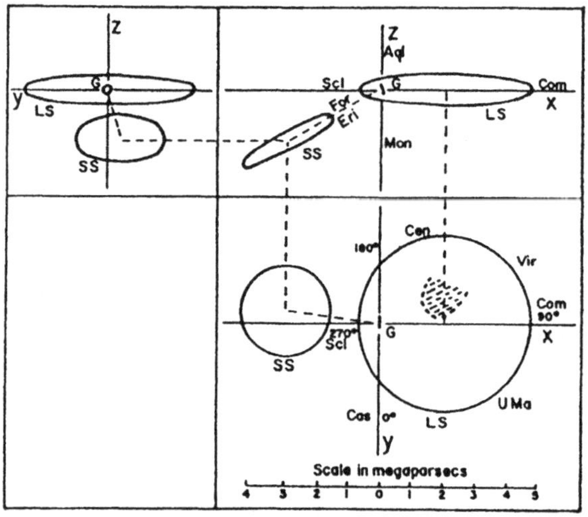

Fig. 12. Presentation of the local supercluster (de Vaucouleurs 1953).

> "Spatial arrangement of the local supergalaxy (LS) and the southern supergalaxy (SS) shown by the projection on the XY plane (that of LS), the XZ plane and the YZ plane (*e.g.*, as seen from the Coma cluster). The galaxy is shown (not to scale) in G; the Virgo cluster is indicated by the interrupted contour on the XY projection. The directions of a few constellations are given for orientation."

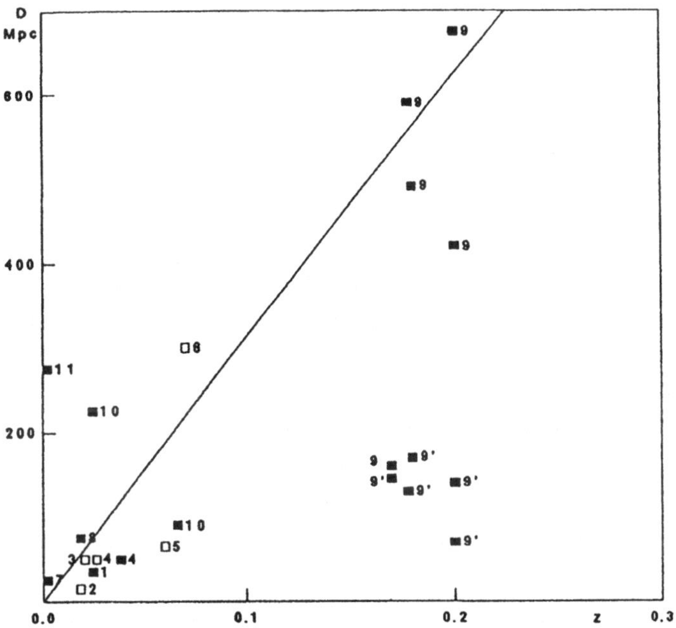

Fig. 13. Increase of observed structure sizes with increasing depth of surveys. Open symbols: voids; filled symbols: clusters and superclusters of galaxies. References indicated by numbers are found in Seitter (1986).

> "Recent studies of the spatial distribution of galaxies and of clusters indicate that practically all clusters and a vast majority of galaxies are concentrated into superclusters. The space between superclusters has no rich clusters and very few galaxies. The whole structure is cellular, with cell walls formed from sheetlike superclusters[2] and the empty cell interiors being huge voids."

A possible definition of superclusters was suggested by Oort (1983):

> "The larger and most conspicuous of these agglomerations may contain several clusters, which explains why they have been given the name 'superclusters'. In their longer dimensions, crossing times exceed the age of the Universe. They are thus unrelaxed. Unrelaxed appearance together with large size might be taken as a definition of superclusters."

Other descriptions of large-scale distributions use the expressions *sponge-like* (Gott *et al.* 1986) or *Voronoi foam*[2] (Icke and van de Weygaert 1987, Icke 1988).

2.1.3 Structures discovered in redshift space

Histograms of z-distributions show peaks suggesting the presence of clusters and empty regions suggesting the presence of voids. In fact, the recognition of voids was possible on the basis of accumulating redshift data. Chincarini (1978) referred to

[2]These terms are not merely descriptive, they are based on definite dynamical theories, see Sect. 2.3.2.

communications since the early 1970s when he wrote:

> "Observations of redshifts on samples complete to a given limiting magnitude have shown that redshifts exist in which for a given region of the sky no galaxies are observed."

A most striking example of a void was presented by Kirshner *et al.* (1981), the Bootes void. The same paper gives references from 1978 onward. A recent review paper by Rood (1988) gives history, present knowledge and future routes concerning voids.

Remarkable structures, so-called bubbles, were found from data of the second CfA (de Lapparent *et al.* 1986). In three-dimensional *real* space these structures are difficult to interpret because of the superposition of two kinds of redshifts: those due to individual motions of the galaxies and those due to their participation in the Hubble flow which is used to position them in real space.

Theory provides explanations for both, the *presence* of bubbles in real space (explosive events during the early history of the universe; Ostriker and Cowie 1981) and the *absence* of bubbles in real space (artefacts in velocity space; Kaiser 1987).

The tendency of structures to 'grow' with increasing depths of surveys is illustrated in *Fig. 13* (Seitter 1986).

2.2 Physical properties of large-scale structures

2.2.1 Masses and the mass to light ratio of galaxies; dark matter

In cosmology, *mass* and *mass distribution* are among the most important physical parameters. The masses of galaxies can be determined directly from their rotation curves and from membership in binary systems and groups and clusters of galaxies, making in each case appropriate assumptions. Gravitional lensing (see Sect. 2.1.1) promises to become an important tool for mass determination in the near future. Masses can be derived indirectly from the *mass to light ratio* for those galaxies which are too distant for direct measurements from rotation curves.

For clusters of galaxies, masses were first determined from the velocity dispersion of the cluster members under the assumption that the virial theorem can be applied. This method was used by Zwicky (1933) to determine the mass of the Coma cluster and by Smith (1936) to derive the mass of the Virgo cluster.

Methods of mass determination for both galaxies and clusters of galaxies were presented by Zwicky (1937c):

> "Present estimates of the masses of nebulae are based on observations of the *luminosities* and *internal rotations* of nebulae. It is shown that both these methods are unreliable; that from the observed luminosities and extragalactic systems only lower limits for the values of their masses can be obtained..., and that from internal rotations alone no determination of the masses of nebulae is possible...
>
> ...three new methods for the determination of nebular masses are discussed, each of which makes use of a different fundamental principle of physics.
>
> Method [1] is based on the *virial theorem* of classical mechanics. The application of this theorem to the Coma cluster leads to a minimum value $\overline{M} = 4.5 \cdot 10^{10}\, M_\odot$ for

the average mass of its member nebulae.
Method [2] calls for the observation among nebulae of certain *gravitational lens* effects.
[Method 3] gives a generalization of the principles of ordinary *statistical mechanics* to the whole system of nebulae, which suggests a new and powerful method which ultimately should enable us to determine the masses of all types of nebulae. This method is very flexible and is capable of many modes of application. It is proposed, in particular, to investigate the distribution of nebulae in individual great clusters."

The last method is one of the early instances of using methods of probability theory to describe galaxy clustering (see Sect. 2.3.1).

The *luminosity function* can be used to derive the total *luminous* mass of a cluster from the mass/light ratio of its individual members.

The earliest determination of the mass to light ratio was made by Öpik (1922) for our own Galaxy in order to be used in his determination of the distance d to the Andromeda nebula. His value is

$$\text{Mass} = 2.6\ \text{Luminosity} \quad \text{(solar units)}$$

(the most recently determined value is 2.7).

With the rotation speed at a given angular radius and the apparent magnitude of the nebula within this radius, the assumption that the centripetal acceleration balances the gravitational acceleration and the ratio M/L from our Galaxy, he had all the data needed to determine d.

Rotation values were first reported by Wolf (M81, 1914) and Slipher (NGC 4594, 1914), and later by Pease (1916, 1918).

The same ratio as employed by Öpik was used by Hubble (1926) to determine the mass of the Andromeda galaxy after he had measured its distance.

Most of the early cosmologists (Lemaître 1931b, Zwicky 1933, Hubble 1934) were aware of the possible existence of dark matter. The *problem* of hidden mass will not be discussed here, though it has become increasingly more debated in cosmology, and in some models is given an important role in the formation of galaxies.

2.2.2 The luminosity function of galaxies

A first attempt to find the luminosity function of extragalactic objects was made by Wirtz in 1926. His tabulated values can be used to construct the diagram shown in *Fig. 14*. His magnitude scale, which is much closer to the modern one than those of other astronomers of his time, is based on diameter measurements. The distribution is Gaussian, as was also found by Lundmark (1927) and Hubble and Humason (1931). Wirtz, using the measurements of Fath (1914), points out the shortcomings, unavoidable at this early stage of investigations:

"The luminosity curve cannot yet be transformed to unit space. Among other things, the limit of the observed space is not known, irrespective of the way of fixing the absolute measuring scale."

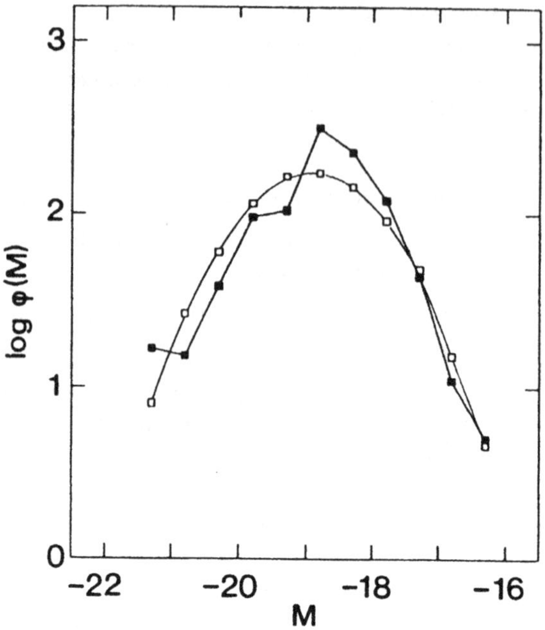

Fig. 14. A reconstruction of the first luminosity function (Wirtz 1926).

"While the distribution function of the apparent total magnitudes shows a considerable scatter and skewness, a much smaller scatter is obtained for the absolute magnitudes and a much closer approximation to the symmetrical curve. The luminosity function would then be to a first approximation a Gaussian error curve."

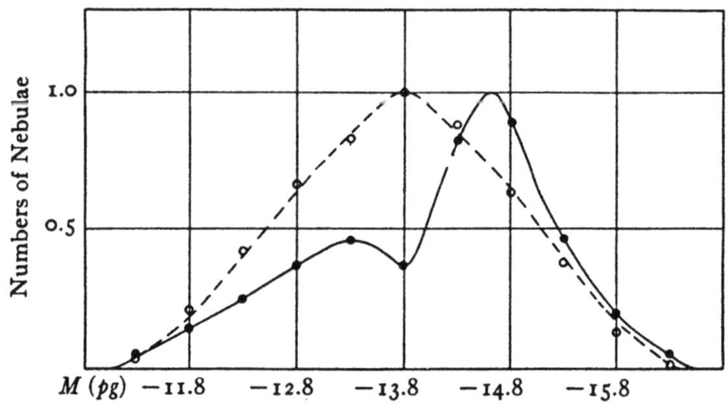

Fig. 15. Hubble and Humason's (1931) luminosity function.

"Frequency distribution of absolute photographic magnitudes among extragalactic nebulae as derived from cluster (circles) and from isolated nebulae (dots). Distances of the isolated nebulae were derived mainly from red-shifts. The range in the two curves is the same. The asymmetry in the curve for isolated nebulae is believed to be due, in part at least, to effects of selection."

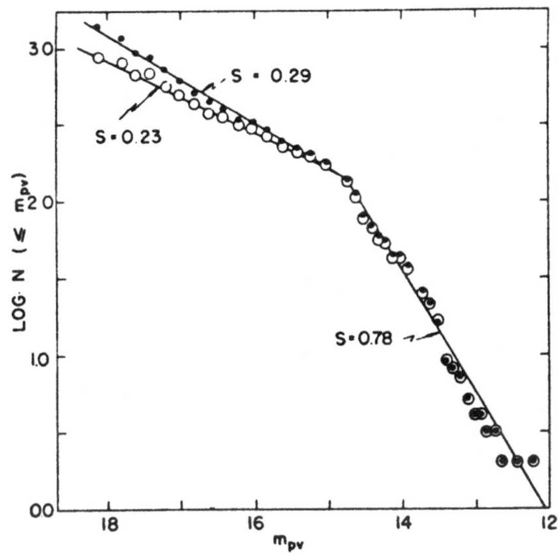

Fig. 16. Abell's (1962) luminosity function.

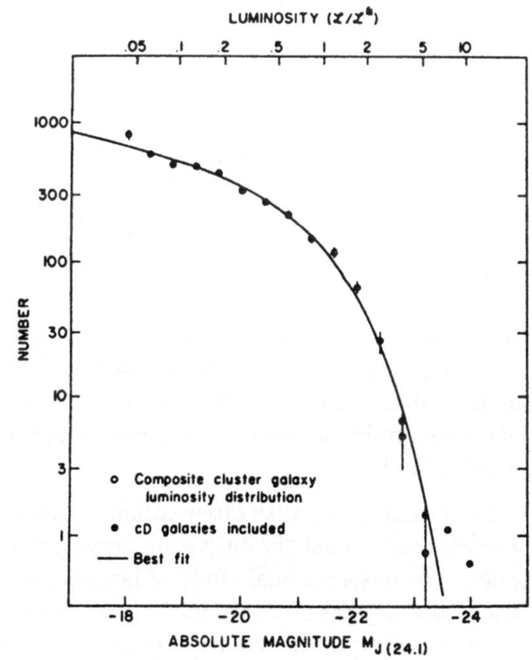

Fig. 17. Schechter's (1976) luminosity function.

"Best fit of analytic expression to observed composite cluster luminosity distribution. Filled circles show the effect of including cD galaxies in composite."

While the luminosity function (LF) determined by Hubble and Humason (1931) is a Gaussian distribution of absolute magnitudes for the cluster nebulae, an asymmetric curve is found for isolated nebulae. The authors consider the latter effect to be at least partly due to selection. The original LF is shown in *Fig. 15*. The LF was also considered Gaussian by numerous subsequent authors (including Zwicky in 1933), at least, it was assumed that the LF has a maximum at a certain luminosity.

An exponential luminosity function was given by Zwicky (1957):

$$N(M) = 0.107 \times 10^{(M+20)/5} \ .$$

Other LFs were to follow, *e.g.* Abell's luminosity function shown in *Fig. 16* (see also references in Schechter, 1976). What was to become known as the *Schechter function* is double exponential when expressed in terms of absolute magnitude (*Fig. 17*):

> "We investigate the expression
>
> $$\phi(L)\, dL = \phi^*(L/L^*)^\alpha e^{-L/L^*}\, d(L/L^*) \ ,$$
>
> where ϕ^*, L^*, and α are parameters to be determined from the data. The parameter ϕ^* is a number per unit volume, and L^* is a 'characteristic luminosity' (with an equivalent 'characteristic magnitude', M^*) at which the luminosity function exhibits a rapid change in the slope in the $(\log\phi, \log L)$-plane. The existence of such a characteristic magnitude has long been stressed by Abell (1962, 1965 – his LF is the approximation of the measured distribution by two lines of different slope, intersecting at M^*), and his notation M^* has been pirated for the present discussion. The dimensionless parameter α gives the slope of the luminosity function in the $(\log\phi, \log L)$-plane when $L << L^*$."

Recently, evidence has been accumulated by Sandage *et al.* (1985), that the Schechter function is the envelope of the individual *Gaussian* distributions of galaxies in different luminosity classes, possibly excepting very faint dwarf galaxies.

2.3 Comparison with theory

Structures *in* the universe are used to deduce the structure and the history *of* the universe. Some methods employed are *statistical* applied to ensembles of particles, assumed to be indistinguishable. Studies of *time-dependent particle properties* are required in order to either ascertain the validity of this assumption or to correct the measured properties appropriately.

World models needed for the comparison with observational results will be described in Sect. 3. The simplest models require that the clustering properties of matter disappear on sufficiently large scales. The observational study of large-scale structures serves to determine the *types*, *scales* and *degrees* of clustering, the possible *causes* of clustering and the *scale size* above which uniform distribution becomes apparent (or not).

2.3.1 Statistics

Statistics appeared in cosmology before the nature of the particles to be used was established. In the first paper (Sect. 3.1.1) which describes the universe in terms of the general theory of relativity (Einstein 1917) it is assumed that the particles filling the universe show *Boltzmann distribution*. Einstein's particles are stars.

Charlier (1922) calculated the number of collisions between galaxies from *Maxwell's equation* in his hierarchical universe which is based on Lambert's (1761) principles but starts with galaxies instead of planetary (or even Jovian) systems.

The universe described by Milne (1933) starts from statistics.

> "We ... impose the condition that this spatio-velocity distribution formula shall actually represent, *statistically*, a concourse of material objects; this is done by a modification of Boltzmann's gas equation.
>
> ... since we expect the clots or sub-systems to be formed in the spatial positions of the singularities in the original swarm, we were let to infer an *exact* correlation between position and velocity ...; this being contrasted with the merely statistical correlation ... It is clear that once we introduce an exact correlation between velocity and position, our system is not a statistical one but a hydrodynamical one."

Milne concludes:

> "Relativity – statistics – hydrodynamics – dynamics – gravitation – such was our course of investigation."

Several points seem noteworthy concerning the earliest applications of statistics to cosmology:

- they include implicitly or explicitly both the distributions of *positions* and *motions*
- no assumption about the distributions are made by Milne; the assumptions made by the other authors are simple statistical or hierarchical distributions.

Another statistic and probabilistic aspect entered extragalactic astronomy with Wirtz (1918) in his search for *parameter correlation*. Covariance functions, now generally called correlation functions, became the standard tools.

Positional distributions became of interest as a consequence of the increasing body of data. The early application of probabilistic methods is well exemplified by the work of Zwicky (1937c, Sect. 2.2):

> "By a bold extrapolation of well-known results of ordinary statistical mechanics we adopt the following working hypothesis... :
>
> 1. The system of extragalactic nebulae throughout the known parts of the universe forms a statistically stationary system.
>
> 2. Every constellation of nebulae is to be endowed with a probability weight $f(\epsilon)$ which is a function of the total energy ϵ of this constellation. Quantitatively the probability P of the occurrence of a certain configuration of nebulae is assumed to be of the type
>
> $$P = A(V/V_0)f(\epsilon/\overline{\epsilon_K}) .$$
>
> Here V is the volume occupied by the configuration or cluster considered, V_0 is the volume to be alloted, on the average, to any individual nebula in the known parts of the universe, and ϵ is the total energy of the cluster in question, while $\overline{\epsilon_K}$ will probably be found to be proportional to the average kinetic energy of individual nebulae. The function $A(V/V_0)$ can be determined a priori. On the other hand, $F(\epsilon/\overline{\epsilon_K})$ presumably will be found to be a monotonously decreasing function in $\epsilon/\overline{\epsilon_K}$,

analogous in type to a Boltzmann factor

$$F = const \cdot e^{-\epsilon/\epsilon_K} .$$

Assuming the principles stated in the preceding to be correct, we may draw the following hypothetical conclusions:

a) The clustering of nebulae is favoured by high values of f and is partially checked by low values of the a priori probability a.

b) If, as would appear to be certain, nebulae are not all of the same mass, nebulae of high mass are favored in the process of clustering, since they contribute most to produce high values of the weight function f.

c) As a consequence of b, we should expect that the frequency with which different types of nebulae occur will not be the same among field nebulae and among cluster nebulae. In other words, clustering is a process which tends to segregate certain types of nebulae from the remaining types. This may contribute toward the correct interpretation of the well-known fact that cluster nebulae are preponderantly of the globular and elliptical types, whereas field nebulae are mostly spirals... It is not necessary as yet to call on evolutionary processes to explain why the representation of nebular types in clusters differs from that in the general field...

d) If cluster nebulae, on the average, are really more massive than field nebulae, the conclusion suggests itself that globular nebulae may, somewhat unexpectedly, be

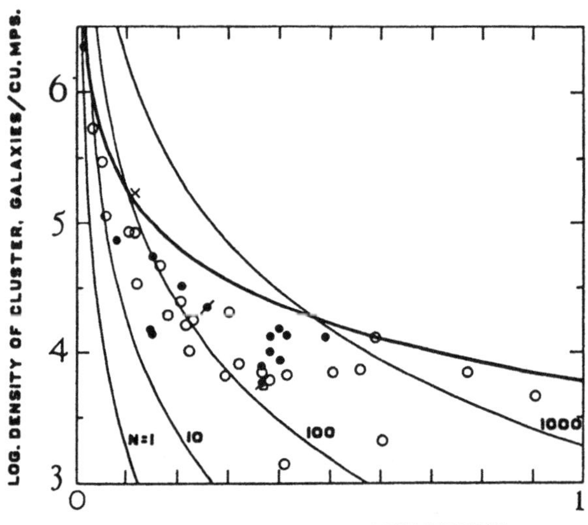

Fig. 18. Density-diameter relation of clusters of galaxies (Carpenter 1938).
"The densities and diameters of 42 clusters of galaxies. The abscissae indicate the diameter, Δ, of the clusters, expressed in megaparsecs. The ordinates indicate the logarithm of the mean density, p, of each cluster, expressed as the number of galaxies per cubic megaparsec. The sources of the data are distinguished as follows: Harvard, open circles; Mount Wilson complete counts, disks; Mount Wilson incomple counts, barred disks; Lundmark, cross; Carpenter, rectangle. The heavy curve is the envelope $p_{max} = 6040\Delta^{-3/2}$. The four lighter curves represent clusters having memberships $N = 1$, 10, 100 and 1000 galaxies."

among the most massive systems. It will be of great interest to check this inference by a search for gravitational lens effects among globular nebulae."

Compare also the work of Holmberg in Sect. 2.3.2.

Another step in analyzing galaxy distributions is the adaption of the correlation function from *parameter* space to *real* space through the definition of *n-point correlation functions for two and three dimensions: the angular and the spatial correlation functions.*

The concept of *clustering* (Sect. 2.1) basic to some of the formal statistical approaches used since the 1950s is presented by Carpenter (1938):

"It is concluded that the density restrictions and the mass restrictions in metagalactic space are real and represent a fundamental property of the distribution of matter in space. It follows as a corollary and from the interpretation of [*Fig. 18*] that there is no basic and essential distinction between the large, rich clusters and the small, loose groups. Rather, the objects commonly recognized as physical clusterings are merely

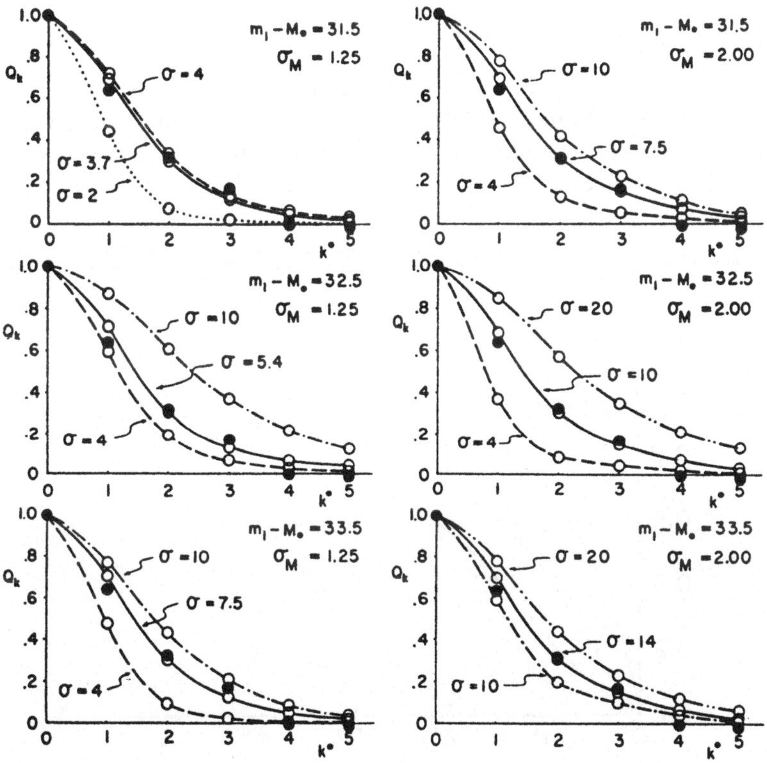

Fig. 19. Quasi-correlations in Lick survey fields (Neyman *et al.* 1953).

"Search for the best-fitting combination of values of the three parameters m_1-M_0, σ_M, and σ. Filled circles show empirical quasi-correlations for the upper right quadrant. All open circles show theoretical quasi-correlations computed by assuming specified values of m_1-M_0, σ_M, and σ..."

the extremes of a nonuniform though not random distribution which is limited by density as well as by population. From this point of view, the term 'supergalaxy' is of questionable propriety, since it implies a distinctive and coherent organic structure inherently of a higher order than the individual galaxies themselves."

Upon the publication of the first data from the Lick catalogue, Neyman and Scott (1952) and Neyman *et al.* (1953) introduced the angular pair correlation functions and Limber (1953, 1954) the angular and the spatial pair correlation function and the integral equation relating the two functions. This initiated the first period of probabilistic studies of the distribution of galaxies. Extensive discussions were given by Neyman *et al.* (1956).

Results of the first applications to the Lick Survey by Neyman *et al.* and Limber are displayed in Figs. 19 and 20.

The full bloom of statistical studies of galaxy clustering using correlation functions began to develop almost two decades after the initial phase. Essential steps leading to it were the publication of the complete Lick catalogue in 1967, its new reduction in 1977 (Seldner *et al.*), the impact of Peebles' books (1971, 1980) and the availability of powerful computing devices.

The importance of *Limber's equation* was realized when more galaxy redshifts became available. Totsuji and Kihara (1969) published an approximate solution for small angles, and the equation was extended to the general relativistic case by Groth and Peebles (1977). More detail is given in an extensive review article covering the 1970s (Fall 1979).

Further statistical techniques are mentioned by Ott (1988). *Fractals* are a new ap-

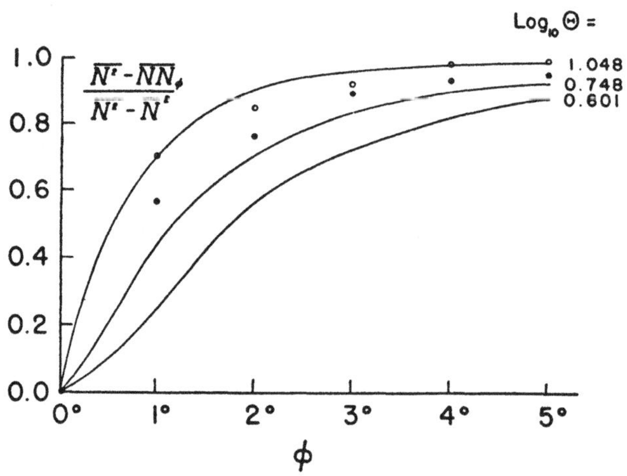

Fig. 20. Correlation functions (Limber 1954).

"Theoretical correlation functions for $\sigma = 2^m$. The open and the filled circles are the points which correspond to the observational material for the high and low latitudes, respectively, for the first approximation to $(\overline{\mathrm{NN}}_\phi - F_\phi)$."

proach to test hierarchical clustering. It was introduced by Mandelbrot and first applied by him to astronomical problems in 1975.

2.3.2 Evolution of clustering and N-body simulation

No generally accepted answer has been found as yet to the question: What causes clustering?

One of the early hypotheses dates back to Lemaître (1934) who suggested a mechanism for the growth of small fluctuations around the mean values of density and velocity in the expanding universe, using the concept that both density and velocity are coupled to 'cosmic repulsion', represented by the cosmological constant (Sect. 4.1). The said fluctuation may locally retard or even reverse the general expansion.

> "For the perturbed motion, *i.e.*, for a distribution of mass and initial velocities somewhat different from the [perfectly homogeneous] idealized model, the motion at some places may be of a completely different type from the notion of the idealized model. The relation between the energy constant h and the mass m may be such that the motion is of the collapsing type...
> Occasionally, we may also have equilibrium regions. The fact that such an equilibrium region is unstable means only that it will occur relatively rarely and that the collapsing regions will be decidedly more frequent than equilibrium-regions.
> The hypothesis we wish to discuss is that collapsing regions must be identified with the extra-galactic nebulae and the equilibrium-regions with the clusters of nebulae."

In the same paper he made another interesting statement not directly connected to the present discussion:

> "The differences of nebulae may be accounted for as a difference of total total angular momentum of the collapsing regions."

Lemaître concluded:

> " We may expect to get a complete theory of all the problems connected with extra-galactic nebulae by applying statistical mechanics to small inhomogeneity in our idealized model. Such an investigation would probably involve only two parameters, one to fix the mean velocity of expansion at the instant of equilibrium, a second one to define the dispersion of the distribution of matter from the idealized model."

Another statistical approach is N-body simulation, a technique which helps to also find evidence concerning the *history of clustering.* The *numerical experiment* consists of three parts: the process to be tested is chosen and presented in mathematical form, numerical values for the relevant parameters are inserted, integration over a predetermined interval is carried out. An early numerical experiment to explain the formation of double and multiple galaxies was performed by Holmberg in 1940. The process is capture, envoked through tidal interaction. The parameters and results are shown in *Fig. 21.*

In a subsequent experiment Holmberg (1941) uses an *analogue device* to test tidal interaction itself. Each one of two galaxies consists of 37 particles.

> "A study of tidal disturbances is greatly facilitated if it can be restricted to two dimensions... In order to reconstruct the orbit described by a certain mass element

belonging to one of the two nebulae, we must first derive as a function of time the x and y components of the total gravitational force acting upon the element. Starting from a certain distribution of mass in the nebulae, we may find the total gravitation by a purely numerical integration. However such an integration is impracticable on account of the large amount of work involved. In the present case a solution has been found by replacing gravitation by light. Every mass element is represented by a small light bulb, the light being proportional to the mass, and the total light along the x and y axes is measured by a combination of a photocell and a galvanometer. The measured values represent the components of the gravitational force. The latter components are obtained by adding up the attractions due to individual mass elements, each multiplied by the cosine of the corresponding projection angle. Consequently the photocell must obey the cosine law as far as the angle of incidence of the light is concerned. If the photocell obeys the cosine law and if the combination of photocell and galvanometer gives a linear relation between light and scale reading, the galvanometer deflection will be proportional to the total gravitational force or, more correctly, to the total acceleration."

For an extensive discussion of Holmberg's extragalactic work, see Rood (1987).

The availability of fast computers solved the problem of tedious integrations and made numerical simulations a standard technique in clustering studies. A recent review article (Sellwood 1987) describes its potential if proper care is taken in the handling and interpretation of simulations with the large numbers of particles which are now within reach. The computation of the evolution of clustering has become a standard tool in cosmology.

There is also a close connection between the early models of Holmberg and the extensive calculations by Toomre and Toomre (1972), which are in turn considered to be the basis for modern concepts of *merging* and induced star formation, *i.e.* *starbursts*

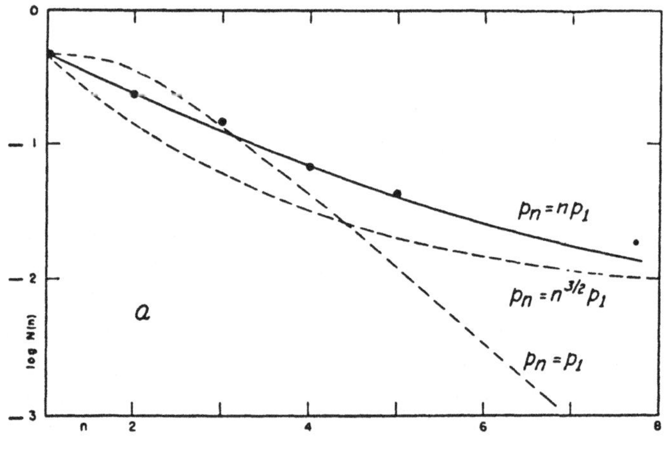

Fig. 21. Holmberg's (1940) results.

"Distribution of nebulae (relative number = $N(n)$) in systems of different order (n). The dots represent observed numbers, whereas the curves give theoretical distributions corresponding to different assumptions for the 'captureability' (p_n) of systems of different order."

in galaxies. The latter has become important especially in the context of infrared emission of heated dust in these systems.

2.3.3 Physical evolution

Physical evolution entered astronomy in the wake of Darwin's biological evolution. Since then it has become the essence of our understanding of stars, galaxies and the universe.

Stellar evolution (starting with Lockyer's "Inorganic Evolution", 1900) lets us expect changing parameters of galaxies. Galaxy evolution is partly due to the evolution of its stellar population (brightness, colour, chemical abundances, gas content), partly due to other internal or external causes (diameter, mass, momentum, angular momentum – see above). The observable properties of the universe are affected by the evolving stars (Olbers Paradox), traceable by our knowledge of the stars (chemical abundances, evolutionary computations), apparently affected by the evolving galaxies (changing galaxy parameters mask and mimic universal evolution) and by its own evolutionary history (which we want to derive).

For the *first cosmological problem*, the Kepler – Lois de Chésaux – Olbers Paradox or the Dark Night Sky Paradox, a number of solutions have been provided throughout the times, the finite life time of stars, *i.e. stellar evolution*, is one of them (Harrison 1981).

Early thoughts about the evolution of galaxies appear in Charlier's (1922) discussion referred to in the previous paragraph: collisions of nebulae are thought to create spiral pattern. Spitzer and Baade (1951) envoke collisions to account for the absence of dust (and thus of the spiral pattern traced by young stars) in S0 galaxies, for merging see above.

Galaxy parameters of unknown time-dependence, making the comparison between near and distant objects illusory were the major concern of Baade, as quoted by Sandage (1987). Tinsley (since 1968, *Fig.* 22) made us see the Hubble diagram not only as a diagram relating universal parameters but in some aspects as a counterpart to the stellar Hertzsprung-Russell diagram: the locus where galaxy evolution can be traced. An early reference to possible evolutionary effects is made by Hubble and Tolman (1935):

> "With regard to the assignment of constant properties to the nebulae over long periods of time, is to be remarked that in the case of the most distant survey, to $m = 21.0$, the 'average' nebula involved must have emitted the light which is now observed at a time of the order of 3×10^8 years in the past. If the recessional explanation of red-shift were adopted, this period would be so nearly comparable with the time of cosmic expansion – possibly of the order of 10^9 to 10^{10} years – that the assignment of constant properties might not appear justified, and in that case we might try to explain excess counts by assuming greater luminosities for the nebulae at earlier times."

The search for traces of evolution is one of the great topics of current observational cosmology, as is the strife for finding evolutionary links between the background radiations, quasars, radio galaxies, and normal galaxies at different redshifts. The *visible* signs of evolution of the universe are the 3 K background radiation, first computed by Alpher and Herman (1948, see also Lemaître 1934, Sect. 4.1) and first measured by Penzias and Wilson (1965), and the abundances of chemical elements, of which a certain percentage can be attributed to the same hot early phase as the background radiation.

In the context of evolution in the expanding universe Lemaître (1931b) opened the discussion on what he later called the primeval atom:

> "Sir Arthur Eddington states that, philosophically, the notion of the beginning of the present order of Nature is repugnant to him. I would rather be inclined to think that the present state of quantum theory suggests a beginning of the world very different from the present order of Nature. Thermodynamical principles from the point of view of quantum theory may be stated as follows: (1) Energy of constant total amount is distributed in discrete quanta. (2) The number of distinct quanta is ever increasing. If we were to go back in the course of time we must find fewer and fewer quanta, until we find all the energy of the universe packed in a few or even in a unique quantum. Now, in atomic processes, the notions of space and time are no more than statistical notions; they fade out when applied to individual phenomena involving but a small number of quanta. If the world has begun with a single quantum, the notions of space and time would altogether fail to have any meaning at the beginning; they would only begin to have a sensible meaning when the original quantum had been devided into a sufficient number of quanta. If this suggestion is correct, the beginning of the world happened a little before the beginning of space and time, I think that such a beginning of the world is far enough from the present order of Nature to be not at all repugnant."

Concerns that not all elements can be made in stars were voiced by Weizsäcker (1938):

> "From theory we must ask a list of testable suggestions when and where in the history of the cosmos the required temperatures and densities could have been realized."

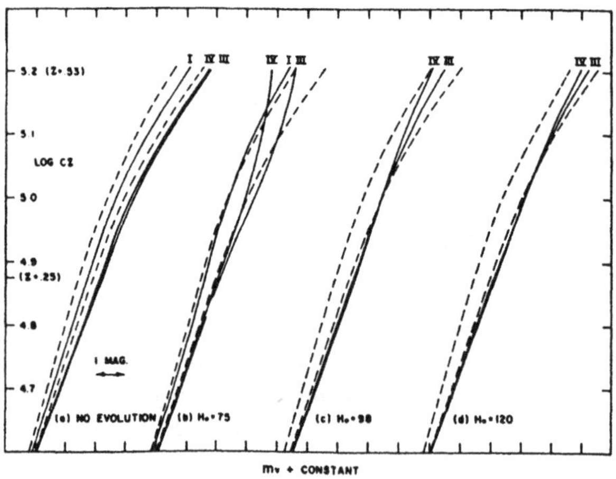

Fig. 22. Tinsley's (1968) evolutionary tracks of galaxies in the Hubble diagram.

In 1946 Gamow wrote:

> "It is generally agreed at present that the relative abundances of various chemical elements were determined by physical conditions existing in the universe during the early stages of expansion, when the temperature and density were sufficiently high to secure appreciable reaction-rates for the light as well as for the heavy nuclei."

Two years later, Gamow (1948) considered element formation and subsequent galaxy formation according to the Jeans criterion in the expanding universe at the time when $\rho_{\rm radiation} = \rho_{\rm matter}$. Because for adiabatic expansion $T \propto 1/R$, the temperature can be obtained at any given time from the integration of dR/dt. Gamow obtained $T = 340$ K for the suggested time of galaxy formation at an age of the universe of $1.3 \cdot 10^8$ years.

Two weeks after Gamow's publication, Alpher and Herman (1948) repeated the computation and carried the temperature determination to a present value of $T = 5$ K.

3 Models of the Universe

3.1 Basic concepts

In view of the complexity of the subject *universe* there seems to be no other approach but the adoption of simple models and attempts at their falsification, in which case other simple models will have to be substituted. So far, tests have not yet *excluded* models based on the

- cosmological principle (isotropy and homogeneity) (from 1916 onwards, see below)
- Einstein's general theory of relativity (1916)
- Friedmann's concept of a *time-dependent* scale factor (1922)
- Friedmann's metric, with the time coordinate *orthogonal* to the three space coordinates, and constant sign of curvature (1922, positive curvature, 1924 negative curvature)
- Heckmann's extension to zero curvature and display of a series of models with $k = +1, 0, -1$ and $\lambda > 0, = 0, < 0$ (1931, 1932)
- Tolman's (1929) and Robertson's (1929) derivation of a Friedmann metric entirely from the assumption of homogeneity and isotropy, and the final formulation of the metric by Walker (1936).

Within the above concepts, cosmology tries to derive numerical values for the

Hubble constant H_0

acceleration parameter q_0

curvature parameter k

matter density ρ_0 or density parameter Ω_0

pressure p_0

age of the universe.

A successful method for the independent derivation of the cosmological constant Λ has not yet been applied[3].

3.1.1 The cosmological principle

Einstein (1917):

> "...The character of space curvature is, depending on the distribution of matter, variable locally and with time, but on a large scale it can be approximated by spherical space. This is at least logically without contradiction and it is the most obvious from the point of view of the general theory of relativity; whether it is tenable from the point of view of our present astronomical knowledge shall not be discussed here. In order to reach this non-contradictory conclusion we had to introduce, however, a new addition[4] to the field equations, which is not justified by our actual knowledge about gravitation. It must, however, be pointed out, that a positive curvature of space also results from the matter contained in it when this additional parameter is not introduced; the latter is only required to ascertain a quasistatical distribution of matter, as corresponds to the fact of small stellar velocities."

de Sitter (1917):

> "Its density (in natural measure) is constant when sufficiently large units of space are used to measure it. Locally its distribution may be very inhomogeneous."

Einstein (1918):

> "In our world matter is, however, not uniformly distributed but concentrated in individual celestial bodies, not at rest, but in slow relative motion (compared to the velocity of light). However, it is well possible, that the mean ("naturally measured") space density of matter taken for spaces which contain very many fixed stars, is a nearly constant quantity."

Friedmann (1922):

> "...by Einstein and also by de Sitter certain assumptions are made concerning the matter tensor, which correspond to the incoherence of matter and its relative rest, that is the velocity of matter is assumed to be sufficiently small compared to the basic velocity – the velocity of light."

Lemaître (1927) - see Sect. 3.2.1.

Robertson (1929):

> "The general theory of relativity attributes the particular metrical properties of the space-time universe... directly to the distribution of matter within it, and has naturally led to speculations concerning the structure of the universe as a whole, in which local irregularities caused by the agglomeration of matter into stars and stellar systems are disregarded. Chief among the resulting relativistic cosmologies are those based on the cylindrical world of Einstein and the spherical world of de Sitter; the line elements on which these interpretations are based have not, however, been derived from the intrinsic properties of homogeneity and isotropy attributable a *priori* to such an idealized universe, but rather are presented as defining manifolds which

---

[3]In the early literature, the designation λ is more frequently used than Λ. When no confusion with wavelength λ is possible, the original designation was kept in the present quotations and discussions.

[4]The cosmological constant discussed below

do possess the desired uniformity. It is the purpose of the present note to formulate explicitely an assumption embodying the uniformity demanded by such a cosmology and deduce all line elements satisfying it...

Space-time shall be spatially homogeneous and isotropic in the sense that it shall admit a transformation which sends an arbitrary configuration in any 3-spaces $t = const, \ldots$ into any other such configuration in the same 3-space in such a way that all intrinsic properties of space-time are left unaltered be the transformation. That is, any such configuration shall be fully equivalent to any other in the same 3-space in the sense that it shall be impossible to distinguish between them by any intrinsic property of space-time ...

If we wish to require in addition that their intrinsic properties be independent of time t, we may amend the above assumption to state that *any configuration*, as there described, *in any 3-space $t = const.$ is fully equivalent to any such in any 3-space of the family.*"

Milne (1933):

"Einstein's [1931] postulate that all places in the universe must be equivalent is modified to read: *The universe must appear the same to all observers.* As the view of any particular observer depends on the space-time frame he adopts, this requires to be made more precise. We therefore posit: *not only the laws of nature, but also the events occurring in nature, the world itself, must appear the same to all observers, wherever they be, provided their space-frames and time-scales are similarly oriented with respect to the events which are the subject of observation.* By 'the world' I do not mean 'the world at an instant' but the totality of the flux of events. This postulate is referred to as the 'extended principle of relativity'."

Here the understanding of the terms *homogeneity* and *isotropy* in cosmology seems to have reached its full meaning. Evolved has what has been termed the *cosmological principle* by Milne.

3.1.2 Curvature and metric in general relativistic universes

The early relativistic world models are static. Einstein (1917) considered positive space curvature with matter acting as the form-giving factor: the *Einstein universe.* When he found that his solution was not stable he introduced a compensating factor, the cosmological constant λ (Sect. 3.2.4).

De Sitter (1917) found all three possible static solutions. For a matter-filled universe it is the solution given by Einstein, the other two solutions pertain to a universe without matter. Assuming $\lambda = 0$, the empty universe is Euclidean and corresponds to an Newtonian universe or to an Einstein universe before the introduction of λ. A projection of the Einsteinian universe into spaces of different curvature requires that time be always and everywhere the same, *i.e.* absolute. With $\lambda \neq 0$ an empty universe of positive curvature can be projected into a Euclidean or hyperbolic space under complete invariance of all four variables.

Observational evidence for a hyperbolical universe is discussed by de Sitter (1917) in the following:

"System A: time is the same everywhere and always, the time coordinate is different from the three space coordinates; system B: there is no universal time, no difference between coordinates in four-dimensional space, no physical meaning of the coordinates.

The three-dimensional space of this system of reference is the space with constant *negative* curvature, or hyperbolical space, or space of Lobatschewski.

In the system B the rays of light are straight lines in hyperbolic space... In space B we have $g_{44} = \cos^2 \chi$. Consequently the frequency of light vibrations diminishes with increasing distance from the origin of co-ordinates. The lines in the spectra of very distant stars or nebulae must therefore be systematically displaced towards the red, giving rise to a spurious positive radial velocity ... Of the following three nebulae, the velocities have been determined by more than one observer:

Andromeda	3 observers	− 311 km/sec.
N.G.C. 1068	3 "	+ 925 "
N.G.C. 4594	2 "	+ 1185 "

... If, however, continued observations should confirm the fact that the spiral nebulae have systematically positive radial velocities, this would certainly be an indication to adopt the hypothesis B in preference of A."

Friedmann (1922) introduced his metric as follows:

"...R is proportional to the curvature radius of space, which then may vary with time.

In the expression for the line element ds^2 the terms g_{14}, g_{24}, g_{34} can be made to vanish for an appropriate choice of the time coordinate, or briefly speaking, time is orthogonal to space. For this second assumption, it seems to me, no physical or philosophical reasons can be given; it serves exclusively the purpose of making the computation simpler...

With the [above] assumptions ds^2 can be brought into the form

$$ds^2 = R^2(dx_1^2 + \sin^2 x_1 dx_2^2 + \sin^2 x_1 \sin^2 x_2\, dx_3^2) + M^2 dx_4^2$$

whereby R is a function of x_4 and M depends in the general case dependent on all four world coordinates."

Friedmann then shows that both the *Einstein universe* and the *de Sitter universe* are special cases of his more general metric. The implicit assumption in Friedmann's metric is that the curvature is positive.

Two years later, Friedmann (1924) extended his models to negative curvature:

"... *we can say that the stationary world with constant negative curvature of space is only possible with disappearing or negative density of matter* ..."

but

"... *the possibility of non-stationary worlds with constant negative curvature of space and positive density* [is given]."

Lemaître (1927) also allowed for both positive and negative curvature.

Tolman (1929) develops his metric with the two following conditions:

"As the first condition to be satisfied by the line element, we shall require it to be compatible for a limited region in space and time with the special theory of relativity ... As the second condition for the line element we shall require the possibility of writing it in a form which is spherically symmetrical in the spatial variables, symmetrical with respect to the past and future time, and static with respect to the time. These requirements, as is well known, lead necessarily to the form

$$ds^2 = -e^{\lambda} dr^2 - r^2 d\theta^2 - r^2 \sin^2\theta + e^{\nu} dt^2$$

where λ and ν are functions of r alone. The requirement of spherical symmetry is an obvious one to impose, since otherwise the universe regarded on a large scale would have different properties in different directions. The requirement of symmetry with respect to past and future time means that the large-scale behaviour of the universe is reversible, and the static form of the line element means that by and large the universe is in a steady state."

Robertson (1929) criticized Friedmann and Tolman on grounds of

"untenable assumptions ... instead of making full use of the intrinsic uniformity of such a space [homogeneous and isotropic] as we do here."

His line element is:

$$ds^2 = dt^2 - e^{2f(t)} \left(\frac{dr^2}{1 - r^2/R^2} + r^2 d\theta^2 + r^2 \sin^2\theta d\phi^2 \right) ,$$

where f is an arbitrary function of time.

While positive and negative curvature had been introduced into expanding relativistic universes by 1924, the Euclidean universe - the only one in Newtonian physics - was not considered until the work of Heckmann, published in Juli 1931:

"The solutions of the Einstein field equations by Friedmann and Lemaître frequently used in recent times, shall in the following be extended. Especially the very simple proof shall be given that besides the assumption of a spherical (or elliptical) closed space, the assumption of a hyperbolical, in a limiting case even Euclidean space, are of entirely equal standing within the framework of the theory of relativity."

It may be interesting to note, that de Sitter in a Harvard lecture series given in October 1931 and published in 1932, discussed already models of different values for λ and the three curvature parameters and included a general sketch for three different cases (For a forerunner of this type of diagram see Sect. 3.1.3)

Heckmann (1932) published the full sequence of possible models with three types of space curvature and three different signs of the cosmological constant λ. All models constitute solutions of the differential equation

$$\dot{R}^2 = \left(\frac{dR}{dt}\right)^2 = \frac{\lambda}{3} R^2 + C + \frac{\kappa}{3R^2} \left[B + A\sqrt{R^2 + \alpha^2} \right] . \tag{1}$$

"(1) has within the framework of the theory of relativity nine different types of solutions, which originate from the combinations of the three cases $C = \begin{cases} +1 \\ 0 \\ -1 \end{cases}$ with the three also possible $\lambda \begin{cases} > \\ = \\ < \end{cases} 0$. These types of solution shall be discussed in the following. The diagrams included for illustration are computed for the special case of

a radiation-filled universe ($A = 0$). Though this case can only command theoretical interest, it is nevertheless typical for the more general cases $A \neq 0$; all characteristic features appear in this model which, however, has the great advantage that it can be expressed in elementary functions...
One can always give (1) the form

$$\left(\frac{dz}{d\tau}\right)^2 = \epsilon z^2 + C + \frac{1}{z^2} M(z) \, ,$$

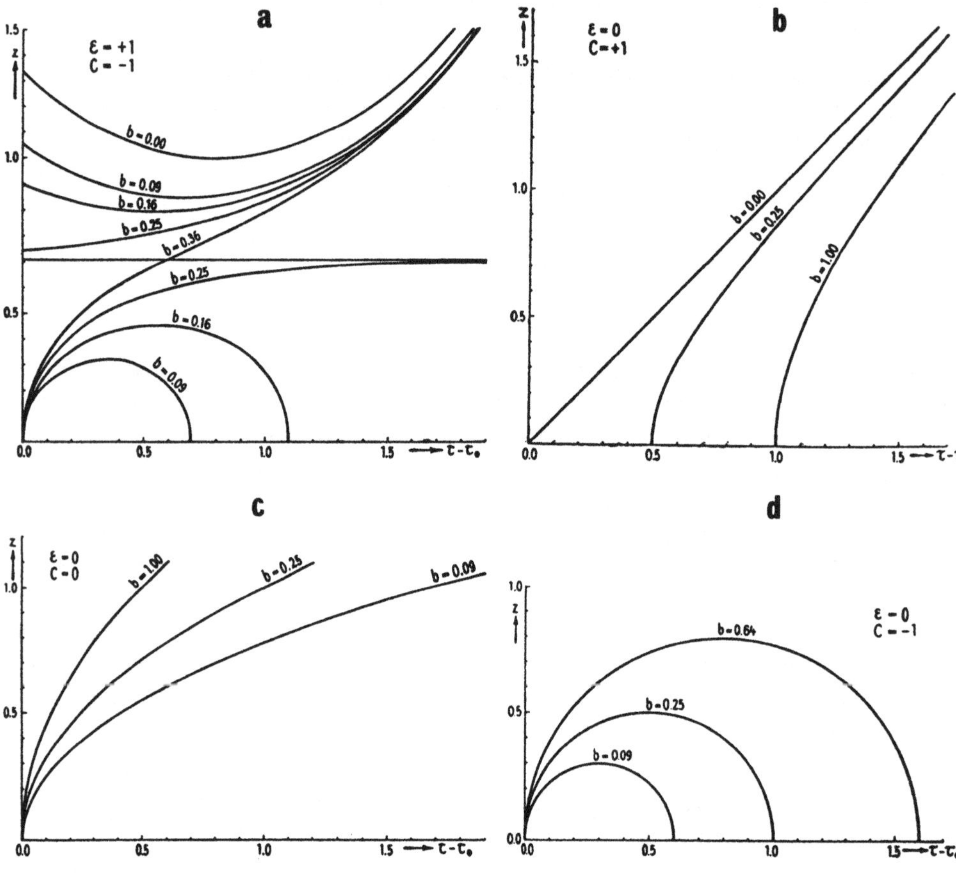

Fig. 23. Four of Heckmann's nine models of the universe (Heckmann 1932).

a "This case is the richest. For $M \equiv 0$ it displays the de Sitter universe..."

b "...Except for vanishing matter term where the solutions become straight lines intersecting the τ-axis under 45°, all solutions rise vertically from the τ-axis to reach inclination 1 asymptotically...at infinity."

c "...With $M \not\equiv 0$ all solutions rise vertically from the τ-axis, their inclination approaching 0 monotonously at infinity..."

d "...for $M \not\equiv 0$ the solutions remain always finite and $(dz/d\tau)^2$ has a zero z_1, so that $z = z_1$ is envelope to the general solutions..."

where z is proportinal to R, τ proportional to t and ϵ assumes the values +1,0,-1. $M(z)$ is defined by

$$M(z) = b + a\sqrt{z^2 + \alpha_1^2}\,; \qquad a > 0,\ b > 0\,,$$

where a, b, and α_1 are proportional to the corresponding constants A, B, and α in (1). C is, as above, +1 for hyperbolic, 0 for Euclidean, −1 for spherical metric."

For the coordinates z and τ compare de Sitter (1930), Sect. 3.1.4. Note that the constant C is the *curvature parameter* k, except that the latter is *positive* for *Euclidean* and *negative* for *hyperbolical space.*

Following Tolman and others, Walker (1936) published the now generally used metric which includes the curvature parameter k:

$$ds^2 = dt^2 - R^2\left(\frac{dr^2}{1-kr^2} - r^2 d\theta^2 + r^2 sin^2\theta d\phi^2\right)\,.$$

Another consequence of Heckmann's paper was an immediately following publication by Einstein and de Sitter (1932) describing the relativistic Euclidean universe, the Einstein-de Sitter universe, which was to become so important within the framework of the inflationary universe (Sect. 4.), opening with the sentence:

"In a recent note in the *Göttinger Nachrichten*, Dr. O. Heckmann has pointed out that the non-static solutions of the field equations of the general theory of relativity with constant density do not necessarily imply a positive curvature of three-dimensional space, but that this curvature may also be negative or zero."

3.1.3 Cosmic time

Time as one coordinate of the four-dimensional world was well established when Friedmann (1922) introduced the concept of the *time-dependence* of cosmic parameters, represented by the first and second time derivatives of the scale factors $R(t)$, and thus the possibility of measuring *cosmic time.*

Well aware of the fact that the cosmological constant λ cannot be determined independently:

"It should be remarked that the 'cosmological' quantity λ remains indetermined in our formulae",

Friedmann considered solutions for t in a world of positive curvature for different values of λ. For this he introduced the constant A, which is directly proportional to the mass M of the universe, and integrated his time-dependent form of Einstein's equations.

Friedmann's "monotonous world of the first kind" is described as follows:

"The time of growth of R from 0 to R_0 we will call *the time since the creation of the world*[5], this time t' is given by

$$t' = \frac{1}{c}\int_0^{R_0}\sqrt{\frac{x}{A - x + \frac{\lambda}{3c^2}x^3}}\,dx\,.$$

[5]"The time since the creation of the world is the time passed since the moment when space was a point ($R = 0$) to the present state ($R = R_0$); this time may also be infinite."

> The time since the creation of the (monotonous) world (of the first kind) considered as function of R_0, A, λ has the following properties:
> 1. it grows with growing R_0; it declines with increasing a, e.g., the mass in space decreases; 3. it decreases with increasing λ."

Friedmann's monotonous world of the second kind, for which it can also be shown that R is an *increasing function of time*, starts from

$$R = x_0' = \vartheta(\lambda, A) .$$

The *"periodic world"* requires that λ lies within the limits $(-\infty, 0)$, with those values excluded which lead to more than infinite period.

> "Our knowledge is completely insufficient to permit number computation and to decide which world is ours ... With $\lambda = 0$ and $M = 5 \cdot 10^{21}$ solar masses, we find that the world period is of the order 10 billion years. But these numbers can, of course, only serve as an illustration for our computations."

The *expansion age* of the universe is given by the inverse Hubble constant (defined below). The *evolutionary age* can be determined from the evolutionary times of its oldest members plus the time from the beginning of the universe to their formation. Two direct methods are currently used:

- age of the oldest solid matter from our vicinity (meteorites, earth, moon) from radioactive dating
- age of the oldest members of the galaxy (globular clusters).

The difference between the observed *expansion age* and the observed *physical age* of the universe has plagued cosmologists almost from the beginning.

3.1.4 Look-back time

Figure 24, taken from de Sitter (1930), shows an early, probably the first, diagram of the relation between world radius R, represented by the dimensionless quantity z, and time t, represented by the dimensionless quantity $\tau - \tau_0$ (see Sect. 3.1.1). De Sitter's definitions are:

$$\frac{z_0 - z}{z} = \frac{R_0 - R}{R} = \frac{\Delta\lambda}{\lambda}$$

and

$$\tau - \tau_0 = c\sqrt{\frac{\Lambda}{3}} \cdot (t - t_0)$$

with λ = wavelength of radiation
c = velocity of light
Λ = Lemaître's cosmological constant in unit length^{-2},
the subscript 0 indicates a fixed time.

With the appropriate choice of R_0, e.g. the present scale factor, an infinite scale factor is reached asymptotically in a finite interval $\tau - \tau_0$, corresponding to infinite time, provided that the cosmological constant is a function of time (see Sect. 3.2.5). The curves labelled I - VII represent different world models.

With reversed scales, taking again the present for reference, one finds that R approaches 0 asymptotically in a finite interval $\tau - \tau_0$. In this form the diagram displays the look-back time for various models. Different scaling permits to display the past behaviour of different models in the form that is frequently used today (*e.g.* Tinsley 1968).

3.2 Basic parameters

3.2.1 $R(t)$ – the scale factor, and its time derivatives

Friedmann (1922) has shown that in a non-stationary world Einstein's field equations, which tie geometry and matter together, relate the time-dependent scale factor $R(t)$ and its first and second derivative to the mass density in the universe, the constant of gravity, the velocity of light and the cosmological constant. His equations

$$\frac{R'^2}{R^2} + \frac{2\,R\,R''}{R^2} + \frac{c^2}{R^2} - \lambda = 0\,.$$

$$\frac{3\,R'^2}{R^2} + \frac{3\,c^2}{R^2} - \lambda = \kappa\, c^2\, \rho\,,$$

with

$$R' = \frac{dR}{dt} \quad \text{and} \quad \kappa = \frac{8\,\pi\, G}{c^2}\,,$$

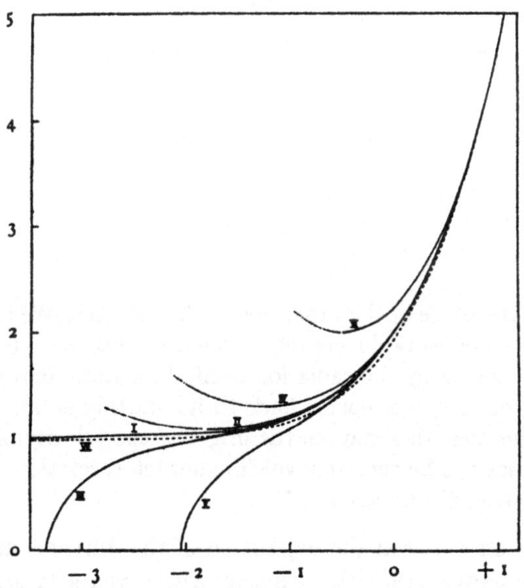

Fig. 24.

"Relation between z and $\tau - \tau_0$. The vertical coordinate is z, the horizontal coordinate is $\tau - \tau_0$." (de Sitter 1930) assume implicitly that the pressure p is zero. Later, these equations were extended to include negative and null curvature (Sect. 3.1.2).

Lemaître (1927) formulated the time dependence of the field:

> "Space is homogeneous and has constant positive curvature; space-time is also homogeneous, for all events are perfectly equivalent. But the partition of space-time into space and time disturbs the homogeneity. The co-ordinates used introduce a centre. A particle at rest at the centre of space described a geodesic in the universe; a particle at rest otherwhere than at the centre does not describe a geodesic. The co-ordinates chosen destroy the homogeneity of the universe and produce the paradoxical results which appear at the so-called 'horizon' of the centre. When we use co-ordinates and corresponding partition of space and time of such a kind as to preserve the homogeneity of the universe, the field is found to be no longer static: the universe becomes of the same form as that of Einstein, with a radius no longer constant but varying with the time according to a particular law."

After introducing R as a function of time and its first and second derivatives, Lemaître arrives at the same expression for the field equations as Friedmann, including, however, the pressure term

$$\frac{R'^2}{R^2} + \frac{2\,R\,R''}{R^2} + \frac{1}{R^2} = \lambda - \kappa p \, ,$$

where κ is again given by Einstein's expression. Lemaître continued:

> "The four identities giving the expression of the conservation of momentum and of energy reduce to
>
> $$\frac{d\rho}{dt} + \frac{3\,R'}{R}(\rho + p) = 0$$
>
> which is the energy equation. This equation can replace [the above equation]. As $V = \pi^2 R^3$, it can be written as
>
> $$d(V\rho) + p\,dV = 0 \, ,$$
>
> showing that *the variation of total energy plus the work done by radiation-pressure in the dilatation of the universe is equal to zero.*"

He summarized:

> "It remains to find the cause of the expansion of the universe. We have seen that the pressure of radiation does work during the expansion. This seems to suggest that the expansion has been set up by the radiation itself. In a static universe light emitted by matter travels round space, comes back to its starting-point, and accumulates indefinitely. It seems that this may be the origin of the velocity of expansion $\dot{R}/R$ which Einstein assumed to be zero and which in our interpretation is observed as the radial velocity of extra-galactic nebulae."

The search for a relation between the expansion of the universe and energy avaliable was continued by Lemaître (1931b). Another three years later, he offered a new physical explanation for the expansion (Sect. 4.1).

In his 1927 paper he also introduced the relation between redshift and variation of the scale factor:

"The relation between radial velocity (redshift) is also given:

$$\frac{v}{c} = \frac{R_2}{R_1} - 1$$

is the apparent Doppler effect due to variation of the radius of the universe. *It equals the ratio of the radii of the universe at the instants of observation and emission diminished by unity.*"

Later it became evident that the first derivative measured in units of R is the increment H in the Hubble diagram or the Hubble parameter while the second derivative measured in units of H^2 is the acceleration parameter q. The symbols

$$H \equiv \frac{\dot{R}}{R}$$

$$-q \equiv \frac{\ddot{R}}{R H^2} .$$

were introduced by Robertson (1955) and Hoyle and Sandage (1956), respectively. The present values [6] are H_0 and q_0.

An attempt to measure the curvature radius of space was first made by Schwarzschild (1900):

"The question shall be discussed how large the curvature radius must at least be chosen... For both cases the elliptical and the hyperbolical space we will now discuss the problem of parallax determination."

The method using parallaxes was employed not only by Schwarzschild but also, in the early days of relativistic cosmology, by de Sitter (1917). In a more general way

$$\text{distance/scale factor} = f(\text{distance measure, metric}) ,$$

the method is still the only one available. It is essentially the measurement of a three-dimensional geometrical figure projected onto the surface of four-dimensional space. The area of the figure relates to the volume of a geodesic sphere and thus to its radius of curvature. The result is obtained through model fitting, so that the figure must be determined theoretically for comparison.

For space with a time-dependent scale factor, the relation between observational and theoretical quantities has been discussed in detail by Heckmann (1942). His *photometric distance* is

$$D = R\, S(\chi)\, (1 + z)$$

with R = scale factor at the time of observation
z = redshift
$S(\chi)$ = distance r.

For small redshifts D is equal to the usual photometric distance. Distance r should be compared to the *spatial distance* defined by Whittaker (1931):

[6]Index 0 was introduced by Friedmann (1922).

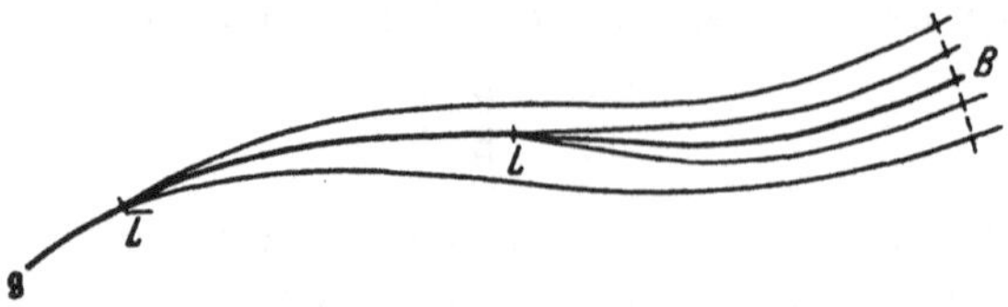

Fig. 25. Light path on a geodesic (Heckmann 1942).

"Concerning the dependence of the apparent magnitude of a light source from its position on the space geodesic which connects it to the observer."

"The spatial distance of two material particles in a general Riemannian space-time may then be thought of as *a relation between two world-points which are on the same null geodesic. It is obviously right that 'spatial distance' should exist only between points which are on the same null geodesic; for it is only then that the particles are in direct physical relation with each other.* This statement brings out into sharp relief the contrast between 'spatial distance and 'interval' defined by

$$ds^2 = \sum_{p,q} g_{pq}\, dx^p\, dx^q \;;$$

for between points on the same null geodesic, the 'interval' is always zero. Thus '*spatial distance exists when, and only when, the 'interval' is zero.*"

A symbolic presentation is given in *Fig. 25* taken from Heckmann (1942).

D is an auxiliary quantity which appears in the series expansions[7]:

$$z = \frac{\dot{R}}{R}D - \frac{1}{2}\left(1 + \frac{\ddot{R}R}{\dot{R}^2}\right)\left(\frac{\dot{R}}{R}D\right)^2 + \ldots$$

$$\log N = 0.6(m - \Delta m_N) - 0.6M + \log n - 3.993\,,$$

$$\Delta m_N = 2.17\left[\frac{\dot{R}}{R}D - \left(1 + \frac{1}{2}\frac{\ddot{R}R}{\dot{R}^2} + \frac{k}{10\dot{R}^2}\right)\left(\frac{\dot{R}}{R}\right)^2 D^2 + \ldots\right].$$

z is directly related to the Hubble parameter H, the deceleration parameter q and the cosmological constant Λ. Δm_N is related to the 'angular' correction of magnitude m for distant objects. Δm_z is the luminosity correction.

For models with $\Lambda = 0$ the dependence of apparent magnitude m on H_0, q_0 and z was given by Mattig (1958, *Fig.* 26).

A recent revival of the idea of measuring the Riemann curvature k/R^2 in a universe with $\Lambda = 0$ or $\neq 0$ is due to Ehlers and Rindler (1987). Through a combination of $m(z)$ and $N(m)$ measurements, the numerical values of the cosmological parameters can be determined, as is apparent from Heckmann's equations.

---

[7]Series expansions were used at least since 1931 (Lemaître) to relate observable quantities and cosmological parameters. Heckmann's presentation is particularly transparent because H and q appear explicitly.

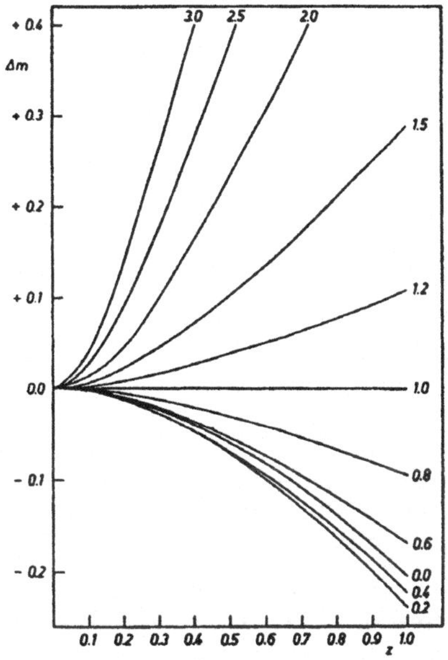

Fig. 26. Mattig's (1958) presentation of magnitude correction Δm as function of z and q_0.

3.2.2 Horizons

Schwarzschild (1916) derived a special solution to the original Einstein equations. The Schwarzschild metric has the line element

$$ds^2 = -\gamma^{-1}dr^2 - r^2 d\theta^2 - r^2 \sin^2\theta d\phi^2 + \gamma dt^2$$

with $\gamma = 1 - \frac{2m}{r}$.

Eddington (1924) showed that the introduction of the cosmological constant led to an extended factor

$$\gamma = 1 - \frac{2m}{r} - \frac{1}{3}\lambda r^2$$

in the line element and thus to a second horizon, the 'world horizon'.

Different usage of the term horizon made standardization desirable. It was provided by Rindler (1956).

> "We shall define a horizon as *a frontier between things observable and things unobservable.* (The vague term *thing* is here used deliberately). There are then two quite different horizon concepts in cosmology which satisfy our definition and to which cosmologists have at different times devoted their attention. The first, which I shall call an *event-horizon*, is exemplified by the de Sitter model-universe. It may be defined as follows: *An event-horizon, for a given fundamental observer A, is a (hyper-) surface in space-time which divides all events into two non-empty classes: those that have been, are or will be observable by A, and those that are forever outside A's possible powers of observation ...*

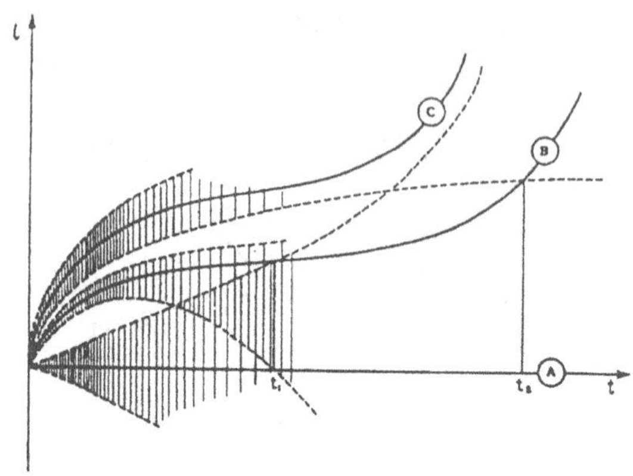

Fig. 27. Light-paths in a model-universe (similar to a Friedmann-Lemaître model) possessing both a particle horizon and an event-horizon (Rindler 1956):

"The origin-observer is denoted by A. B is an observer on a typical particle which becomes visible to A at creation-time t_1 (when A and B enter each other's creation-light-cones) and which passes beyond A's event-horizon at time t_2, so that events at B after t_2 are outside A's possible powers of observation. C is the critical particle which becomes visible to A only at $t = \infty$. C's creation-light-track towards A is that of the unique photon which reaches A at $t = \infty$, and which we have already identified with A's event-horizon. And in the same way that A approaches asymptotically the boundary of C's creation-light-cone, so C approaches that of A's creation-light-cone. Evidently all particles beyond C are entirely outside A's cognizance. In the diagram only the positions near the vertices of the creation-light-cones have been shaded in. We may note that the existence of a critical particle with properties analogous to those of C above, one on each line of vision of each fundamental observer, is a general feature of all models possessing both types of horizon."

The other type of horizon, which I shall call a *particle-horizon*[8], is exemplified by the Einstein-de Sitter model-universe. It may be defined as follows: *A particle-horizon, for any given fundamental observer A and cosmic instant t_0 is a surface in the instantaneous 3-space $t = t_0$, which divides all fundamental particles into two non-empty classes: Those that have already been observable by A at time t_0 and those that have not.*"

Fig. 27 gives an illustration of horizons.

3.2.3 The current matter density of the universe ρ_0 and its normalized value Ω_0

The matter density in space is of particular interest in relativistic universes, because it is closely related to the scale factor.

With the basic family of models introduced above (general relativity, Friedmann-Robertson-Walker metric) and the assumption $\Lambda = 0$, and a presently matter domi-

[8] "It will be understood that whenever we speak of particles in this context we always mean fundamental particles, *i.e.* the representations of the nebulae in the world-model."

nated universe ($p_0 = 0$), ρ_0 relates in simple ways scale factor, deceleration parameter and Riemannian space curvature.

$$H_0^2\, q_0 = \frac{4\pi G \rho_0}{3}$$

$$H_0^2\,(2q_0 - 1) = \frac{kc^2}{R_0^2}$$

The value of k is either $+1, 0, -1$ according to whether

$$\rho_0 \; \begin{matrix} > \\ = \\ < \end{matrix} \; \frac{3H_0^2}{8\pi G}$$

It was from this formula, with $k = 0$, using the then accepted Hubble constant of $500\,\mathrm{km\,sec^{-1}\,Mpc^{-1}}$ that Einstein and de Sitter derived the value

$$\rho_0 = 4 \cdot 10^{-28}\,\mathrm{g\,cm^{-3}}$$

in their famous paper of 1932.

Hubble (1934) determined the density of luminous matter in space from his counts of 44 000 nebulae with

$$\log \mathrm{Mass} = \log b + 0.4\,(5.7 - M)$$

with b = mass/luminosity ratio
M = absolute magnitude.

$\log b$ is estimated from nearby systems to be of order unity, M is estimated as -13.8 to - 14.5. The result is given as

$$\log \rho_0 = -29.8 \quad \text{or} \quad -29.9$$

with an uncertainty probably less than 0.5 (units of ρ_0 are $\mathrm{g\,cm^{-3}}$):

> "The discussion, of course, ignores the existence of internebular matter, the density of which, even in an optimal form, might be several thousand times greater without introducing appreciable absorption. Since absorption depends upon the state of material (the density for large meteorites, for instance, might surpass that of the galactic system without introducing appreciable obscuration), upper limits can be assigned to the density of internebular space only from dynamical considerations."

An earlier value (Hubble 1926) is

$$\rho_0 = 1.5 \cdot 10^{-31}\,\mathrm{g\,cm^{-3}}\;.$$

The matter density ρ_0 is frequently replaced by the dimensionless quantity Ω_0 which measures ρ_0 in terms of the Hubble constant, according to the above formula:

$$\Omega_0 = \rho_0 \cdot \frac{3H_0^2}{8\pi G}\;.$$

For $\Lambda = 0$, $\Omega_0 = 1$, corresponding to the Euclidean universe (*e.g.* Börner 1988).

Ω_0 can be determined directly from dynamical evidence, including the 'cosmic virial theorem', and from the abundance of elements formed in the early universe.

3.2.4 The cosmological constant Λ

The cosmological constant λ was introduced by Einstein (1917). It relates to the *matter-density* ρ and is measured in units time^{-2}. The cosmological constant used by Lemaître (1927) is related to *matter-energy-density* ρc^2 and measured in units length^{-2}, it is the same as the cosmological constant used today.

Einstein wrote:

> "If it would then be certain, that the field equations which I have used so far, are the only ones in agreement with the postulate of the general theory of relativity, we must well conclude, that the theory of relativity does not admit the hypothesis of a spatial closure of the world.
>
> ...We can, however, on the left-hand side of the field equations add the fundamental tensor $g_{\mu\nu}$, multiplied with a presently unknown universal constant λ, without disturbing the general covariance; we put in lieu of the field equations
>
> $$G_{\mu\nu} - \lambda g_{\mu\nu} = \kappa \left(T_{\mu\nu} - \frac{1}{2} g_{\mu\nu} T\right) \ldots$$
>
> The newly introduced universal constant λ thus determines the average distribution density ρ, which can remain in equilibrium, as well as the radius R of spherical space and its volume $2\pi^2 R^3$."

Einstein (1918):

> "The G-field is determined entirely by the masses of the bodies. Since mass and energy are the same sccording to the results of the special theory of relativity and the energy is formally described by the symmetrical energy tensor ($T_{\mu\nu}$) this means, that the G-field be entirely caused and expressed by the energy tensor of matter."

Einstein calls this postulate 'Mach's principle':

> "I have chosen the name Mach's principle because this principle is a generalization of Mach's requirement that inertia must be interpreted as the interaction of bodies.
>
> According to [the originally proposed field equations] a G-field would be possible without any generating matter, contrary to Mach's principle.
>
> But the postulate is fulfilled – according to my present understanding – by field equations which are obtained by adding the λ-term ...
>
> According to this equation a space-time continuum free from singularities with an everywhere disappearing energy tensor of matter seems not possible."

4 The modern universe

4.1 The role of Λ for inflation

A decade later when Einstein and others wanted to drop the cosmological constant, because they no longer considered it necessary in an expanding universe, the *astrophysicist* among cosmologists, Lemaître, offered physical interpretations for Λ. He suggested:

1. the cosmological constant to be related to the driving mechanism of expansion
2. a *physical* process to be responsible for the existence of Λ.

While earlier explanations did not survive time, the concept proposed three years later has won surprising actuality in connection with the *inflationary universe.*
Lemaître (1934):

> "The problem of the universe is essentially an application of the law of gravitation to a region of extremely low density. The mean density of matter up to a distance of some ten millions of light years from us is of the order of $10^{-30}\,\mathrm{g\,cm^{-3}}$; if all the atoms of the stars were equally distributed through space there would be about one atom per cubic yard, or the total energy would be that of an equilibrium radiation at the temperature of liquid hydrogen [12 K]. The theory of relativity points out the possibility of a modification of the law of gravitation under such extreme conditions. It suggests that, when we identify gravitational mass and energy, we have to introduce a constant. Everything happens as though the energy in vacuo would be different from zero. In order that absolute motion, *i.e.*, motion relative to vacuum, may not be detected, we must associate pressure $p = \rho c^2$ to the density of energy ρc^2 of vacuum. This is essentially the meaning of the cosmological constant λ which corresponds to a negative density of vacuum ρ_0 according to
>
> $$\rho_0 = \left(\frac{\lambda c^2}{4\pi G}\right) \approx 10^{-27} g\, cm^{-3} ."$$

From the equations given later in the text:

$$r = R(t) \sin \chi$$

$$\frac{dr}{dt} = r\sqrt{\frac{\lambda c^2}{3}} = r\sqrt{\frac{4\pi}{3} G\rho_0}$$

with χ defined in the usual way as varying between 0 and 2π over the complete map of the universe, he obtains for the location of the observer

$$\frac{\dot{R}}{R} = \sqrt{\frac{4\pi}{3} G\rho_0}$$

with the solution

$$R(t) = \exp\sqrt{\frac{4\pi}{3} G\rho_0}\; t = \exp\sqrt{\frac{\lambda c^2}{3}}\; t\,.$$

With the then accepted value for the Hubble constant, $H_0 = 500\,\mathrm{km\,s^{-1}\,Mpc^{-1}}$, Lemaître obtained the numerical value quoted for the negative vacuum density.

It can also be deduced that Lemaître might have been aware of the fact that Λ can be obtained from observations and that the data available to him give

$$\Lambda = 10^{-54}\; cm^{-2}\,.$$

The connection with the Hubble constant appeared indirectly already in de Sitter's paper (1930, Sect. 3.1.4). The relation

$$R(t) = exp\,(\Lambda\, t)\;.$$

is required for the inflationary universe, where in the current models a rapid *increase* of $R(t)$ occurs because of the large amount of false vacuum energy, proportional to Λ, suddenly becoming available.

4.2 The link to particle physics

In the context of particle physics, a recent paper by Straumann (1987) illustrates the role of Weyl in introducing gauge theory into cosmology. Weyl (1918) discusses the then only known other interaction besides gravity, the electromagnetic interaction, in the frame of cosmology.

> "*A true near-geometry, however, may know only one principle of tranferring a length from one point to an infinitely neighbouring one*, and then one has as little reason to assume that the problem of transferring a length from one point to another at finite distance is integrable, as the problem of transferring directions is found to be integrable. In removing this said inconsequence, one obtains a geometry which, surprisingly, applied to the world, *explains not only the effects of gravity, but also those of the electromagnetic field.* According to the theory thus formulated, both come from the same source, even more, *in general one cannot separate in an arbitrary way gravity and electricity*. In this theory *all physical properties have a world geometrical meaning; in particular, the quantity of action appears from first principles as a pure number. It leads to an essentially uniquely defined World Law; it even permits in a certain sense to understand why the world is four-dimensional...*
> The occuring formulae must correspondingly show a double invariance: 1. they must be *invariant with respect to arbitrary continuous coordinate transformations*, 2. they must remain unchanged, *when the* g_{ik} *are substituted by* λg_{ik}, where λ is an arbitrary continuous function of position. The appearance of a second invariance property is characteristic for our theory...
> ...in the same way as according to investigations by Hilbert, Lorentz, Einstein, Klein and the author *the four conservation laws of matter* (of the energy-momentum-vector) *are connected to the invariance of action*, which contains four arbitrary functions, *against coordinate transformations, is the here newly appearing 'gauge invariancc'* [Maßstab-Invarianz], which introduces an arbitrary fifth function, *connected with the conservation of electricity.*"

5 Basic data

Three directly measurable quantities are available in observational cosmology:

- apparent flux f_ν, integrated flux or apparent magnitude m
  (broadband, narrowband, colours)
- angle θ – position and extent
- redshift z
  (frequency-measurement)
- combinations, such as surface brightness B (flux/solid angle)

Polarization measurements are employed only implicitly through the astrophysical properties of the objects used in cosmological studies. The fifth measureable quantity of astronomy, time – here the age of the universe – is obtained indirectly (Sect. 3.1.3).

5.1 The role of data in cosmology

The cosmological principle permits only *global* actions of uniformly distributed matter in order that a unique scale factor $R(t)$ of the universe may exist (Sect. 3.1.1). Einstein's equations, derived under these assumptions, include general expansion and combined gravitational attraction, *i.e.* coordinated motion and acceleration. According to Lemaître, the cosmological constant is linked to expansion; acceleration is governed by the presence of matter. Einstein's equations, in the form first presented by Friedmann, include only the first and second time derivatives of $R(t)$.

Severe problems arise for the comparision of theory with observational data. The simple theory based on the cosmological principle (Friedmann–Lemaître–Robertson–Walker metric) does not offer an algorithm which provides the general inclusion of local effects; and observers have largely contented themselves with simple corrections to data obtained from comparatively simple structures (such as rich clusters of galaxies). Now, we begin again to realize, how difficult it is, to derive *global* information using data which are (or at least may be) coarsely affected by *local* phenomena on all scales reached so far.

The situation is well summarized by Ehlers (1988):

> "The actual universe is inhomogeneous. In theory, one tries to account for this by adding perturbations to the homogeneous isotropic background and matter variables (density, velocity,...).
> Observational data refer, of course, to the real inhomogeneous universe. In order to derive from these data the global parameters of cosmological models, one has to apply corrections to account for local effects. This procedure is perhaps not yet well understood.
> For a detailed discussion see, *e.g.*, 'Relativistic cosmology: its nature, aims and problems' (Ellis 1984)."

In practice, it requires assumptions to be made of the numerical values of the cosmological parameters, *i.e.* of the world model, which is to be derived from the observations.

Strict recurrence to the data is advocated by Milne (1934):

> "Now what Dr. Hubble, Dr. Shapley and their co-workers actually observe may be described as follows. A certain area on a photographic plate is taken, representing a certain solid angle in the sky, and attention is fixed on a number of small nebulous patches and their spectra. For each patch its Doppler shift s and apparent brightness b are measured, and the patches are counted. Independent of all conventions as to distance, velocity, etc., the observations give firstly the behaviour of b as a function of s: secondly the number of patches $n(s)ds$ with Doppler shifts lying between s and $s+ds$: thirdly, some idea of the stage of evolution of the patch of Doppler shift s – all at a given epoch of observation. Fourthly, in principle, if observations could be carried out through long intervals of time, they would give the variation of s with epoch of observation t for a given nebulous patch, $s = f(s_0, t)$, and the functional dependence of $b(s)$ and $n(s)$ on t, which we may accordingly write as $b(s,t)$ and $n(s,t)$. Every solution of the cosmological problem, every world-model, predicts in principle the smoothed-out values to be expected for $f(s_0, t)$ for a given patch, and the brightness

and distribution functions $b(s,t)$ and $n(s,t)$ for different patches. Two theories differ when their predictions of these functions differ. This method of comparison avoids all reference to distance-assignments, world-geometry, schemes of projection or the like."

5.2 Redshifts and distances

Light and particles (galaxies) move along (different) geodesics. When, in an expanding universe, light travels along a smooth geodesic from the source towards the observer, its frequency changes with the changing scale factor according to $z_{cosm} = \frac{\Delta\lambda}{\lambda} = \frac{\Delta R}{R}$ (Sect. 3.2.1). This is the *global* effect.

Redshift z_{cosm} is related to the *distance* r of the object, given by the invariable fraction of the scale length, and the present scale factor R_0. When distance is *measured* by a distance-dependent object property, such as apparent brightness or angular extent, the deduced *distance value* depends not only on the cosmological parameters: Hubble constant H_0, deceleration parameter q_0 and curvature constant k, but also on the *measuring process*. If the process is brightness measurement the resulting distance is luminosity distance r_M; angular diameters give angular distances r_θ, parallax measurements parallax distances r_p, *etc.* The differences occur because the scale factor R_0 enters differently into the measured quantities.

McCrea (1935, following Tolman, 1930, Walker, 1933, and others) gives an extensive discussion of distance determinations which he introduces:

> "...any specific astronomical measurement of 'distance' ...carried out in any relativity model of space-time must lead to a result which depends on the particular operations of measurement."

For small distances from the observer, z can be approximated by the relativistic or the classical Doppler formula, and the distance r is determined from v/H_0.

Contributions to the *observed* redshift z_{total} result from the warping of the smooth geodesic due to local mass concentrations:

Local effects on the light path contribute z_l. *Local effects*, which can be described as *peculiar motions of the emitting particle* (galaxy) and the *observer* along particle geodesics, contribute z_m. When they are sufficiently small, z and r can again be approximated by the classical or special relativistic Doppler formula. The superposition $z_{total} = z_{cosm} + z_l + z_m$ makes it observationally difficult to separate the cosmological and the local contributions.

5.3 Magnitudes (fluxes)

5.3.1 Effects due to space curvature

The *global* effect is the geometrical dilution of light along the light path (see *Fig. 25*) from light *emission* (scale factor R_1) to light *reception* (scale factor R_0). The global effect depends on z and the cosmological parameters and applies to the *total amount of light*, measured e.g. as bolometric magnitude.

Light is generally measured in limited frequency intervals, *e.g.* the blue wavelength region (B-magnitude). Due to the redshift, different wavelength regions are moved into the frequency range measured, and a *K-correction* has to be applied. The correction is a function of the spectral energy distribution.

Local distortions of the light path may lead to considerable changes of magnitude, *e.g.* enhancement through gravitational lensing.

Local effects of particle motion due to the presence of mass concentrations influence magnitude measurements in the same way as the global effects. This holds for both the motion of the object and the motion of the observer. In the special relativistic approximation, dimming or brightening due to local motion can be described as aberration: the change in the amount of light received because of the concentration of radiation into a cone extending into the direction of particle motion. For peculiar galaxy velocities, which are generally much smaller than c, the effect can be neglected.

A $(1+z)$ correction to the magnitude was introduced by Hubble and Humason (1931):

> "A redshift, by redistributing the radiation to correspond with a lower temperature and hence with a later spectral type, introduces an increment to the photographic magnitude..."

In a paper by Hubble and Tolman (1935) the correct $(1+z)^2$ correction is used:

> "...it has been derived in such a way as to make proper allowance, first, for the double effect of nebular recession in reducing both the individual energy and the rate of arrival of the photons, and then for the further circumstance that a change in spectral distribution of the energy that does arrive will lead to changes in its photographic effectiveness."

5.3.2 Absorption corrections

Frequency dependent (selective) and frequency independent (neutral) absorption – atmospheric absorption corrected for – can be present in:

- our galaxy
- the observed galaxy
- intergalactic space.

Selective absorption in our Galaxy is well observed; it is strongly dependent on galactic latitude. Absorption in galaxies depends on morphological type and galaxy orientation. Intergalactic absorption – general or in localized clouds (Rudnicki 1988) – is still a matter of debate.

Suggestions for absorption go back to the interpretation of the earliest data, summarized in the NGC and Index catalogues. Charlier (1922) comments:

> "A remarkable property of the image is that the nebulae seem to be piled up in clouds (as also the stars in the Milky Way). Such a clouding of the nebulae may be a real phenomenon, but it may also be an accidental effect caused by *dark matter* in the space or declared by condensation of the observations in singular points of the sky."

The effects of galactic absorption were well understood by Hubble (1934), who presented the law of absorption essentially in its presently used form:

"Systematic *variations in longitude* are appreciable only in the lower latitudes, where obscuration appears to be conspicuously greater in the direction of the galactic center than in the opposite direction. There is a definite variation with *latitude*, which from the poles to about $\beta = 15°$ is represented by the *cosecant formula*

$$\log N_m = C - 0.15 \operatorname{cosec} \beta ,$$

indicating an obscuration of $0\overset{\mathrm{m}}{.}5$ from pole to pole, with *no appreciable difference* between the two hemispheres."

Stebbins and Whitford (1937) found evidence of absorption in extragalactic nebulae from their extensive measurements of magnitudes and colours:

"The colors of the first three types E, Sa, Sb indicate an excess..., possibly because of selective absorption within the nebulae. There is no indication of selective absorption in internebular space."

Intergalactic absorption was claimed by Wirtz (1924):

"While all computations of the correlation coefficient for the nebular characteristics, surface brightness and diameter, give very low absolute values, uncertain in each individual case, one finds, however, from different material always the same negative sign, *i.e.* the intensity [surface brightness] increases with increasing apparent diameter. If one takes the apparent diameter as a measure of distance, one can interpret this effect as resulting from a general cosmic absorption. If one positions the nebulae, distant Milky Way systems, at distances which correspond to the parallaxes of K. Lundmark ... one can determine the amount of absorption. One finds it, with the help of the regression line relating magnitude to the logarithm of the large axis, to be extremely small, of the order of 10^{-5} magnitudes per 1000 parsec."

5.3.3 Aperture and orientation corrections

The measurement of total galaxy magnitudes requires that contributions from all parts of a galaxy can be measured. This is not possible when

- parts of the galaxy are not included by the measuring process
- parts of the galaxy are hidden.

Magnitude measurements may require corrections for the limited size of the measuring aperture or uncertain sky correction in the case of apertures which are too large.

The aperture effect is mentioned implicitly by Whitford (1936) in the first report about photoelectric magnitude measurements of galaxies. The more detailed paper by Stebbins and Whitford (1937) lists magnitude measurements made with different apertures. Form their tabulated values the increase in total brightness with increasing aperture size is clearly apparent.

Hidden light due to intervening matter is discussed above. Light is also hidden as a function of orientation: inclinations of the plane of the galaxy relative to the plane of the sky near 90° require the largest corrections. Regions of low surface brightness may be lost in the sky background.

Hubble (1932) presented an interesting effect which is related to the above phenomenon: objects of low surface brightness are lost in the background as a function of surface brightness (*Fig. 28*).

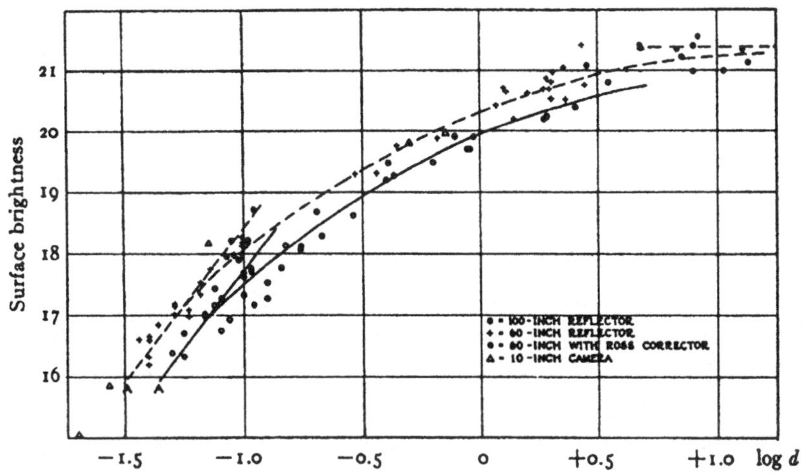

Fig. 28. "Relation between surface brightness and diameter of threshold images." (Hubble 1932).

5.4 Angles

5.4.1 Effects due to space curvature

Globally, the angular extent of an object seen at a given distance r depends on redshift z and on the cosmological parameters.

Local effects on the light path may become apparent in gravitational lensing, where the angular sizes can grow up to 2π. Other local effects can generally be neglected.

5.4.2 Corrections

The corrections mentioned in Sect. 5.3 also apply to angular measurements, where aperture effects may be present and central absorption discs at large inclinations may mask the outer parts of the galaxies.

5.5 Combined measurements

When angles, defined by isophotes, are measured, or when surface brightness, defined by the light contained in a given solid angle, is determined, all the effects described above contribute jointly to the measured value.

5.6 Selection

Selection effects arise because the observer sees different populations at different distances. With increasing distance only the absolutely brighter and larger galaxies are registered. The 'Scott' effect (Scott 1957, 1962) states the same phenomenon with respect to cluster properties: at larger distances only rich and highly concentrated clusters appear above the sky background (*Fig. 29*). The combined effects of magnitude or diameter *limits* and the *distribution functions* of magnitude or diameter result in *selection functions.*

Selection effects are critical for sampling, especially when large population gradients exist across the sampling volume and when the measuring errors are large. The effect is referred to as Malmquist bias (1920). Malmquist introduced it originally in the context of stellar statistics, where he assumed the error distribution to be Gaussian. The Malmquist correction is used to compensate for the effect.

All correlations derived for magnitude- and diameter-limited samples suffer from severe Malmquist biases. Volume-limited samples are less influenced by population gradients. Corrections for the effects of error distribution must always be applied.

Other selection effects are introduced by the measuring procedures. They result from the fact that most methods are more sensitive to certain parameter values than to others. An example is the preferential discovery of Ly α-quasars (*e.g.* Gericke 1988) which is a combined effect of favourable wavelength region and emission characteristics of the line.

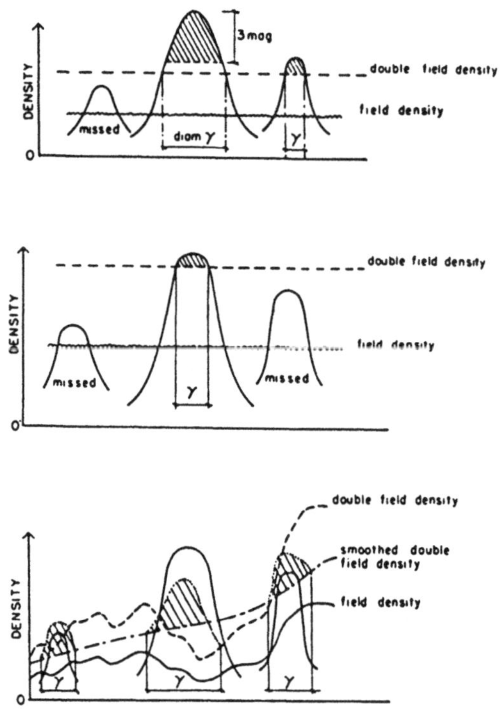

Fig. 29. Visibility of clusters (Scott 1962).

"Effects of varying cluster richness, varying compactness, and varying field density on the delineation of the cluster contour, on the estimated diameter and on the estimated number of cluster galaxies."

5.7 Evolutionary effects

Physical evolution (Sect. 2.3.3) may require magnitude, colour, and angular corrections. Examples are the apparently larger numbers of blue galaxies in distant clusters (Butcher and Oemler 1984) and light, colour, and angular changes in merging (starburst) galaxies.

6 Results expected from the data

A brief history of the connection between *observation* and *theory* of the "particles" in the universe was given in Sect. 2.3. The final comment here concerns the observational tests for the theory discussed in Sects. 3 and 4.

McCrea (1935) summarized:

> "This paper has sought to make clear what particular set of assumptions are being tested when comparison is made with any particular set of observational results. The *broad* conclusions to which we are led are these:
>
> If we compare a relation like that connecting apparent size, apparent brightness, and redshift we are testing merely the correctness of our interpretation of the observed quantities, and not any particular theory of them.
>
> If we make all possible observations on the "distances", red-shifts, and numbers of extra-galactic nebulae we are testing the possibility of representing them as fundamental particles in a universe of the type (1) [with Friedmann-Lemaître-Robertson-Walker metric], and the correctness of the derivation of 'world-pictures' in such a model.

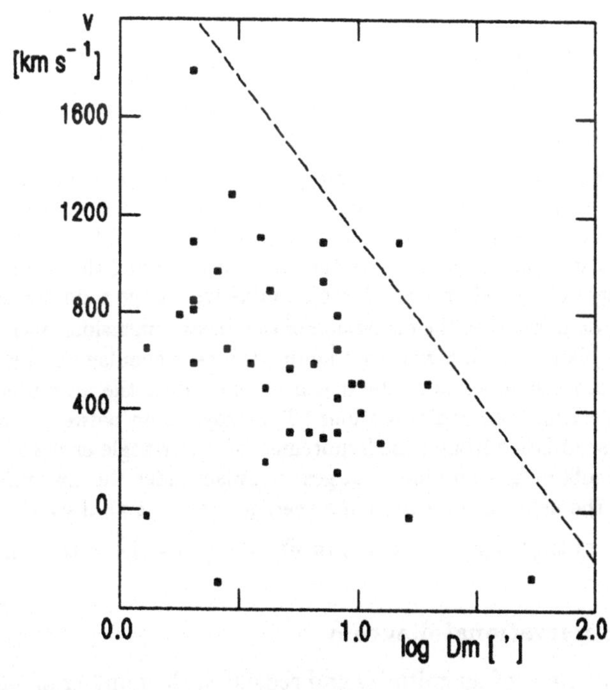

Fig. 30. Reconstruction of Wirtz's original diagram.

Table 1: Parameter values for Fig. 31.

Curve	H_0	q_0	Λ	M	K
solid	100	0.5	0	−22	**2.0 z**
solid	100	0.5	0	−22	**4.0 z**
solid	**100**	0.5	0	−21	$4.14\,z - 0.44\,z^2$
solid	**50**	0.5	0	−21	$4.14\,z - 0.44\,z^2$
dotted	100	**0.1**	0	−21	$4.14\,z - 0.44\,z^2$
dotted	100	**1.0**	0	−21	$4.14\,z - 0.44\,z^2$
dashed	75	0.5	0	**−21.5**	$4.14\,z - 0.44\,z^2$
dashed	75	0.5	0	**−19.5**	$4.14\,z - 0.44\,z^2$
dashed	75	0.5	0	**−18.5**	$4.14\,z - 0.44\,z^2$

In order however to choose between such models and classical ones it will in general be necesssary to test... terms in k in expressions involving the numbers of nebulae ...".

Thus the connection is made to the most recent suggestion concerning k/R^2 measurements (Sect. 3.2.1).

One diagram of each type of relation mentioned by McCrea will be shown in the following.

6.1 The $\log\theta(z)$-diagram

The first diagram used in observational cosmology appears to be the $\log\theta(z)$-diagram. In ist original version, it was not displayed, but is described in detail by Wirtz (1924):

"When the different quantities v [velocity] versus lg Dm [diameter] are represented graphically, the diagram shows a V-shaped or triangular form from which one deduces the following facts: for the apparently small nebulae one finds smallest and largest v, the apparently largest nebulae have the smallest v, among the nebulae with small v one finds large and small objects, large nebulae with large v do not exist.

From this one deduces, that the dispersion of the linear dimensions of the nebulae fills the triangular plane in such a way, that among the near nebulae absolutely small and large objects are visible while in the depth of space only the absolutely largest are subject to observing their radial motions. The progression between v and apparent Dm is reproduced best through the hypotenuse of the triangle enclosing all observed points, which follows the absolutely largest nebulae under the assumption that the giants among the nebulae have the same average extent at all distances."

A reconstruction of the diagram is given in *Fig. 30*, using the data from papers quoted by Wirtz.

6.2 A basic observational diagram

Measurements are made of magnitudes and redshifts; the number of galaxies per magnitude and redshift interval is counted. The three-dimensional version of the diagram

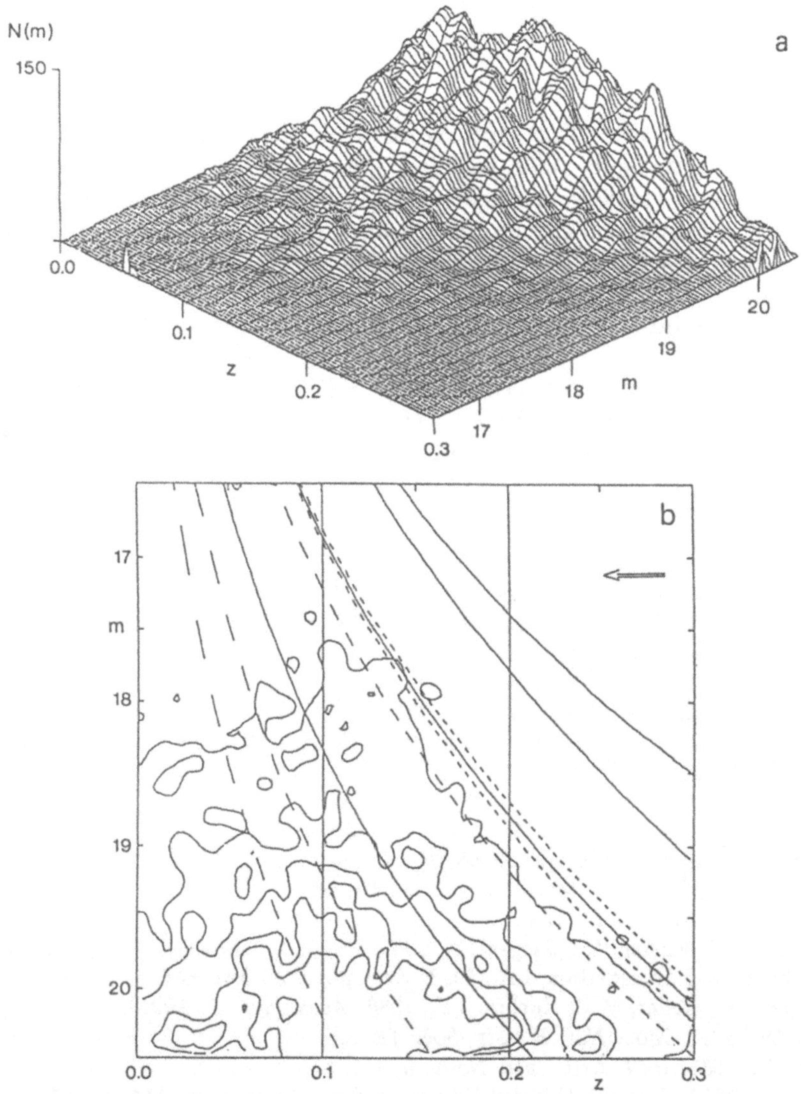

Fig. 31. Basic observational diagram: $m, z, N(m, z)$ (Seitter *et al.* 1988).
Data from ESO/SRC field No. 411. Going along the direction of the arrow (upper right of *Fig. 31b*), groups of two or three curves are reached in the order given in Table 1. Within each group one parameter (given in bold face in Table 1) is varied; all parameters are listed. Isopleths are drawn at levels $N = 50, 100, 200, 400, 800$ per intervals of 0.5 magnitudes and 0.03 z-values.

The two vertical lines are the projections of curves following the $N(m)$-'landscape'. They correspond to observational luminosity functions before normalization.

H_0 is given in $\mathrm{km\,s^{-1}\,Mpc^{-1}}$.

is shown in *Fig. 31a*. The bias towards certain redshift-magnitude combinations is clearly visible. A two-dimensional projection is shown in *Fig. 31b*. Various computed curves are inserted in the diagram; their parameters are given in Table 1.

7 Conclusions

The introduction to these proceedings is written in the form of historical annotations to the three topics: *large scales*, including the largest one conceivable: the scale of the universe; *large numbers*, in a field which – on the observational side – has to rely on extensive accumulation of data; *large efforts*, traced as a tribute to the minds to whom we owe many of the thoughts and methods used in our present work.

Acknowledgements

Sincere thanks go to G. Börner and J. Ehlers, MPI Astrophysik Garching, for critical comments on parts of the manuscript. The active support of the MRSP group and of H.E. Nager of the Universitätsbibliothek der WWU Münster is gratefully acknowledged.

References

Abell, G., 1958. *Astrophys. J. Suppl.*, **3**, 211.
Abell, G., 1962. In *IAU Symp. 15, Problems of Extra-Galactic Research*, ed. McVittie, G.C., MacMillan, New York, p. 213.
Abell, G., 1965. *Ann. Rev. Astr. Astrophys.*, **3**, 1.
Alpher, R.A., Herman, R., 1948. *Nature*, **162**, 774.
Baum, W.A., 1962. In *IAU Symp. 15, Problems of Extra-Galactic Research*, ed. McVittie, G.C., MacMillan, New York, p. 390.
Börner, G., 1988. *The Early Universe*, Springer, Berlin.
Butcher, H.R., Oemler, A., 1984. *Astrophys. J.*, **285**, 426.
Carpenter, E.F., 1938. *Astrophys. J.*, **88**, 344.
Charlier, C.V.L., 1922. *Ark. Math. Astr. Fys.*, **16**, No. 22.
Chincarini, G., 1978. *Nature*, **272**, 515.
da Costa, L.N., Pellegrini, P.S., Sargent, W.L.W., Tonry, J., Davis, M., Meiksin, A., Latham, D.W., Menzies, J.W., Poulson, I.A., 1988. *Astrophys. J.*, **327**, 544.
de Lapparent V., Geller, M.J., Huchra, J.P., 1986. *Astrophys. J.*, **302**, L1
de Sitter, W., 1917. *Mon. Not. R. astr. Soc.*, **78**, 3.
de Sitter, W., 1930. *Bull. Astr. Inst. Neth.*, **5**, 157.
de Sitter, W., 1932. *Kosmos*, Harvard University Press, Cambridge, MA, p. 103.
de Vaucouleurs, G. 1953. *Astr. J.*, **58**, 30.
Dreyer, J.L.E., 1888. *A New General Catalogue of Nebulae and Clusters of Stars, Mem. R. astr. Soc.*, **49**, 1.
Dreyer, J.L.E., 1895. *Index Catalogue of Nebulae, Mem. R. astr. Soc.*, **51**, 185.
Dreyer, J.L.E., 1908. *Mem. R. astr. Soc.*, **59**, 105.
Eddington, A.S., 1924. *The Mathematical Theory of Relativity*, Cambridge University Press, Cambridge.
Edge, D.O., Shakeshaft, J.R., McAdam, W.B., Baldwin, J.E., Archer, S., 1959. *Mem. R. astr. Soc.*, **68**, 37.
Ehlers, J., 1988. Private communication.

Ehlers, J., Rindler, W., 1987. *Astr. Astrophys.*, **174**, 1.

Einasto, J., Corwin Jr., H.G., Huchra, J., Miller, R.H., Tarenghi, M., 1982. In *Highlights of Astronomy Vol. 6*, ed. West, R.M., p. 757.

Einstein, A., 1916. *Ann. Phys.*, **49**, 769.

Einstein, A., 1917. *Sitz.-Ber. Kgl. Preuß. Akad. Wiss., phys.-math. Kl.*, 1917, p. 142.

Einstein, A., 1918. *Ann. Phys.*, **53**, 130.

Einstein, A., 1931. *Sitz.-Ber. Kgl. Preuß. Akad. Wiss., phys.-math. Kl.*, 1931, p. 235.

Einstein, A., 1936. *Science*, **84**, 506.

Einstein, A., de Sitter, W., 1932. *Proc. Nat. Acad. Sci. Washington*, **18**, 213.

Ellis, G.F.R., 1984. In *General Relativity and Gravitation, 10th International Conference*, eds. Bertotti, B., de Felice, F., Pascolini, A., Reidel, Dordrecht, p. 215.

Fall, S.M., 1979. *Rev. Mod. Phys.*, **51**, 21.

Fath, E.A., 1914. *Astr. J.*, **28**, 75.

Flin, P., 1988. *Acta Cosmologica*, **15**, 25 and 43.

Ford, H.C., Harms, R.J., Ciardullo, R., Bartko, F., 1981. *Astrophys. J.*, **238**, L113.

Forman, W., Janes, C., Cominsky, L., Julien, P., Murray, S., Peters, G., Tananbaum, H., Giacconi, R., 1978. *Astrophys. J. Suppl.*, **38**, 357.

Friedmann, A. 1922. *Z. Phys.*, **10**, 377.

Friedmann, A. 1924. *Z. Phys.*, **12**, 326.

Gamow, G., 1946. *Phys. Rev.*, **70**, 572.

Gamow, G., 1948. *Nature*, **162**, 680.

Geller, M.J., Huchra, J.P., de Lapparent, V., 1987. In *IAU Symp. 124, Observational Cosmology*, eds. Hewitt, A., Burbidge, G., Fang, L.Z., Reidel, Dordrecht, p. 301.

Gericke, V., 1988. *These proceedings*, p. 235.

Gott III, J.R., Melott, A., Dickinson, M., 1986. *Astrophys. J.*, **306**, 341.

Groth, E.J., Peebles, P.J.E., 1977. *Astrophys. J.*, **217**, 385.

Harrison, E.R. 1981. *Cosmology*, Cambridge University Press, Cambridge.

Heckmann, O. 1931. *Veröff. Univ. Sternw. Göttingen*, Heft 17.

Heckmann, O. 1932. *Veröff. Univ. Sternw. Göttingen*, Heft 23.

Heckmann, O. 1942. *Theorien der Kosmologie*, Springer, Berlin; 2nd ed., 1968.

Herschel, J.F.W., 1864. *Catalogue of Nebulae and Clusters of Stars, Phil. Trans.*, **154**, 1.

Herschel, W., 1811. *Phil. Trans.*, **101**, 269.

Holmberg, E., 1940. *Astrophys. J.*, **92**, 200.

Holmberg, E., 1941. *Astrophys. J.*, **94**, 385.

Hoyle, F., Sandage, A. 1956. *Publ. astr. Soc. Pacific*, **68**, 301.

Hubble. E., 1926. *Astrophys. J.*, **64**, 362.

Hubble, E., 1932. *Astrophys. J.*, **76**, 106.

Hubble, E., 1934. *Astrophys. J.*, **79**, 8.

Hubble, E., 1936. *Astrophys. J.*, **84**, 270.

Hubble, E., Humason, M.L., 1931. *Astrophys. J.*, **74**, 43.

Hubble, E., Tolman, R.C., 1935. *Astrophys. J.*, **82**, 302.

Huchra, J.P., 1988. In *IAU Symp. 130, The Structure of the Universe*, eds. Audouze J., Szalay, A., Kluwer, Dordrecht, in press.

Huchra, J., Davis, M., Latham, D., Tonry, J., 1983. *Astrophys. J. Suppl.*, **52**, 39.

Humason, M.L., Mayall, N.U., Sandage, A.R., 1956. *Astr. J.*, **61**, 97.

Icke, V., 1988. These proceedings, p. 259.

Icke, V., van de Weygaert, R., 1987. *Astr. Astrophys.*, **184**, 16.

Kaiser, N., 1987. *Mon. Not. R. astr. Soc.*, **227**, 1.

Kirshner, R.P., Oemler, Jr., A., Schechter P.L., Shectman S.A., 1981. *Astrophys. J.*, **248**, 157.

Koo, D.C., 1985. *Astr. J.*, **90**, 418.

Kron, R.G., 1988. In *3rd ESO/CERN Symp. on Cosmology, Astronomy and Fundamental Physics*, Kluwer, Dordrecht (in press).

Lambert, J.H., 1761. *Cosmologische Briefe über die Einrichtung des Weltbaues*, E. Kletts Wwe., Augspurg; engl. edit. and transl. by Jaki, S.L., 1976, Scottish Academic Press, Edinburgh.

Lemaître, G., 1927. *Ann. Soc. Sci. Bruxelles*, **47**, Ser. A, 103; engl. transl.: 1931, *Mon. Not. R. astr. Soc.*, **91**, 483.

Lemaître, G., 1931a. *Nature*, **128**, 704.

Lemaître, G., 1931b. *Mon. Not. R. astr. Soc.*, **91**, 490.

Lemaître, G., 1934. *Proc. Nat. Acad. Sci. Washington*, **20**, 12.

Limber, D.N., 1953. *Astrophys. J.*, **117**, 134

Limber, D.N., 1954. *Astrophys. J.*, **119**, 665.

Lockyer, J.N., 1900. *Inorganic Evolution*, MacMillan, London.

Loh, E.D., 1988a. In *Towards Understanding Galaxies at High Redshift*, eds. Kron, R.G., Renzini, A., Kluwer, Dordrecht.

Loh, E.D., 1988b. *These proceedings*, p. 267.

Lundmark, K., 1925. *Mon. Not. R. astr. Soc*, **85**, 865.

Lundmark, K., 1927. *Nova Acta R. Soc. Sci. Upsal. = Medd. Astr. Obs. Upsala*, **No. 30**

Malmquist, K.G., 1920. *Medd. Lund Obs.*, Ser. 2, No. 22.

Mandelbrot, B., 1975. *C. R. Acad. Sci. Paris, Ser. A*, **280**, 1551.

Mattig, W., 1958. *Astr. Nachr.*, **284**, 109.

McCrea, W.H., 1935. *Z. Astrophys.*, **9**, 290.

Mills, B.Y., 1952. *Austr. J. Sci. Res., Ser. A.*, **5**, 266.

Milne, E.A., 1933. *Z. Astrophys.*, **6**, 1.

Milne, E.A., 1934. *Observatory*, **56**, 24.

Neyman, J., Scott, E.L., 1952. *Astrophys. J.*, **116**, 144.

Neyman, J., Scott, E.L., Shane, C.D., 1953. *Astrophys. J.*, **117**, 92.

Neyman, J., Scott, E.L., Shane, C.D., 1956. In *Proc. 3rd Berkeley Symp. on Math. Statistics and Probability*, **III**, Univ. California Press, Berkeley, p. 75.

Nilson, P., 1973. *Uppsala General Catalogue of Galaxies, Nova Acta R. Soc. Sci. Ups. Ser. V:A*, **I** = *Uppsala Astr. Obs. Ann.*, **6**.

Öpik, E., 1922. *Astrophys. J.*, **55**, 406.

Olowin, R.P., 1987. In *IAU Symp. 124, Observational Cosmology*, eds. Hewitt, A., Burbidge, G., Fang, L.Z., Reidel, Dordrecht, p. 1.

Oort, J.H., 1983. *Ann. Rev. Astr. Astrophys.*, **21**, 373.

Ostriker, J.P., Cowie, L.L., 1981. *Astrophys. J.*, **243**, L127.

Ott, H.-A., 1988. *These proceedings*, p. 125

Palumbo, G., Tanzella-Nitti, G., Vettolani, G. 1983. In *IAU Symp. No. 104, Early Evolution of the Universe and Its Present Structure*, eds. Abell, G.O., Chincarini. G., Reidel, Dordrecht, p. 177.

Partridge, R.B., 1988. In *3rd ESO/CERN Symp. on Cosmology, Astronomy and Fundamental Physics*, Kluwer, Dordrecht (in press).

Pawsey, J.L., 1957. In *IAU Symp. No. 4, Radio Astronomy*, ed. van de Hulst, H.C., Cambridge Univ. Press, Cambridge, p. 228.

Pease, F.G., 1916. *Proc. Nat. Acad. Sci. Washington*, **2**, 517.

Pease, F.G., 1918. *Proc. Nat. Acad. Sci. Washington*, **4**, 21.

Peebles, P.J.E, 1971. *Physical Cosmology*, Princeton Univ. Press, Princeton.
Peebles, P.J.E., 1980. *The Large-Scale Structure of the Universe*, Princeton Univ. Press, Princeton.
Penzias, A.A., Wilson, R.W., 1965. *Astrophys. J.*, **142**, 419.
Pinkau, K., 1979. *Nature*, **277**, 17.
Reinmuth, K. 1916. *Veröff. Sternw. Heidelberg-Königstuhl*, **7**, No. 8, 175.
Reinmuth, K., 1940. *Veröff. Sternw. Heidelberg-Königstuhl*, **12**, No. 6, 27.
Rindler, W., 1956. *Mon. Not. R. astr. Soc.*, **116**, 662.
Robertson, H.P., 1929. *Proc. Nat. Acad. Sci. Washington*, **15**, 822.
Robertson, H.P., 1955. *Publ. astr. Soc. Pacific*, **67**, 82.
Rood, H., 1987. *Publ. astr. Soc. Pacific*, **99**, 921.
Rood, H., 1988. *Ann. Rev. Astr. Astrophys.*, in press.
Rubin, V., 1951. *Astr. J.*, **56**, 47.
Rudnicki, K., 1988. *These proceedings*, p. 225
Ryle, M., Smith, F.G., Elsmore, B., 1950. *Mon. Not. R. astr. Soc.*, **110**, 508.
Sandage, A.R., 1987. In *IAU Symp. 124, Observational Cosmology*, eds. Hewitt, A., Burbidge, G., Fang, L.Z., Reidel, Dordrecht, p. 1.
Sandage, A.R., Binggeli, B., Tammann, G.A., 1985. *Astr. J.*, **90**, 1759.
Schechter, P., 1976. *Astrophys. J.*, **203**, 297.
Schuecker, P., 1988. *These proceedings*, p. 142.
Schwarzschild, K., 1900. *Vierteljahrsschr. Astr. Ges.*, **35**, 337.
Schwarzschild, K., 1916. *Sitz.-Ber. Kgl. Preuß. Akad. Wiss., phys.-math. Kl.*, 1916, p. 189.
Scott, E.L., 1957. *Astr. J.*, **62**, 248.
Scott, E.L., 1962. In *IAU Symp. 15, Problems of Extra-Galactic Research*, ed. McVittie, G.C., MacMillan, New York, p. 269.
Seares, F.H., 1925. *Astrophys. J.*, **62**, 168.
Seitter, W.C., 1986. *Mitt. Astr. Ges.*, **67**, 401.
Seitter, W.C., Schuecker, P., Duemmler, R., 1988. In preparation.
Seldner, M., Siebers, B., Groth, E.J., Peebles, P.J.E., 1977. *Astr. J.*, **82**, 249.
Sellwood, J.A., 1987. *Ann. Rev. Astr. Astrophys.*, **25**, 151.
Shakeshaft, J.R., Ryle, M., Baldwin, J.E., Elsmore, B., Thomson, J.H., 1955. *Mem. R. astr. Soc.*, **67**, 106.
Shane, C.D., Wirtanen, C.A., 1950. *Proc. American Phil. Soc.*, **94**, 13.
Shane, C.D., Wirtanen, C.A., 1954. *Astr. J.*, **59**, 285.
Shane, C.D., Wirtanen, C.A., 1967. *Publ. Lick Obs.*, **22**, Pt. 1.
Shapley, H., 1934. *Mon. Not. R. astr. Soc.*, **94**, 791.
Shapley, H., 1938. *Proc. Nat. Acad. Sci. Washington*, **24**, 282 and 527.
Shapley, H. 1957. *The inner Metagalaxy*, Yale Univ. Press, New Haven.
Silk, J., 1970. *Space Sci. Rev.*, **11**, 671.
Slipher, V.M., 1913. *Lowell Obs. Bull.*, **2**, No. 58, 56.
Slipher, V.M., 1914. *Lowell Obs. Bull.*, **2**, No. 62, 65.
Smith, S. 1936. *Astrophys. J.*, **83**, 23.
Soucail, G., Mellier, Y., Fort, B., Hammer, F., Mathez, G., 1987. *Astr. Astrophys.*, **184**, L7.
Spitzer, L., Baade, W., 1951. *Astrophys. J.*, **113**, 413.
Stebbins, J., Whitford, A.E., 1937. *Astrophys. J.*, **86**, 247.
Straumann, N., 1987. *Phys. Bl.*, **43**, 414.
Tinsley, B.M., 1968. *Astrophys. J.*, **151**, 547, and unpublished Ph. D. thesis, Univ. of Texas, 1967.

Tolman, R.C., 1929. *Proc. Nat. Acad. Sci. Washington*, **15**, No. 4, 297. and *Astrophys. J.*, **69**, 245.
Tolman, R.C., 1930. *Proc. Nat. Acad. Sci. Washington*, **16**, 511.
Toomre, A., Toomre, J., 1972. *Astrophys. J.*, **178**, 623.
Totsuji, H., Kihara, T., 1969. *Publ. Astr. Soc. Japan*, **21**, 438.
Tully, R.B., 1987. *Astrophys. J.*, **323**, 1.
Walker, A.G., 1933. *Mon. Not. R. astr. Soc.*, **9**, 159.
Walker, A.G., 1936. *Proc. London Math. Soc. 2nd Ser.*, **42**, 90.
Weizsäcker, C.F. von, 1938. *Phys. Zs.*, **39**, 633.
Weyl, H., 1918. *Sitz.-Ber. Kgl. Preuß. Akad. Wiss., phys.-math. Kl.*, 1918, p. 465.
Whitford, A.E., 1936. *Astrophys. J.*, **83**, 424.
Whittaker, E.T., 1931. *Proc. R. Soc. London, Ser. A*, **133**, 93.
Wirtz, C., 1918. *Astr. Nachr.*, **206**, 109.
Wirtz, C., 1922. *Astr. Nachr.*, **215**, 349.
Wirtz, C., 1924. *Astr. Nachr.*, **222**, 21.
Wirtz, C., 1926. *Astr. Nachr.*, **228**, 40.
Wolf, M., 1901. *Publ. astrophys. Obs. Königstuhl-Heidelberg*, **1**, 11.
Wolf, M., 1914. *Vierteljahrsschr. Astr. Ges.*, **49**, 162.
Wolf, M., 1916. *Veröff. Sternw. Heidelberg-Königstuhl*, **7**, No. 7, 169.
Wright, Th., 1750. *An Original Theory or New Hypothesis of the Universe*, London (repr. 1971, intr. by Hoskin, M.A., Macdonald, London).
Zwicky, F., 1933. *Helv. Phys. Acta*, **6**, 110.
Zwicky, F., 1937a. *Phys. Rev.*, **51**, 290.
Zwicky, F., 1937b. *Phys. Rev.*, **51**, 679.
Zwicky, F., 1937c. *Astrophys. J.*, **86**, 217.
Zwicky, F., 1938. *Publ. Astr. Soc. Pacific*, **50**, 218.
Zwicky, F., 1957. *Morphological Astronomy*, Springer, Berlin.
Zwicky, F., 1959. In *Handbuch der Physik*, **LIII**, ed. Flügge, S., Springer, Berlin.
Zwicky, F., Herzog, E., Wild, P. Karpowicz, M., Kowal, C., 1961-1968. *Catalog of Galaxies and Clusters of Galaxies*, **I-VI**, California Institute of Technology, Pasadena.

Part I

Two-Dimensional Distribution of Galaxies

Heinz Horstmann

Neil Heydon–Dumbleton

Joseph Hollósi

Steve Maddox

The Edinburgh/Durham Galaxy Survey

N.H. Heydon–Dumbleton and **C.A. Collins**
University of Edinburgh, Dept. of Astronomy
Blackford Hill, Edinburgh, Scotland, UK

H.T. MacGillivray
Royal Observatory, Blackford Hill
Edinburgh, Scotland, UK

Abstract

We are presently producing the Edinburgh/Durham galaxy catalogue, based on digitized scans of UK Schmidt plates by the COSMOS measuring machine. Our initial catalogue covers some 60 plates and contains approximately 10^6 galaxies to $b_J = 20$. For each galaxy we store 27 image parameters, including image magnitude, RA, DEC, area, and position angle, enabling the database to be used for a wide range of astronomical applications. Here we discuss our data reduction techniques for both image classification and image photometry. Preliminary results are presented, including maps of galaxy distribution, together with number-magnitude counts and the galaxy–galaxy correlation function.

1 Introduction

This paper describes the construction of the Edinburgh/Durham Southern Sky Galaxy Survey. The aim of this project is to produce a homogeneous galaxy catalogue of a large part of the southern sky ($\simeq 1$ steradian) to a limiting magnitude of $b_j = 20$. The raw data for this work are COSMOS scans of glass copies of SERC J Survey plates. Previous surveys of the distribution of galaxies, carried out 'by eye', have been plagued by systematic variations in completeness and limiting magnitude. The advantage of using a microdensitometer such as COSMOS, is not only the speed at which it can detect and store image data, but also how, by experimentation, theory, and careful data reduction, it is possible to quantify and reduce such errors. This survey will therefore allow us to make an unparalleled study of the large-scale galaxy distribution, as well as providing a database that will be available for use in a wide range of astronomical applications.

This paper details our work in four sections. First we discuss the motivation for a machine based galaxy catalogue and the limits this sets on the data quality. Secondly we present brief details of the COSMOS machine and the plate material we use. The mechanics of the catalogue production are detailed in Sect. 4. These include a summary of the problems we have encountered, the data reduction techniques employed and the results of the quality assurance tests we have applied. Finally we present some preliminary results from analysis of the catalogue data. These are presented to indicate the type of work that can be carried out using this catalogue and will be discussed in detail elsewhere.

2 Motivation

There have been several previous surveys of the large-scale distribution of galaxies. Two notable examples are the Shane and Wirtanen counts (1967), hereafter SW, and the Abell cluster catalogue (1958). These surveys have been used extensively to study the distribution and formation of galaxies in the north celestial hemisphere. Our data provides the basis for comparative studies in the southern hemisphere.

Recent work has cast doubt on the results from these visually selected or 'eyeball' surveys because of their heavily subjective nature. In particular both the break in the two point galaxy angular correlation function, $w_{gg}(\theta)$ at $10h^{-1}$Mpc (Groth and Peebles 1977) and the filamentary appearance of the galaxy sky distribution, may be the result of plate boundaries and observer bias (Geller *et al.* 1984; de Lapparent *et al.* 1986; Postman *et al.* 1986). Variations in counting efficiency or survey limiting magnitude introduce spurious galaxy number density gradients across plates or discontinuities at plate boundaries. For a new survey to address the reality of these features and investigate the general properties of the large-scale distribution of galaxies it is vital to reduce such systematic errors. Geller *et al.* (1984) estimate that for surveys of comparable depth to the Lick survey, systematic errors in photometric calibration must be kept within $0\overset{m}{.}05$ *rms*.

Technology is now available to repeat these 'eyeball' surveys using computer based image detection and classification. Plate measuring machines such as COSMOS are able to detect and parameterise all images on a photographic plate in a matter of hours. Computer analysis of these image parameters enables the reduction and quantification of systematic errors. The improvement in photographic emulsions (IIIa-J) and telescope design (large field Schmidt telescopes) have improved the quality of image data on photographic plates and enable new surveys to go deeper.

There are several fundamental questions in astrophysics that need to be addressed by a new survey. As well as doubts about the break in $w_{gg}(\theta)$, there exist similar doubts about the amplitude and extent of the cluster–cluster correlation function, $w_{cc}(\theta)$. Using various cluster detection algorithms we will produce an objective cluster catalogue from which we can determine clustering power on large scales. This will provide a test of various galaxy formation scenarios, while the ratio of the amplitudes of the galaxy–galaxy and cluster–cluster correlation functions provides a measure of the bias required in models where light does not trace mass.

3 Data

3.1 COSMOS

COSMOS is a high speed, flying spot scanning microdensitometer, which can digitise a square area of size 287 × 287 mm of a photographic plate ($5\overset{\circ}{.}35 \times 5\overset{\circ}{.}35$ on a UKST plate). The basic information stored for each pixel is a 14-bit transmission value. This

Table 1. COSMOS Image Analysis Mode parameters used in the Edinburgh/Durham Galaxy Catalogue.

Field.	Units.	Description.
RA	hours	right ascension (1950).
DEC	degrees	declination (1950).
XMIN	0.1 μm	x-minimum of the image.
XMAX	0.1 μm	x-maximum of the image.
YMIN	0.1 μm	y-minimum of the image.
YMAX	0.1 μm	y-maximum of the image.
AREA	pixels	area of the image.
IMAX	intensity	maximum intensity above sky.
COSMAGCAL	magnitude	COSMOS image magnitude.
ISKY	intensity	sky intensity at centroid.
IXCEN	0.1 μm	intensity weighted x-centroid.
IYCEN	0.1 μm	intensity weighted y-centroid.
UMAJAX	0.1 μm	unit weighted semi-major axis.
UMINAX	0.1 μm	unit weighted semi-minor axis.
UTHETA	degrees	orientation on plate.
IMAJAX	0.1 μm	intensity weighted semi-major axis.
IMINAX	0.1 μm	intensity weighted semi-minor axis.
POSANGLE	degrees	position angle on sky.
CORMAGCAL	magnitude	field effect corrected magnitude.
SIGMA	pixels$^{\frac{1}{2}}$	Gaussian fit parameter.
IDSEQ	integer	sequential identification number.
LOGAREA	real	log(number of pixels).
GEOM	real	geometric filling factor.
GEOMLOG	real	GEOM × LOGAREA.
FIELDNO	integer	ESO/SERC J Survey field number.
BJMAG	magnitude	image magnitude in b_J

provides a measure of the emulsion density, which is converted to an intensity using a Baker density curve obtained from the 16 densitometer spots or the step wedge at the edge of each plate (MacGillivray and Stobie 1984). Details of the COSMOS machine can be found in MacGillivray and Stobie (1984), but briefly in threshold mapping mode (TM), an intensity cut above the local sky background is applied and then a pattern analyser (Thanisch *et al.* 1984) calculates a set of image moments for connected pixels with intensities above the threshold.

These moments are used to provide information on the structure of each image (Stobie 1980) through the COSMOS parameters (Table 1).

3.2 UKST J Survey

The plate material being used to construct the Edinburgh/Durham galaxy catalogue are glass copies of grade A IIIa-J SERC Southern Sky Survey plates taken with the 1.2m UKST. Each plate covers $6^\circ\!.4\times6^\circ\!.4$ and has a limiting sky background of typically $22^{\rm m}\!.0$ arcsec^{-2} in b_j. The survey consists of plates from 890 fields, covering $\delta=0^\circ$ to the south celestial pole, with $\simeq \frac{3}{4}^{\circ}$ overlap between plates. The principal advantage of using glass copies is their availability. Copy plates are produced with the density level of the plate background lower than that of the original plates. This has the advantage of reducing the noise in the measuring process resulting from a combination of machine noise and emulsion grain clumps. Comparisons of image quality and noise characteristics on original and copy plates show the degradation and information loss, in the copying process, are minimal (Brück and Waldron 1984, Stobie *et al.* 1984). The copying process does reduce the dynamic range and therefore there is a cut off in useable information at high density levels, however in practice the useful density range is not set by the photographic plate, but by COSMOS (MacGillivray and Stobie 1984).

The J survey is homogeneous in that all plates are taken on the same telescope, pointing to zero hour angle, in good seeing conditions (< 3 arcsec), using IIIa-J emulsion, with a nominal 60 minute exposure adjusted for variation in emulsion sensitivity and hypersensitisation. There are several changes to the telescope and plate material, that have been implemented during the survey (UKSTU Handbook). These include, the fitting of an achromatic corrector (1977), hypersensitisation (1975) and nitrogen flushing (1982). Some of these changes affect the image structure and the background density variation across the plate (Dawe and Metcalfe 1982) and can therefore effect the efficiency of image classification.

4 Mechanics

In a machine based survey of this type there are two primary sources of completeness variation as a function of position. The first is the variation in star–galaxy classification efficiency and the second is the variation in survey limiting magnitude due to systematic errors in image photometry. We examine these separately.

4.1 Image Classification

The systematic error in survey limiting magnitude allowed by Geller *et al.* (1984) also provides a constraint on the allowed variation in classification efficiency. The change in differential galaxy number density due to a $0\overset{m}{.}05$ calibration error is 7% ($\frac{\Delta N}{N} = \ln(10) \cdot \gamma \cdot \Delta m$, $\gamma = 0.6$, this paper). Correspondingly, a 7% difference in the number of images classified as galaxies is equivalent to a $0\overset{m}{.}05$ calibration error.

For the galaxy survey, classification techniques must be more rigorous than for a stellar survey. Since stars out-number galaxies for magnitudes brighter than $\simeq 19\overset{m}{.}5$, the raw data effectively consists of a complete star catalogue with a small galaxy contamination, from which we must construct a complete galaxy catalogue with a small stellar contamination. The aim of the project is a catalogue of maximum completeness. We find that this necessitates some stellar contamination, so by experiment we have designed a classification procedure to reduce contamination and the variation in classification efficiency to a minimum.

4.1.1 Basic parameters

In general star-galaxy image classification of digitised data from photographic plates, can be performed, because the two populations occupy different regions of some classification parameter *vs.* magnitude parameter plane (classification plane). Since most image information is contained in the parameters formed from the zero and second order moments (*i.e.* area, major and minor axis, *etc.*), which also suffer least from machine noise, it is these parameters or combinations of them, that are commonly used for image classification. We concentrate on classifying images using three COSMOS classification parameters together with a magnitude parameter. The COSMOS magnitude parameter is defined below and the 'field effect corrected' image magnitude, CORMAGCAL, is discussed in Sect. 4.1.2. *Fig. 1* shows the distribution of objects in each of these three classification planes.

For relatively bright images we use image geometry information carried in the parameter, GEOM, which is defined below. This parameter measures how effectively an image fills the ellipse fitted to its major and minor axes. For stars nominally brighter than $b_j \simeq 15.5$ diffraction spikes increasingly dominate the fitted ellipse so that GEOM < 0.9, while for galaxies GEOM $\simeq 1.0$. Intermediate magnitude galaxies have a lower surface brightness than point spread function stellar images, so in the LOGAREA *vs.* magnitude plane, galaxies lie above a tight stellar locus. For fainter images the SIGMA parameter provides a means of classification. This is defined as the width of a Gaussian fit to the image area and maximum intensity. For stars the Gaussian width is set by the point spread function and therefore stars again occupy a tight locus below galaxies in the classification plane.

The GEOM and SIGMA classification parameters are defined by

$$\mathrm{GEOM} = \frac{\pi\, a\, b}{\mathrm{AREA}} \tag{1}$$

$$I_{th} = \mathrm{IMAX}\, \exp\left(\frac{-\mathrm{AREA}}{2\pi\ \mathrm{SIGMA}}\right) \tag{2}$$

and the COSMOS magnitude is

$$\mathrm{COSMAGCAL} = -2.5 \log\left(\sum_{pix}\left(\frac{I_{pix} - I_p}{I_p/A_{pix}}\right)\right) \tag{3}$$

where $I_{th} =$ intensity of the threshold, IMAX $=$ maximum image intensity, a and b are the semi-axes, I_p is the local background intensity (mags arcsec^{-2}) and A_{pix} is the pixel size in arcsec^2.

We also use the parameter GEOMLOG which is a combination of two basic classification parameters. It is defined as

$$\mathrm{GEOMLOG} = \mathrm{GEOM} \times \mathrm{LOGAREA} \tag{4}$$

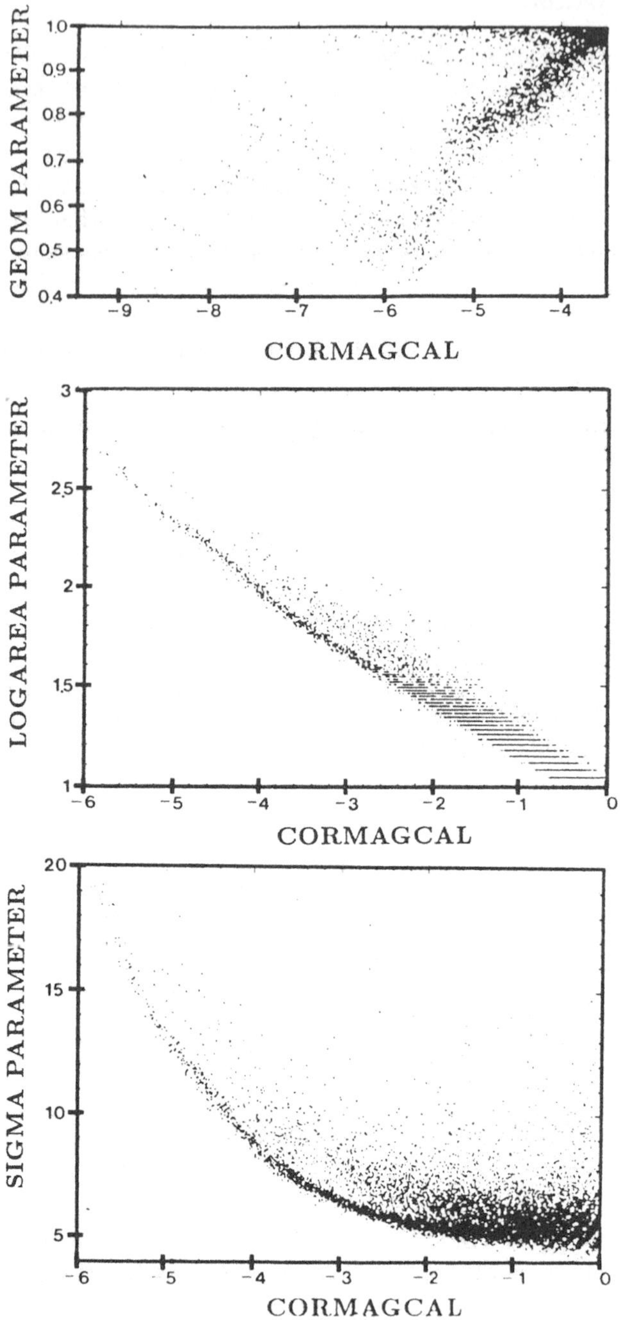

Fig. 1. Distribution of images in the 3 main classification planes: (a) GEOM *vs.* CORMAGCAL, (b) LOGAREA *vs.* CORMAGCAL, (c) SIGMA *vs.* CORMAGCAL.

4.1.2 Field effects

The major aspect of the J survey, which makes uniform star–galaxy classification difficult is the variation in the background density across the plate. This leads to a variation in the measured properties of point images as a function of position on the plate, referred to as field effects.

For unsaturated images COSMAGCAL $= b_j - m_{sky}$, where m_{sky} is the sky background in magnitudes per square arcsec (Eqn. 6). For saturated images of a given b_j, COSMAGCAL varies as a function of plate position.

The primary cause of this variation is the change in background density across plates. This is due to vignetting and differential de-hypersensitisation, which causes a further decrease in the background density over that of the standard vignetting curve (Dawe and Metcalfe 1982). Differential desensitisation is caused by the varying amount of moist air in the curved plate holder (Dawe *et al.* 1984) and is particularly acute on hypersensitised plates taken prior to nitrogen flushing (plates # 1541 to # 8271). We find an average variation in COSMAGCAL of $0\overset{m}{.}2$, though worse case variations are of order $1\overset{m}{.}0$. Our estimates from COSMOS scans agree with those of Dawe and Metcalfe (1982).

Field effects for saturated stellar images (b_j brighter than $\simeq 19.0$) cause COSMAGCAL values at the edge of the plate to be 'brighter' than those at the centre. *Fig. 2a* shows how there is no one optimum discrimination locus for images from all areas of the plate. Field effects wash out any discrimination between stars and galaxies and must be corrected for.

To solve this problem we define a correction to COSMAGCAL for each image. This correction is a function of image area and position on the plate and is based on the difference in magnitude between a saturated image at the plate centre and at the image position. *Fig. 2b* shows that when this corrected magnitude is used, we can define a single discrimination locus for all images regardless of position. We stress that this corrected magnitude is only used for the purposes of image classification. As galaxy images are mostly unsaturated (Sect. 4.2.1), using the corrected magnitude would introduce a position dependence into galaxy photometry.

4.1.3 Objective classification

In many previous papers on image classification the discrimination boundary has been fit to the data 'by eye' (Sebok 1986, Kron 1980). In experiments to measure possible sources of systematic errors we found variations of 20 % between plate data sets classified in this manner. Subjective discrimination is unacceptable for large catalogues as it is observer bias, unrepeatable and systematic errors are unquantifiable. Because of these large variations it is also impossible to optimise the classification procedure and quantify the efficiency of the classification parameters as a function of magnitude.

For the purposes of our catalogue we have designed an objective classification algorithm, that classifies individual objects based on the distribution of all images in the classification plane. It is assumed that at any given magnitude the images below the

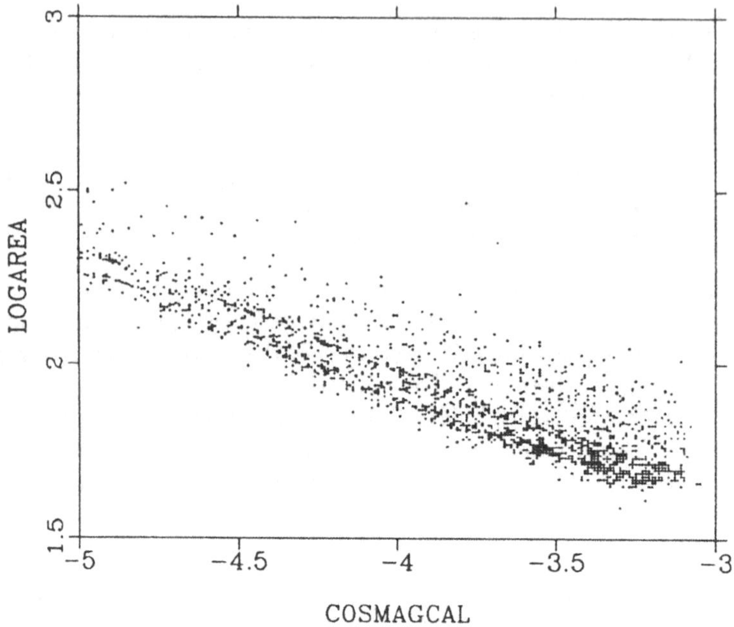

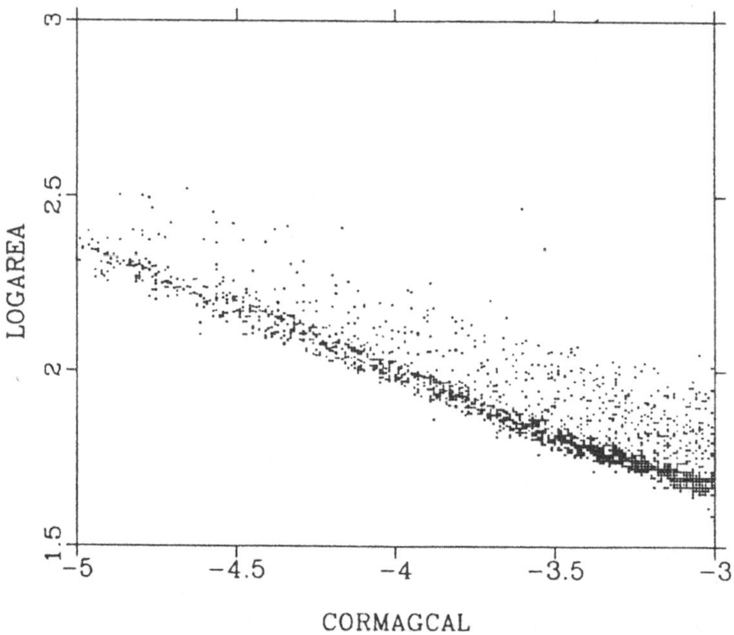

Fig. 2. Field effects on UKST J plates: (a) Data from 0.5° strips at the western edge and centre of the plate, plotted in the LOGAREA *vs.* COSMAGCAL plane. The stellar locus for data from the edge lies below that for data from the centre. (b) The same data after field effect correction plotted in the LOGAREA *vs.* CORMAGCAL plane.

mode of the classification parameter distribution are stars, while above it the likelihood that an image is a galaxy increases with distance from the mode. The algorithm makes no assumption about the shape of the stellar locus (*i.e.* parameter noise characteristics), other than it is symmetrical about the mode. For each image magnitude a 'separation' value of the classification parameter is found by estimating the position of the 95th percentile of the stellar image distribution. The smoothed locus of these values as a function of magnitude provides the discrimination boundary.

This algorithm provides consistent classification over plates of differing quality. It is repeatable and using it we have been able to investigate the variation in efficiency, with magnitude of each of the classification parameters.

We found that there is an optimum magnitude range over which each classification parameter should be used. If the parameter is used outside this range the resultant catalogue is very incomplete and/or very contaminated. The optimum magnitudes for changing parameters vary from plate to plate and we have confined our effort to designing an objective procedure for predicting these changeover magnitudes.

The details of this procedure are described elsewhere (Heydon–Dumbleton *et al.* 1988), but it is based on identifying features in the variation of image geometry with magnitude. There is some leeway ($\simeq 0^{\mathrm{m}}\!.5$) in the LOGAREA to SIGMA parameter changeover ($b_j \simeq 19$) and we find $< 1\%$ variation in the number of images classified as galaxies by either parameter in this region. The GEOM to LOGAREA parameter changeover is less precise due to the gradual onset of diffraction spikes. From stellar $b_j \simeq 12$ to 11.5 the diffraction spikes are long enough to lower the measured stellar surface brightness, but not long enough to reduce GEOM significantly below 1.0. This results in substantial misclassifcation of galaxies as stars if the LOGAREA classifier is used alone. The fact that the eye can detect two separate populations in the LOGAREA classification plane at these magnitudes is insufficient to ensure completeness.

Figure 3 shows how galaxies, selected as 'stars' by LOGAREA classification, can be 'sieved' out in the GEOMLOG classification plane. The same sieve could be applied in the GEOM classification plane, but requires a curved 'sieve' boundary rather than a linear boundary used in the GEOMLOG plane. Without this procedure the catalogue would be up to 50 % incomplete between galaxy $b_j \simeq 16.5$ and 17.

Despite this small iteration we conclude that, for our data, there is little need for classification in multiparameter space. For most magnitudes there is either only one acceptable classification parameter or no further improvement in efficiency is gained using other parameters.

4.1.4 Residual systematics

Table 2 shows the results of comparing our automatically selected galaxy samples with visually classified samples. The plate material tested is representative as it covers a wide range of plate qualities and epochs. From these results we can estimate residual plate to plate variation in number density due to systematic errors in our classification procedure. The Edinburgh/Durham galaxy catalogue will be $> 95\%$

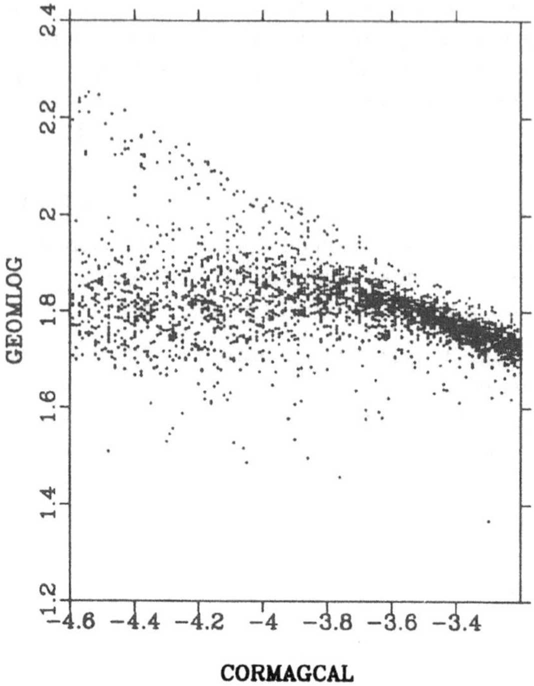

Fig. 3. Images selected as 'stars' by surface brightness criteria, plotted in the GEOMLOG *vs.* CORMAGCAL plane. Misclassified galaxies lie above the 'true star' locus.

Table 2. Visual estimates of galaxy catalogue contamination (*ctm*) and completeness (*cmp*). Figures in brackets are estimates given by the objective classification algorithm.

FIELDS:	J5304		J3579		J8046		J2693		J6124	
classification parameter(s)	*ctm*	*cmp*	*ctm*	*cmp*	*ctm*	*cmp*	*ctm*	*cmp*	*ctm*	*cmp*
LOGAREA + GEOMLOG	2%	95%	5%	96%	11%	100%	2%	100%	2%	100%
LOGAREA	10% (6%)	98%	2% (4%)	96%	11% (12%)	100%	11% (6%)	100%	7% (5%)	95%
SIGMA	7% (6%)	96%	8% (3%)	98%	11% (12%)	100%	9% (5%)	96%	11% (6%)	98%

complete and have < 10% stellar contamination with < 5% variation in the number of objects classed as galaxies. The requirement to maintain these statistics sets the limiting magnitude of our survey at $b_j = 20$. Below COSMAGCAL = −2.0 there is insufficient information in any of the available COSMOS parameters to classify images within the 95 % completeness limit (Heydon-Dumbleton *et al.* 1988).

Figure 4 shows an estimate of the systematic number density variation in a W – E direction across an average plate. This was obtained by stacking, *i.e.* summing binned number density counts, for 40 plates to remove real structure. The *rms* variation < 3 % *rms*, compared with 7 % *rms* in the raw SW counts.

We have performed an independent test of our classification efficiency using the differential galaxy number-magnitude counts. Taking a theoretical slope of 0.6, we have added a Monte Carlo stellar contamination. As shown in *Fig. 9* (insert) a contaminated catalogue would have number-magnitude counts with a flatter slope. Number-magnitude counts for our galaxy catalogue show a 0.6 slope over a wide range of magnitudes, $b_j \simeq 13$ to 19. From the uncertainty in the slopes we are able to rule out a 10 % stellar contamination of our catalogue at the $3\,\sigma$ level, and 20 % at the $5\,\sigma$ level.

4.2 Photometry

Systematic errors in image photometry provide the second source of catalogue completeness variation. Here we describe our approach to ensuring that our final calibra-

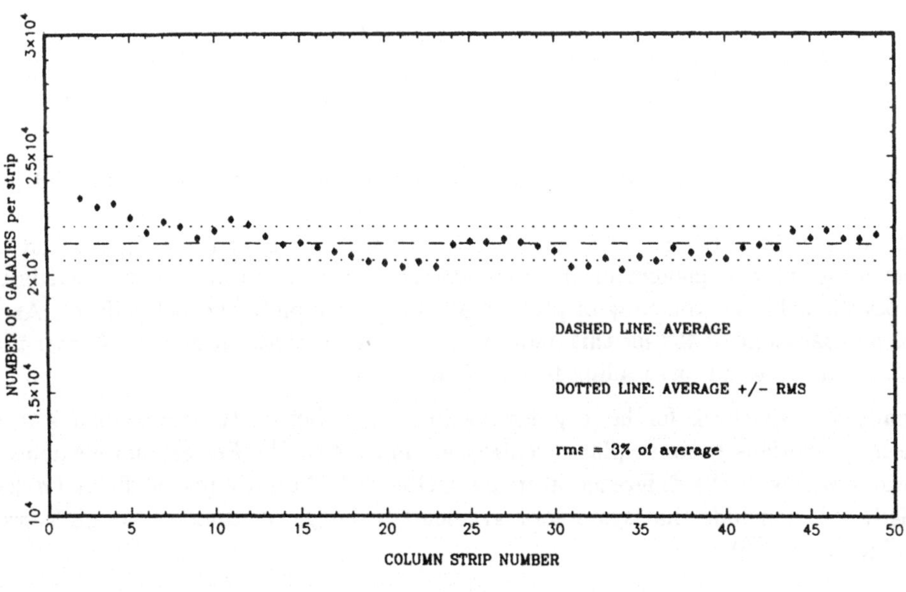

Fig. 4. The variation of binned number density counts summed over 40 plates. The dotted line shows the *rms* variation which is 3 %. This estimates the systematic number density gradient across an average plate.

tion error lies within the limit set by Geller *et al.* (Sect. 2). We stress that we are dealing with galaxy and not stellar image photometry. Different reduction techniques are required for each image type. A technique which produces consistent stellar magnitudes will produce widely inaccurate galaxy magnitudes (*cf.* Hawkins 1988). We will discuss our photometric reduction procedure in terms of the consistency of image photometry across a particular plate dataset ('intra-plate' photometry) and from plate to plate ('inter-plate' photometry).

4.2.1 Intra-plate photometry

The photometry of individual images is based on the definition of the COSMOS magnitude COSMAGCAL. This magnitude, defined in Eqn. 3, is constructed from the ratio of summed image intensity to the local plate background intensity. Since only pixels with intensities above a given threshold are included in the image, COSMOS magnitudes are isophotal magnitudes. It is found, however, that because this threshold is on average only 8 % above the sky, they are effectively total magnitudes (Shanks 1984). The internal measurement accuracy of a COSMOS magnitude for an individual image is $0\overset{m}{.}03$ (MacGillivray and Stobie 1984).

In comparison to vignetting and de-hypersensitisation, intrinsic sky background variation, zodiacal light, plate fogging and stray light are all negligible. The COSMOS magnitude for unsaturated images is then independent of plate position and related to b_j by

$$\text{COSMAGCAL} = b_j - m_{sky} \tag{5}$$

with

$$m_{sky} = -2.5 \log \left(\frac{I_{sky}}{A_{pix}} \right) \tag{6}$$

where I_{sky} is the intrinsic sky background intensity in b_j (typically $22\overset{m}{.}0$ arcsec^{-2}).

If these assumptions are correct, a plot of COSMAGCAL *vs.* b_j for COSMOS data should produce a linear graph with a 45° gradient. *Fig. 5a* shows such a plot for data from field 411 using the photoelectric data of Kirshner *et al.* (1978) and the electronographic sequences of Hawkins (1981). These data are widely distributed across the field and the copy of plate # 4606 has extremely bad field effects. As we find a linear slope of 45° for this data, we are confident that, for galaxies fainter than $b_j \simeq 15$ the assumptions leading to Eqn. 5 are valid.

Finally we can check for large gradients in image photometry across plates using the large overlaps between plates centred around RA = 0^{h}. *Fig. 6a* shows smoothed contour values of the difference in image COSMAGCAL on adjacent fields (# 298, #348). We conclude that systematic gradients in image photometry are on average confined to $< 0\overset{m}{.}03$.

4.2.2 Inter-plate photometry

COSMOS magnitudes are defined relative to the intrinsic sky background intensity. This has to be calibrated for each plate to ensure a uniform limiting magnitude. We have used the $\frac{2}{3}^\circ$ overlap between scanned regions of adjacent fields to investigate the variation of m_{sky}. *Fig. 6b* shows a plot of COSMAGCAL on one plate *vs.* COSMAGCAL on an adjacent plate. The 45° slope is as predicted from Eqn. 5. The *rms* variation in m_{sky} from the nominal value of $22^{\mathrm{m}}.0$ is $0^{\mathrm{m}}.2$, although there is a worst case difference of $0^{\mathrm{m}}.7$. Accurate m_{sky} calibration is therefore vital.

As part of the above investigation we have examined the residual magnitude shift after using the overlap regions of adjacent plates to calibrate closed loops of 4 or more plates. We found on average a residual of $0^{\mathrm{m}}.04$, but there are cases where the residual was much higher, $\simeq 0^{\mathrm{m}}.1$. Investigations reveal that there are a few overlaps where the COSMAGCAL *vs.* COSMAGCAL calibration shows a deviation from 45° slope. We do not believe this invalidates our conclusion on intra-plate photometry as the remaining 6 overlaps of these plates, with surrounding plates, generally show 45° slopes. The non–45° calibration curves are randomly distributed through the survey and are not correlated with regions of bad vignetting. The use of overlap regions to calibrate over a large number of plates will therefore introduce calibration errors larger than the $0^{\mathrm{m}}.05$ limit. In addition to this any matching procedure will propagate calibration errors as is the case in the SW counts (Geller *et al.* 1984).

Our approach to ensure inter-plate photometric consistency has been to obtain CCD sequences for every second field. For each calibration field we have obtained obser-

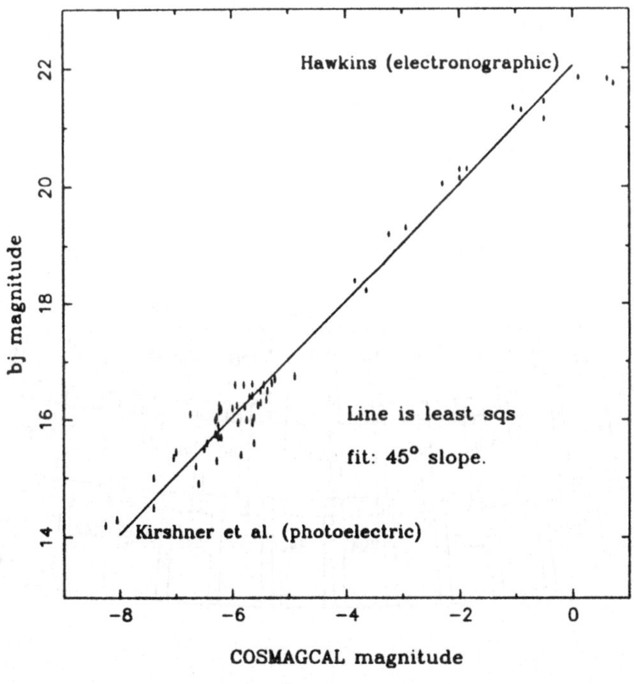

Fig. 5. Calibration of photographic magnitudes in field 411. Galaxy b_j magnitudes from photoelectric (Kirshner *et al.* 1978) and electronographic (Hawkins 1981) sequences plotted *vs.* COSMAGCAL. The 45° slope shows the validity of Eqn. 5.

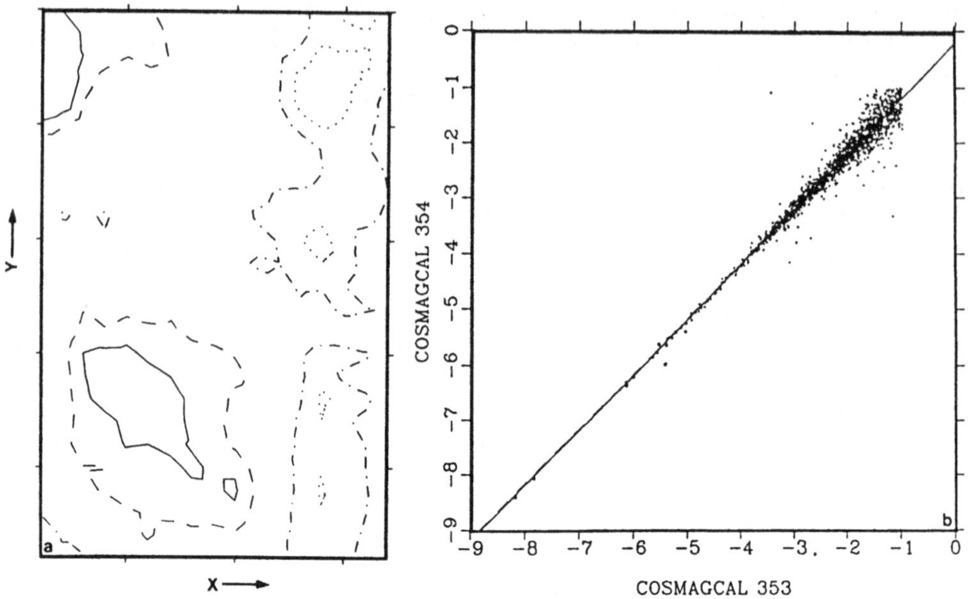

Fig. 6. (a) Smoothed contour values of the difference in COSMAGCAL values on adjacent fields (# 293, # 348), centred around RA = 0^h. Contours range from $-0\overset{m}{.}04$ to $+0\overset{m}{.}04$ in $0\overset{m}{.}02$ intervals. (b) Inter-plate calibration using COSMAGCAL values for images in the overlap region of fields # 353 and # 354. The line is a least squares fit to the data and has a gradient of 1.0 ± 0.005

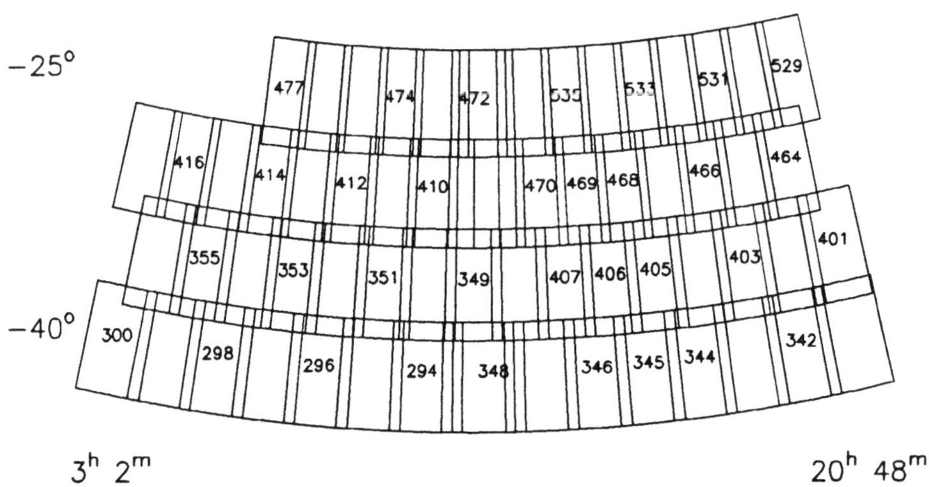

Fig. 7. The fields covered by the Edinburgh/Durham Galaxy Survey and the distribution of CCD sequences obtained for inter-plate calibration.

vations in B and V for approximately 15 galaxies covering a magnitude range of $B = 20.0$ to 14.0. This will ensure accurate calibration of m_{sky} and colour transformation to b_j. The distribution of our sequences shown in *Fig. 7*, enables us to detect any erroneous overlaps and monitor the accuracy of our calibration.

Using sequences on adjacent plates we estimate an inter-plate limiting magnitude variation of less than $0\overset{m}{.}04$.

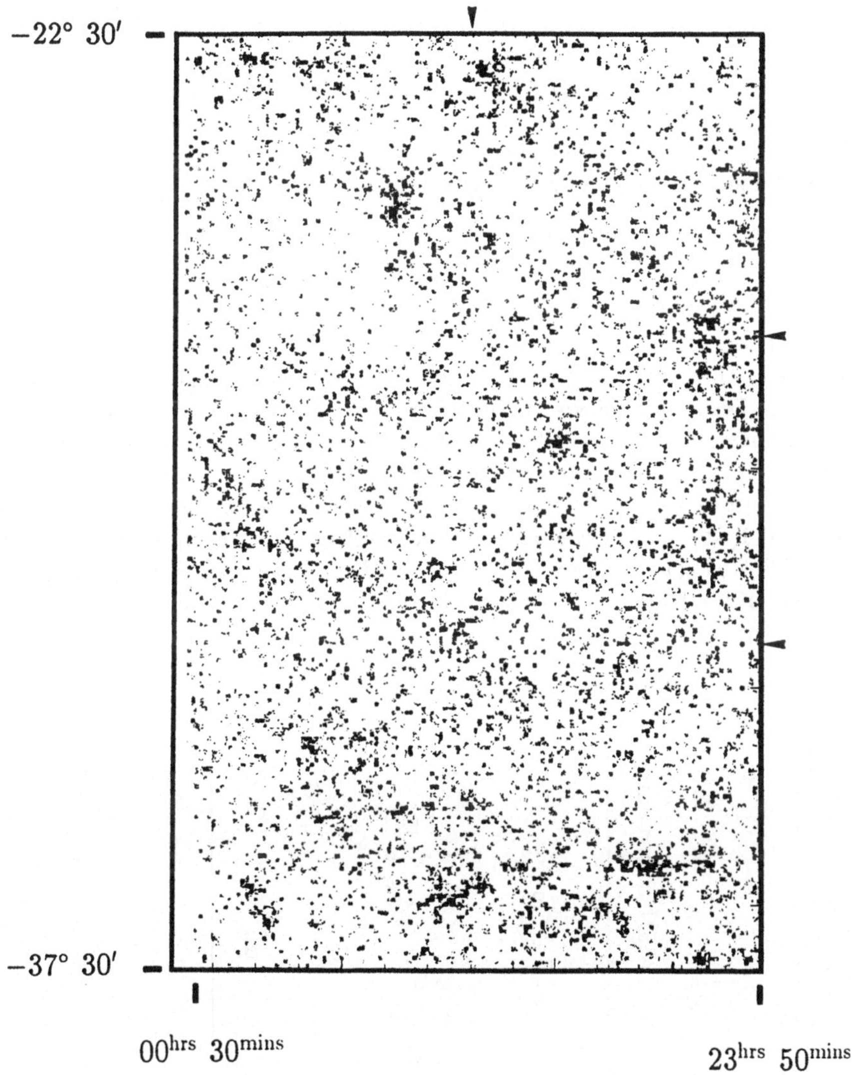

Fig. 8. A greyscale plot of binned data from a 6 plate mosaic covering fields 349, 350, 409, 410, 472, 473 (Plate boundaries are indicated by the arrows). The region subtends $\simeq 100$ degrees^2 and the average number of galaxies per bin for a random distribution is 1. Grey bins contain 3 or more galaxies, while black bins have > 10 galaxies per bin.

5 Applications

The preliminary work on the Edinburgh/Durham Galaxy Catalogue is nearing completion. The intial catalogue consists of a mosaic of 60 plates around the South Galactic Pole complete to a limiting magnitude of $b_j = 20$. The quality assurance we carry out during the production of the mosaic ensures that only objects from one plate in the overlap regions are included and that spurious objects have been removed from the catalogue. These include objects produced by the breakup of bright stars ($b_j < 16$), star clusters, satellite trails and defocused regions. The full catalogue contains all the image parameters for some 10^6 galaxies. As well as the full catalogue we will produce a density map of galaxy counts (*Fig. 8*) and a cluster catalogue based on various detection cluster detection algorithms (Turner and Gott 1976, Dodd and MacGillivray 1986). *Fig. 9* shows a plot using the image parameter data for the members of a small compact cluster detected in Field 414.

Finally we present some preliminary results of interest. *Fig. 10* shows number-magnitude counts and the the two point angular correlation function from the 6 plate region shown in *Fig. 8*.

The number-magnitude counts show a 0.6 slope over a wide range of magnitudes, $b_j \simeq 13$ to 19. They agree well with the counts of Shanks *et al.* (1980), but differ from the counts of Tyson and Jarvis (1979) which show a 0.4 slope fainter than $J \simeq 16.5$. We are confident that our result is not due to stellar contamination (Sect. 4.1.4) or errors in our magnitude scale (Sect. 4.2.1). Our result indicates that the assumptions of an homogeneous galaxy distribution, and flat space are not the 'oversimplification' suggested by Tyson and Jarvis and may be valid to a depth of $600\,h^{-1}$ Mpc (using $M^* = -19.8 + 5\log h$; Phillips and Shanks 1987). Some of the implications of the number-magnitude counts are discussed in Heyden-Dumbleton *et al.* (1988).

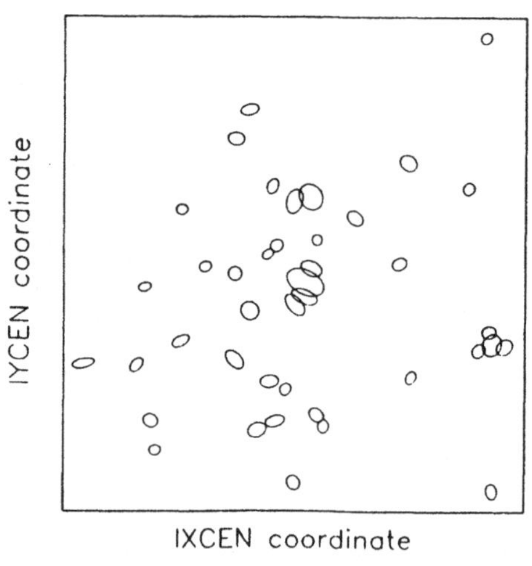

Fig. 9. A plot of COSMOS image data from a compact cluster at RA: $1^h\,49^m\,59\overset{s}{.}8$, DEC: $-32°\,9'\,10\overset{''}{.}0$.

Our preliminary two-point angular correlation function (*Fig. 11*) shows a break at around 2° from a power law of slope −0.7. Scaling of our correlation function, which is at a depth of $b_j = 19.5$, to the SW depth of $18^{\rm m}\!.6$ moves the break out to an angular separation of 2.5 to 3°. This is in agreement with the break found by Groth and Peebles (1977) and is indicative of a break in the spatial correlation function $\xi(r)$, at $10\,h^{-1}$Mpc. Our results suggest that the observer bias and limiting magnitude variations in the SW counts found by Geller *et al.*, though well founded, do not effect the correlation analysis of Groth and Peebles. Further analysis of the correlation function is presented in Collins *et al.* (1988).

NHD would like to thank the organisers of this workshop for the opportunity to present this work. We would like to thank the UKST unit for the use of the plate material and the COSMOS group for providing the scan data. For further information on the Edinburgh/Durham Galaxy Survey please contact NHD or CAC (@UK.AC.ROE.STAR). For details on and requests to use COSMOS please contact HMG (@UK.AC.ROE .STAR).

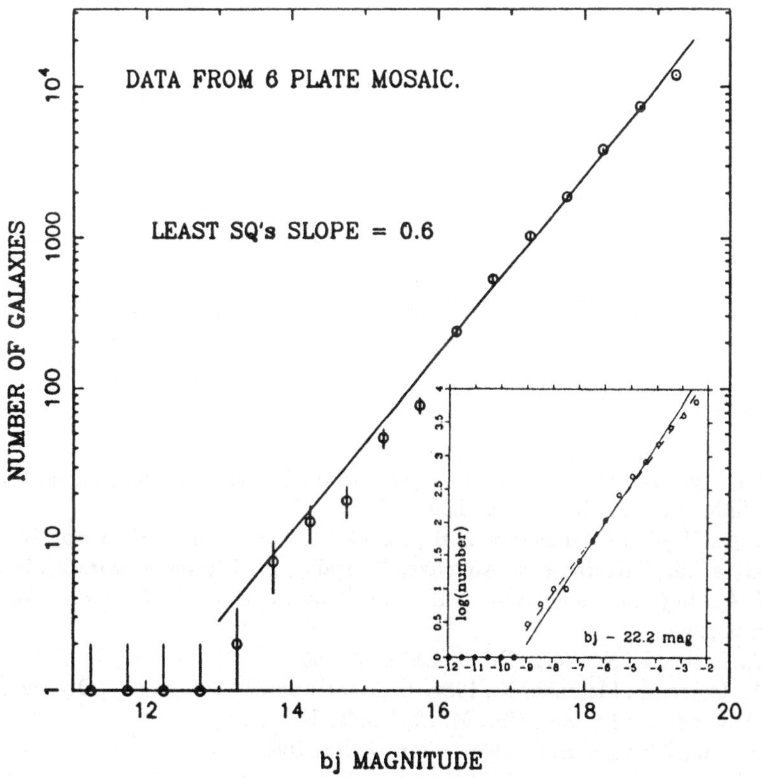

Fig. 10. Number-magnitude counts for galaxies from the Edinburgh/Durham catalogue. The insert shows number-magnitude counts for a galaxy catalogue with 20 % stellar contamination. The slope for the contaminated catalogue is significantly ($> 3\sigma$) flatter than that for the uncontaminated catalogue.

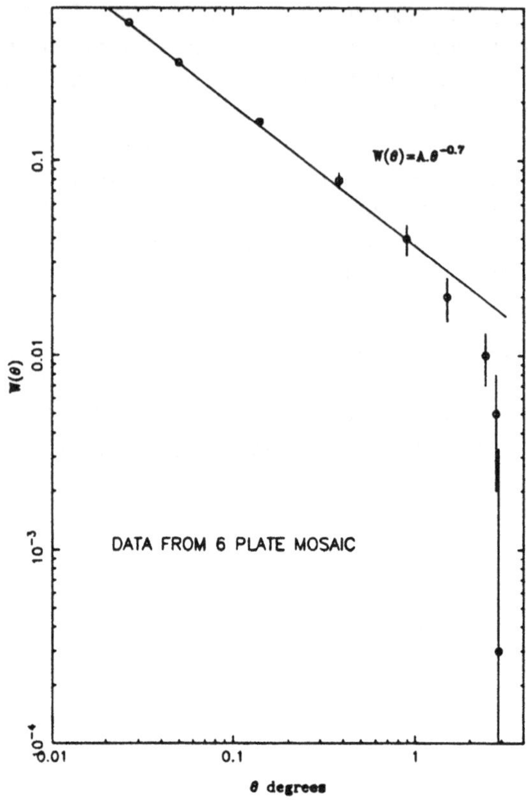

Fig. 11. The two-point angular correlation function at a depth of $b_j = 19.5$ for the 6 plate mosaic shown in *Fig. 8*. There is a break from the -0.7 power law slope at 2°.

References

Abell, G.O., 1958. *Astrophys. J. Suppl.*, **3**, 211.

Brück, M.T., Waldron, J.D., 1984. In *Astronomical Photography*, Occasional Reports Royal Obs. Edinb., **14**, eds. Sim, M. E., Ishida, K., p. 185.

Collins, C.A., Heydon–Dumbleton, N.H., Shanks, T., 1987. In *IAU Symp. No. 130, The Structure of the Universe*, eds. Audouze, J., Szalay, A., Kluwer, Dordrecht, in press.

Collins, C.A., Heydon–Dumbleton, N.H., MacGillivray, H.T., 1988. *Mon. Not. R. astr. Soc.*, submitted.

Dawe, J.A., Metcalfe, N., 1982. *Proc. A.S.A.*, **4**, 466.

Dawe, J.A., Coyte, E., Metcalfe, N., 1984. In *Astronomical Photography*, Occasional Reports Royal Obs. Edinb., **14**, eds. Sim, M. E., Ishida, K., p. 59.

Dodd, R.J., MacGillivray, H.T., 1986. *Astr. J.*, **92**, 706.

Fong, R., Jones, L.R., Shanks, T., Stevenson, P.R.F., Strong, A.W., Dawe, J.A., Murray, J.D., 1987. *Mon. Not. R. astr. Soc.*, **224**, 1059.

Geller, M.J., de Lapparent, V., Kurtz, M.J., 1984. *Astrophys. J.*, **287**, L55.

Groth, E.J., Peebles, P.J.E., 1977. *Astrophys. J.*, **217**, 384.

Hawkins, M.R.S., 1981. *Mon. Not. R. astr. Soc.*, **194**, 1013.

Hawkins, M.R.S., 1988. *These proceedings*, p. 285.

Heydon–Dumbleton, N.H., Collins, C.A., MacGillivray, H.T., 1988. *Mon. Not. R. astr. Soc.*, submitted.

Kron, R.G., 1980. *Astrophys. J. Suppl.*, **43**, 305.

Kirshner, R.P., Oemler Jr., A., Schechter, P.L., 1978. *Astr. J.*, **83**, 1549.

de Lapparent, V., Kurtz, M.J., Geller, M.J., 1986. *Astrophys. J.*, **304**, 585.

MacGillivray, H.T., Stobie, R.S., 1984. *Vistas Astr.*, **27**, 433.

Phillips, S., Shanks, T., 1980. *Mon. Not. R. astr. Soc.*, **227**, 115.

Postman, M., Geller, M.J., Huchra, J.P., 1986. *Astr. J.*, **91**, 1267.

Sebok, W.L., 1986. *Astrophys. J. Suppl.*, **62**, 301.

Shane, C.D., Wirtanen, C.A., 1967. *Publ. Lick Obs.*, Vol XXII, Part 1.

Shanks, T., Fong, R., Ellis, R.S., MacGillivray, H.T., 1980. *Mon. Not. R. astr. Soc.*, **192**, 209.

Shanks, T., Stevenson, P.R.F., Fong, R., MacGillivray, H.T., 1984. *Mon. Not. R. astr. Soc.*, **206**, 767.

Stobie, R.S., 1980. *J. Brit. Interplanetary Soc.*, **33**, 323.

Stobie, R.S., Okamura, S., Davenhall, A.C., MacGillivray, H.T., 1984. In *Astronomical Photography*, Occasional Reports Royal Obs. Edinb., **14**, eds. Sim, M. E., Ishida, K., p. 219.

Turner, E.L., Gott III, J.R., 1976. *Astrophys. J. Suppl.*, **32**, 409.

Thanisch, P., McNally, B.V., Robin, A., 1984. *Image & Vision Computing*, **2**, 4.

Tyson, J.A., Jarvis, J.F., 1979. *Astrophys. J.*, **230**, L153.

UKSTU Handbook, United Kingdom Schmidt Telescope Unit, Royal Observatory, Edinburgh.

The APM Galaxy Survey: Some Data Reduction Techniques

S.J. Maddox, J. Loveday,
W.J. Sutherland and **G. Efstathiou**
Institute of Astronomy
Cambridge, England, UK

Abstract

We have used the Automatic Plate Measuring machine in Cambridge to scan 176 photographic survey plates from the UK Schmidt Telescope and have compiled a galaxy catalogue covering 4400 degrees^2 and containing over four million galaxies with $B_J < 21$. We give a brief description of the plate material, measuring procedure and data reduction techniques. The star–galaxy separation technique and tests of its reliability are described. CCD observations of groups of faint galaxies have been used to construct 66 photometric sequences of galaxies in the range $B_J = 17 - 21$.

1 Introduction

Wide-field surveys of galaxies and clusters are an indispensable tool for studying large scale structure in the universe. The Abell catalogue (Abell 1958), Zwicky catalogue (Zwicky *et al.* 1961-1968), and the Lick survey (Shane and Wirtanen 1967, Seldner *et al.* 1977) have provided many statistical results of key importance to our understanding of galaxy formation and clustering (see *e.g.* Peebles 1980). However, these surveys were constructed more than 20 years ago. Since then, there have been major technological developments in photographic emulsions, automatic scanning machines and computers. It is therefore possible to improve significantly on earlier surveys by generating deep galaxy catalogues with high photometric precision and uniformity over wide areas of sky. Over the last four years, we have taken advantage of these developments to construct a new survey of several million galaxies.

The photographic material that we have used is based on the $6° \times 6°$ SERC IIIa-J plates taken by the UK Schmidt Telescope Unit (UKSTU). The UKSTU plates cover the entire sky south of $\delta = -20°$. Our limiting magnitude corresponds to $B_J \approx 21$, more than 2 magnitudes deeper than the Lick survey. The European Southern Observatory (ESO) has supplied us with high quality glass copies of the original plates (West 1978) which we have scanned in Cambridge.

The SERC Automatic Plate Measuring (APM) machine in Cambridge is a high speed automated densitometer with on-line sky subtraction and image analysis (Kibblewhite *et al.* 1984). It can scan the central $5\overset{\circ}{.}8 \times 5\overset{\circ}{.}8$ of a UKSTU plate in about 14 hours giving accurate measurements of the position, magnitude and shape parameters for each image above a fixed detection threshold. We have used the APM machine to scan 176 of the 190 UKSTU fields within the area $b < -40°$ and $\delta < -20°$. The remaining

14 plates have yet to be supplied by ESO. Over the 4400 square degrees covered by our survey, we have detected about 3.6 million galaxies at a magnitude limit of $B_J = 21$. *Fig. 1* shows the field boundaries in an equal area projection centred on the south galactic pole. For statistical studies of the large-scale structure, it is important that the galaxy selection function is uniform over the whole survey. We have therefore used a combination of photometric CCD calibrations and comparisons between plate overlaps to ensure that the magnitude limit and star-galaxy classification are uniform. The positions of the calibration sequences are shown by the dots in *Fig. 1*.

2 The APM measurements

An outline of the technical details of the APM system is given by Kibblewhite *et al.* (1984). The microdensitometer uses a scanning laser to sample the transmission of the emulsion on a grid of positions with $\approx 8\,\mu$m separation. The image of a star or galaxy forms a group of pixels that are brighter than the local average sky and during the measurement the APM system locates all the images which contain more than a certain number of pixels brighter than a set threshold above sky.

To estimate the local sky brightness at each point, the APM carries out a preliminary scan of the plate. During the scan the pixel values are analysed in 64×64 pixel groups, corresponding to 0.5 mm squares on the plate. For each square a frequency histogram of the number of times each pixel value occurs is calculated. Then the sky brightness is estimated by fitting to the peak of the histogram. The sky values from all of the squares over the plate form a 640×640 pixel map of the background sky brightness. The map is filtered with a two dimensional non-linear filter and used to give an estimate of sky at the position of each pixel.

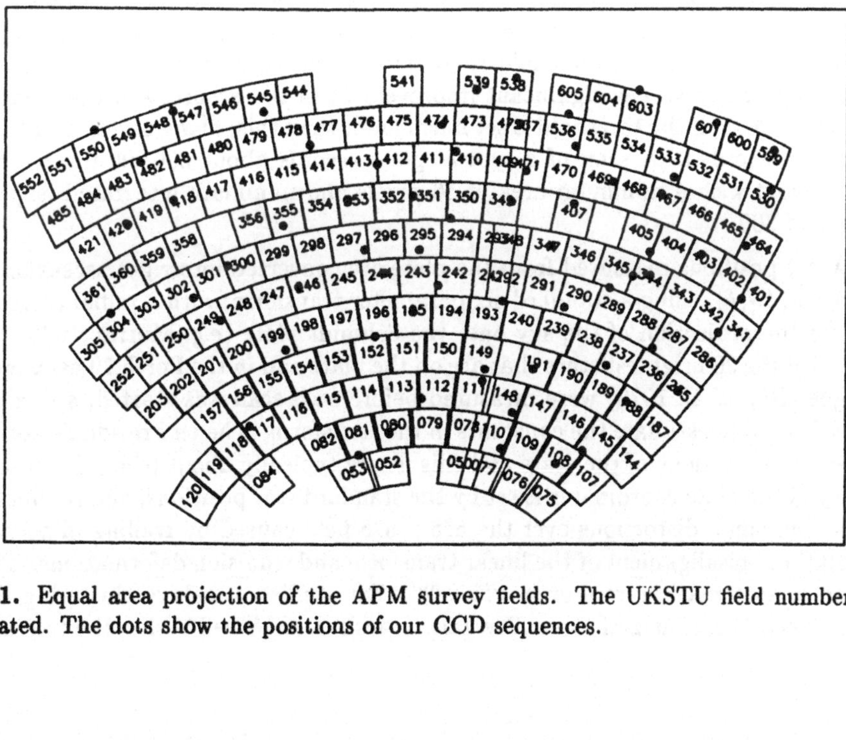

Fig. 1. Equal area projection of the APM survey fields. The UKSTU field numbers are indicated. The dots show the positions of our CCD sequences.

The pixel values input to the image processor after sky subtraction are truncated to 8 bits. For each detected image containing more than a certain minimum number of pixels above the threshold the image processor calculates parameters from the sky subtracted densities,

$$D = D_o - D_s ,$$

where D_o is the density of an object pixel, and D_s is the local sky value.

The relative brightness of each image is measured by the summed density above sky for pixels above the threshold, t,

$$I = \sum_{D_o > D_t} (D_o - D_s) = \sum_{D>t} D , \tag{1}$$

where $D_t = D_s + t$. The position of each image is calculated as the density weighted mean of the X and Y of the pixels contained in it, so $X_0 = \langle XD \rangle / I$, and $Y_0 = \langle YD \rangle / I$. The shape of each image is parameterised by $\langle x^2 D \rangle / I$, $\langle xyD \rangle / I$, $\langle y^2 D \rangle / I$, where $x = X - X_0$, and $y = Y - Y_0$. These parameters can be used to calculate the radius of gyration, the eccentricity and orientation of an equivalent bivariate Gaussian image. The surface brightness profile of each image is given by the peak density above threshold, $D_p - D_t$, and the areal profile. The areal profile is the area of the image at 8 increasing surface brightness levels. The levels that are used are $D \geq D_t + 1, D_t + 2, D_t + 4, D_t + 8, D_t + 16, D_t + 32, D_t + 64, D_t + 128$. This combination of parameters gives all the useful information contained in the faint images and simultaneously gives a good representation of the brighter images.

The threshold set for each scan was twice the *rms* noise in the measured sky value. For the UKSTU J copies the noise is typically between 5 and 6 density units. The thresholds correspond to between 8 % and 11 % of the sky surface brightness, or $\approx$ 24.5 to 24 B_J magnitudes arcsec^{-2}. Measures of the sky subtracted density for pixel values lower than this threshold are dominated by the grain noise of the emulsion and so would not give any significant improvement in the accuracy of the measured parameters. Also, if the threshold is set much lower than this, many useless small noise images are recorded. A scan of a typical plate records about 300 000 images using this threshold with a minimum area of 16 pixels and minimum integrated isophotal intensity of 300.

The (X, Y) positions measured from the plate are converted into right ascension and declination, (α, δ), using a standard six parameter transform, with a radial correction factor for the projection of the sky onto the Schmidt plate (*e.g.* Murray 1983). The positions of the 20 brightest stars that are on the plate and in the Perth 70 astrometric catalogue (Høg *et al.* 1970) were measured before each scan was started, and a least squares fit used to estimate the constants in the transform. The *rms* residuals between the measured positions of the stars and this fit are typically about $0''.5$. The absolute accuracy of the plate coordinates is set by the standard star positions, and is about $1''$. We also find small distortions over the $5^\circ.8 \times 5^\circ.8$ field caused by trailing, differential field rotation, misalignment of the linear transform and emulsion deformations. These give rise to systematic errors of $\approx 1''.5 - 2''$. For studies of galaxy clustering these positional errors are insignificant. The galaxy positions are well within the tolerances required for multi-object fibre spectroscopy at large telescopes. For example, Colless and Hewett (1987) have used the APM survey to generate fibre masks at the Anglo-Australian Telescope for their study of the dynamics of rich clusters.

Figure 2 shows an image surface density map in the same projection as *Fig. 1* for 20 368 362 images before making any corrections for uniformity. The magnitude limits vary between plates, reflecting differences in plate quality and detection threshold. These variations can be seen as changes in the number density of images across the field boundaries. However, even with no corrections the survey is fairly uniform and interesting features can be seen. For example, the fall in the stellar density away from the galactic plane and away from the galactic centre is clearly visible. Also the dwarf galaxies in Sculptor and Fornax, and a few globular clusters are partly resolved and show up as small areas of high image density. Many other intermediate scale features from galaxy clustering can be seen.

3 Measurements of the APM density *vs.* flux relation

The APM parameters are calculated using density measurements rather than flux measurements. Therefore the relation between the measured intensities and standard photometric magnitudes needs to be investigated. The APM initially measures the transmission for each pixel on a 12 bit scale using a photomultiplier to measure the intensity of the stabilised laser beam after it has gone through the plate. The transmission values are converted to a 10 bit density using a look-up table. The look-up table was chosen so that on an average copy plate the density range from the

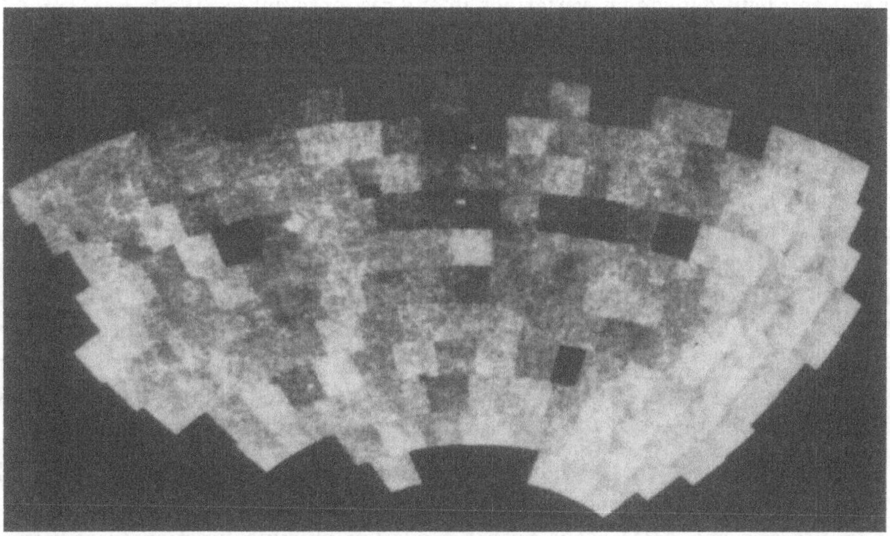

Fig. 2. Equal area projection showing the surface density of images in the unmatched APM survey scans at a magnitude limit of $M = 8$, corresponding to $B_J \approx 21.5$. There are 20.4×10^6 images in the $120° \times 60°$ area shown. The grey levels are set so that black = 0 image density and white = 7.3×10^3 images per square degree.

minimum sky value to the saturation level of the copies roughly fitted in the 10 bit density scale. After sky subtraction, only the most significant 8 bits are retained and so the effective range of the densities used in the calculation of the image parameters is 0 to 255. The optical density range measured on a copy plate is roughly 0 to 2.5. Therefore the APM density used to calculate the image parameters is related to the standard optical density, O,

$$D = \frac{256}{2.5} \log \left(\frac{4096}{T} \right) \approx 102.4\, O + const\,. \tag{2}$$

The density *vs.* flux relation can be measured using the step wedges that the UKSTU exposed on the edge of each plate during the main exposure. Unfortunately the emulsion is desensitised near the edges of the plates (Dawe and Metcalfe 1982), so the response of the emulsion for the central area of plate is different to that at the location of step wedges. The halo around the laser spot also alters the effective density *vs.* flux relation. The halo makes the density measurements at a point dependent on the surrounding density values. For a pixel in the wedge, the surrounding pixels have the same density, but for an image pixel most of the surroundings are close to the local sky. Therefore the density measured for a high flux pixel in an image is different to the density measured for the same flux in the step wedge.

A more direct way to measure the density *vs.* flux relation for images is to compare a raster scan from the APM with a CCD frame of the same area (Cawson *et al.* 1987). This gives a direct measurement of the flux corresponding to the density measured for each image pixel. These comparisons include the adjacency effects of the spot halo and any field dependent variations in the response curve. We have many CCD frames which were taken with the 1.0 m telescope at the South African Astronomical Observatory to provide photometric calibration. The area of plate covered by the CCD frames for field 344 was raster scanned with the APM. The size of pixels in the raster scan is the same as that used for the image mode scans, but is different to the CCD frames. The CCD frames need to be accurately rescaled and positioned to match with the APM scan so that the image pixels match up exactly. If there is a mismatch between APM pixels and CCD pixels, the calibration curve is spread out into a series of loops corresponding to the difference between the image cross sections in the two pixel sets. We used an image detection algorithm similar to that in the APM to locate and measure the image positions in the data sets. Then we paired up each image and did a least squares fit to give 6 constants in a linear transform between the two coordinate systems. Each B and V frame was transformed so that the image positions matched the images in the corresponding raster scan.

The seeing on the survey plate and the two frames are different, so the image profiles are not exactly the same, even when correctly matched up. We smoothed both the CCD frames and the APM scans with a Gaussian blurring function to reduce the relative difference. The smoothing also reduces the random noise in the curves.

If a B or V band CCD frame is used alone, each image gives a slightly different curve. This is because the images have different colours, and so need different colour correction terms to convert from the B or V to the J band of the UKSTU plates. The frames were added using standard colour equations (Blair and Gilmore 1982, Walker 1984) to give a composite J frame which was compared pixel by pixel to the raster scan. *Fig. 3* shows a point for each pixel pair. The lowest pixel values are those for the sky, which is well above the fog level for these frames.

These measurements show that the APM density D below the emulsion saturation is fairly well approximated by

$$D = \alpha f + \beta \,, \tag{3}$$

where α, β are constants, and f is the incident flux. This approximation is plotted as the solid line in *Fig. 3.* The more conventional way of relating optical density O to incident flux is to assume

$$O = \gamma \log_{10}(f) + c \,, \tag{4}$$

but for lower densities this is a poor approximation to the measured curve. A better approximation uses the Baker density (Baker 1925), defined as

$$B = \log_{10}(10^{O} - 10^{O_{fog}}) \,, \tag{5}$$

so that

$$B = \gamma \log_{10}(f) + c \,. \tag{6}$$

This fits both the low and high density parts of the curves very well below the emulsion saturation. The linear approximation is just as good as Baker density before the copy

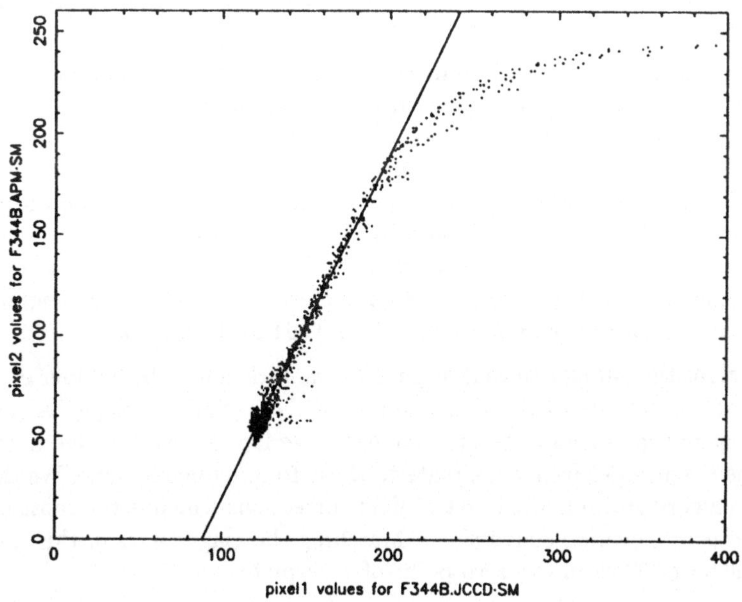

Fig. 3. Comparison of APM density measurements with CCD flux measurements in field 344. The density of each pixel from a $2' \times 3'$ raster scan of field 344 is plotted against the CCD flux measurement for the same piece of sky. The solid line is the $D = \alpha f + \beta$ approximation, with $\alpha = 1.7$ and $\beta = -150$.

emulsion saturates and is much simpler. Since the density is linearly related to the flux below saturation in this approximation, using densities throughout the image analysis is equivalent to using fluxes. Therefore the APM magnitude for galaxies is simply related to the total magnitude, $M = 2.5\log_{10}(I) \approx z - B_J$, where z is the zero point for the plate. We have constructed 66 photometric sequences of faint galaxies from the CCD observations, and used them to measure zero points for the fields. They typically give $z \approx 29 - 30$.

Multiple scans of the same plate show that the measured magnitudes are repeatable to 5 %, even for images close to the plate limit. However, comparisons between different plates of the same field show that variations in sensitivity and grain noise lead to large magnitude errors for images fainter than $B_J \approx 21$. Also, it becomes very difficult to distinguish galaxies from stars near the plate limit (see Sect. 6). For the final catalogue we have therefore aimed at uniformity for objects brighter than $B_J = 20.5$. At this limit we find about 17 000 galaxies per plate.

4 Plate overlaps and matching

The field centres are separated by 5° so there is a generous overlap area of about 6° × 1° between each plate. The parameters of matched pairs of images in the overlaps provide our primary means of ensuring uniformity in the selection function over the survey area.

A good demonstration of our photometric accuracy and the extent to which we can remove systematic field effects is provided by plate pairs near $\alpha = 0$. Unlike the majority of plates in the survey, some of the plates at $\alpha = 0$ have large overlaps of up to 3° × 6°. Thus we can compare measurements made at one plate centre with those made at another plate edge. Any systematic errors in the magnitudes on each plate will show up as a positional variation in the mean difference between the two sets of measurements. *Fig. 4* shows a contour map of the average difference between the magnitudes of images measured from one plate at $\alpha = 0$ and its neighbour. The differences over most of the area are less than $0\overset{\text{m}}{.}04$ peak-to-peak.

We use an algorithm similar to that applied to the Lick counts by Seldner *et al.* (1977) to determine corrections to the magnitudes from the plate overlaps. A polynomial fit to the magnitude-magnitude plot for each overlap is used to give a conversion between the magnitudes from each plate to those from its neighbours. We then apply an iterative algorithm to find the set of field corrections which is the most consistent with the measured overlap conversions. Matching the magnitudes in this way gives a residual scatter of $0\overset{\text{m}}{.}02$ in the zero point of each plate.

The edge matching procedure is sensitive to residual field effects which introduce small systematic drifts in the plate zero points. Without any additional checks, these would cause large-scale gradients in the survey in the same way as that described for the Lick survey by Groth and Peebles (1986). To prevent this evidently undesirable effect, we use the CCD sequences as tie points in the plate matching algorithm.

5 Star-galaxy separation

The APM measures parameters for all images on each field, including stars, galaxies and dust, emulsion defects and other noise. Many techniques to distinguish stars from galaxies using parameterised shape data have been developed by several groups including Godwin *et al.* (1983), Jarvis and Tyson (1981), Sebok (1979) and MacGillivray *et al.* (1976). All of the techniques rely on the fact that the profile and shape of a stellar image is determined only by the instrumental response and seeing conditions during the observation, whereas a galaxy image has a different intrinsic profile. Therefore the image parameters of stars lie in a well defined region of the parameter space and can be excluded to leave a sample of galaxies.

The first attempt to define a galaxy sample from the APM survey used the standard APM automated classification program STATS. Pairs of the measured parameters are plotted against each other, and the stars fall along a well defined line. The program locates the peak position and measures the *rms* scatter about the stellar locus in four parameter plots. The plots are shown for a typical field in *Fig. 5.* The images which are more than $2\,\sigma$ away from the stellar locus are flagged as non-stellar. In the plot of peak density *vs.* total intensity an attempt is made to get a reasonable measure of the unsaturated value of the peak by extrapolating the areal profile to the centre of the image. Due to measuring noise and emulsion defects some images have parameters that are impossible for real images and these are excluded from the plots and flagged as noise images.

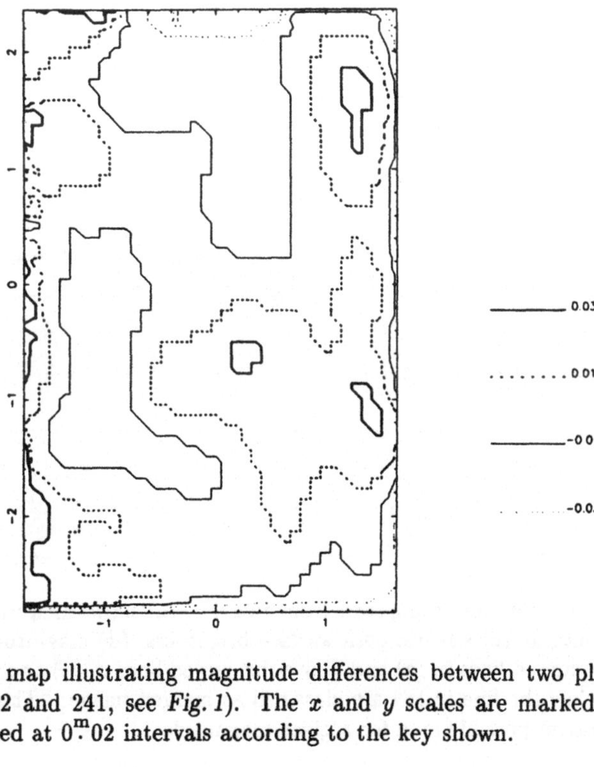

Fig. 4. Contour map illustrating magnitude differences between two plates with a large overlap (fields 292 and 241, see *Fig. 1*). The x and y scales are marked in degrees. The contours are spaced at $0\overset{m}{.}02$ intervals according to the key shown.

The positions of each image on the four plots are combined by adding the distances of each image from the stellar lines in quadrature. Assuming that the differences from the stellar line are caused by Gaussian errors, this combination of distances is effectively a maximum likelihood estimate of whether the image is a star. Finally, each image is flagged as a star, eccentric star, galaxy, merged image or noise image, depending on the combined estimate and also the individual estimates. The eccentricity of each image is also calculated, and if an otherwise stellar image has a very different eccentricity than the average for stars, it is flagged as an eccentric star and grouped with the

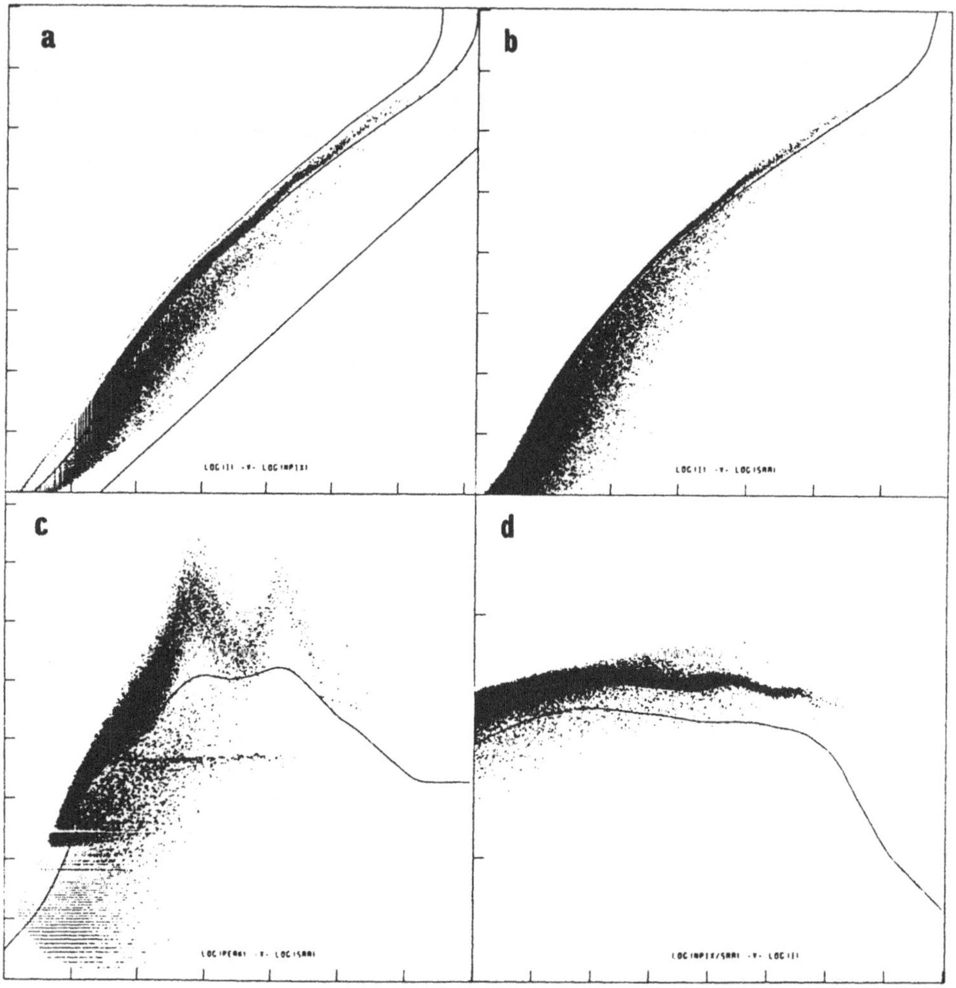

Fig. 5. Parameters used in separating stars from galaxies. (a) shows area *vs.* magnitude, (b) mean size *vs.* magnitude, (c) size *vs.* peak surface brightness, (d) magnitude *vs.* area/size. Points between the curved lines in (a) and above the curved lines in (b) and (c) are classified as stars. Points below the line in (d) are classified as merged images. The straight line in (a) shows the minimum possible area for a given magnitude.

merged images. The eccentricity of stars need not be zero because small errors in the tracking during the exposure can lead to all stellar images having a small eccentricity.

Using this classification algorithm, the main residual source of error in our matched galaxy maps is caused by variations in star-galaxy separation. Nevertheless some correlation analyses can be carried out as described in Maddox *et al.* (1988). Over the survey area the stellar density varies by a factor of ~ 3 from the galactic pole to low galactic latitudes. Most of the residual variation is caused by an increase in stellar contamination in the directions of the galactic centre and anticentre. Therefore we have developed a more sophisticated classification algorithm which uses all of the APM parameters.

6 Profile classification

The areal profile gives 8 measurements of the surface brightness profile for each image. Some typical stellar profiles are shown in *Fig. 6* together with some galaxy profiles. For a large range of magnitudes, the stellar profiles are very similar to each other, and the galaxy profiles are more varied with much lower surface brightnesses. Stars with $M < 10$ are faint enough to peak below the emulsion saturation and the profiles are linear in the $\log_{10}(D)$ *vs.* N_{pix} plot. Therefore their surface brightness profile is given by $f = Pe^{-(r^2/2\sigma^2)}$. The galaxy profiles are not very different to those of the stars for $M < 9$, so at faint magnitudes the classifications are not very reliable. The profiles of stars $13 > M > 10$ are affected by saturation, but the stellar profiles are still easily distinguishable from galaxies. Stars with $M > 13$ are bright enough that the halo affects the profile, and so the stellar profile becomes indistinguishable from the galaxy profiles.

A simple way of using all of the APM parameters is to treat each of the levels and the peak density as different measurements of the surface brightness profile, and use nine separate pairwise plots against magnitude. The radius of gyration is included as a tenth plot against magnitude. The distance of each image from the peak of the median stellar line is measured from each plot. Then for each image the residuals from the plots are summed in quadrature with a set of weighting factors, to give the final profile residual, ψ, which is equivalent to the integrated difference between the particular image profile and the stellar profile at the same magnitude,

$$\psi = \sum_{levels} \frac{(P(M) - P_{locus}(M))^2}{\sigma^2_{locus}(M)} . \tag{7}$$

Some fields show significant variation in the position of the stellar locus on the plots as a function of position over the plate. Therefore the position of the peak of the stellar locus for each paramater is calculated separately for each cell in a 8×8 array over the field. Then for every image the residual in each parameter is calculated from the peak position interpolated from the neighbouring cell centres.

The residuals are weighted by $\frac{1}{\sigma^2_{locus}}$ and so, assuming the errors are approximately Gaussian, the residual sum for stars has a χ^2 distribution. The σ_{locus} is calculated

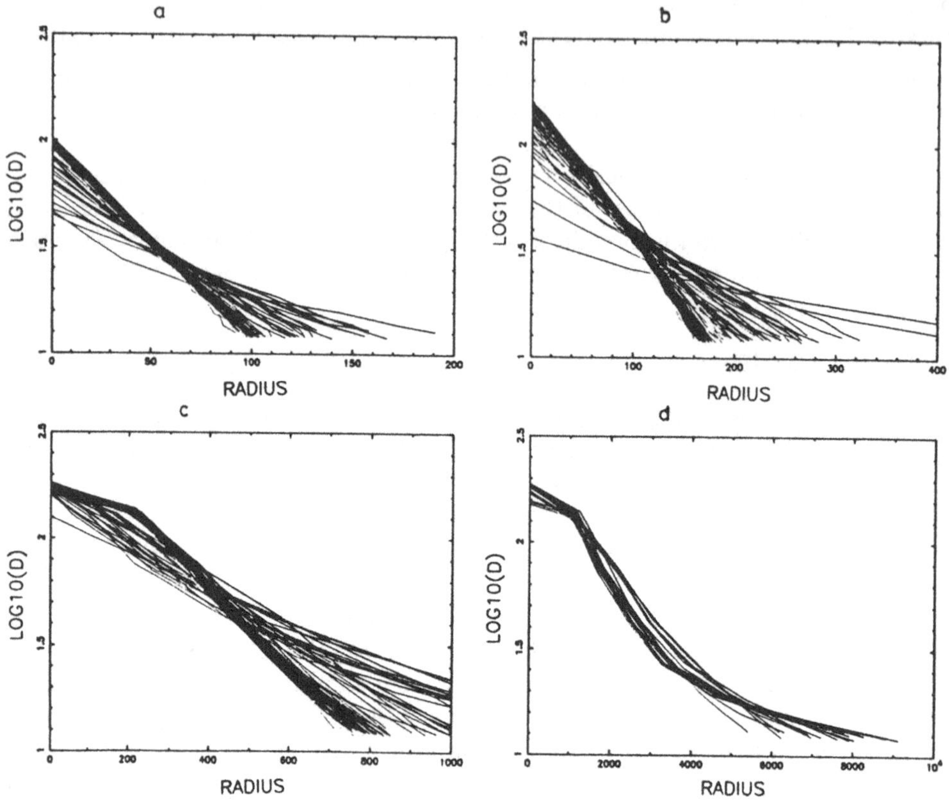

Fig. 8. Areal profiles of images at different magnitude slices. The density of each level is $D_n = D_t + 2^{n-1}$, and $\log(D_n)$ is plotted against N_n, the number of image pixels with $D > D_n$. For the fainter two slices, (a), $M = 9.0 - 9.1$, and (b), $M = 10.0 - 10.1$, the stellar profiles are all straight lines with the same slope. The galaxies are lower surface brightness, and more varied. For (c), $M = 12.0 - 12.1$, the stellar profiles are saturated near their peak, but are still easily distinguishable from galaxies. Brighter than $M = 14.0$, (d), the stars are saturated except for the halo, and so their profiles are indistinguishable from galaxies.

as a function of magnitude in each of the plots, so a fixed limit of ψ gives a sample of images which have a fixed minimum confidence level that they are significanctly non-stellar.

The σ_{locus} can be measured internally from the widths of the stellar locus in each of the plots, as in the STATS algorithm. Unfortunately, the variation in σ_{locus} for different plates makes the confidence levels different for each field and so it would be neccessary to adjust the classification boundary for each field to obtain a uniform sample of galaxies. In order to keep the scale of ψ the same for each field and also maintain uniformity as a function of magnitude, we assume that the σ_{locus} has the same magnitude dependence in each field. We have used the overlap between two fields to estimate the *rms* error in the measurements of each parameter and hence the σ_{locus} as a function of magnitude. The comparison of the measurements of each of the profile areas from two neighbouring fields show the *rms* errors are fairly well approximated by $\sqrt{N_{pix}}$, equivalent to Poisson noise in the count of pixels above each density level.

The distribution of $\log\psi$ with these weightings, is shown for images from a typical field in *Fig. 7*. The stellar images fainter than $M = 12.5$ have a distribution which is well approximated by a constant χ^2 for all magnitudes. The galaxy images are well separated from the stars for $12.5 > M > 9.5$. For images fainter than this, the seeing and increased noise make it impossible to distinguish a stellar image from a galaxy, and the confidence level that can be given to any image being non-stellar is smaller. For images with $M < 9.0$, no reliable distinction can be made.

In order to measure the reliability of this classification algorithm we have inspected 6 areas on each of 16 plates in a test region and compared with the parameter classification. Each area is about $1 \times 1\,\mathrm{cm}^2$ and contains about 100 images brighter than $B_J \approx 21$, which we classified as stars, galaxies and merged objects by eye. We agreed with each other for about 95 % of the objects, and the other 5 % of the images were too faint to be sure about. Histograms of the number of images as a function of the $\log\psi$ classification parameter are plotted for the visually checked samples of stars and galaxies in *Fig. 8*. Each plot shows images in a different magnitude range.

For images in the faintest range the star and galaxy distributions have a large overlap.

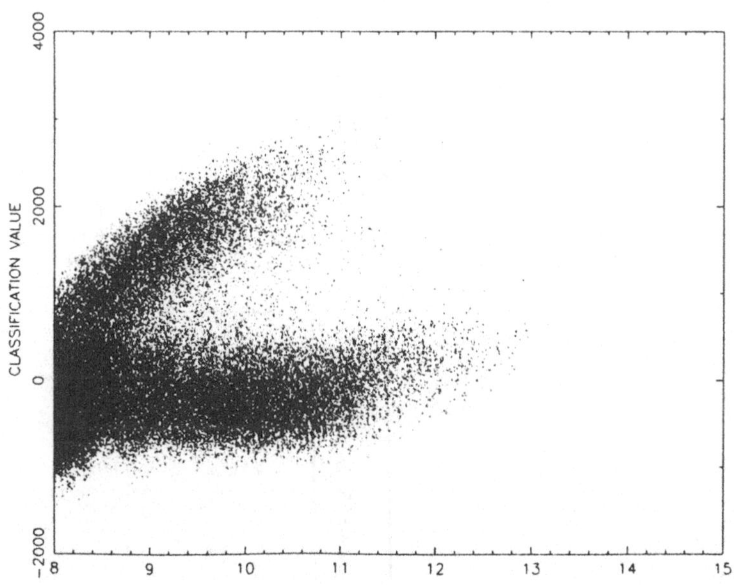

Fig. 7. The distribution of the ψ classification value for field 78. The stars fall along a well defined ridge with $\psi \approx 0$. The galaxies with $9 > M > 13$ fall in a well separated ridge with $\psi > 1000$. Near the plate limit, the atmospheric seeing blurs any image structure so the two ridges merge together and no distinction can be made for images fainter than $M \approx 8.5\,(B_J \approx 21)$.

These images are close to the plate limit, and their shapes are determined mainly by the the seeing profile. This means that it is very hard to tell the difference between stars and galaxies by eye, or from the measured parameters, and so it is not possible to select a complete and uncontaminated sample of galaxies. For the brighter ranges where the intrinsic galaxy profiles are visible, most stars fall in a narrow peak near zero as expected. The galaxies form a wider distribution extending to higher values because the intrinsic profiles are very different from the median stellar profile. Some high surface brightness galaxies have profiles which are very close to stellar, and so the galaxy distribution extends into the stellar distribution.

For images $M > 13$ the parametrisation of the shape does not contain all of the information available from the plate. The haloes around bright stars, and saturation of both stars and galaxies make the parameters for all objects very similar, and also increase the noise in the measurements. Therefore classifications based on the image parameters become unreliable and it is neccessary to visually check each bright image. We have visually classified galaxies brighter than $B_J = 16.3$ to give a sample of 11 963 galaxies with morphological types.

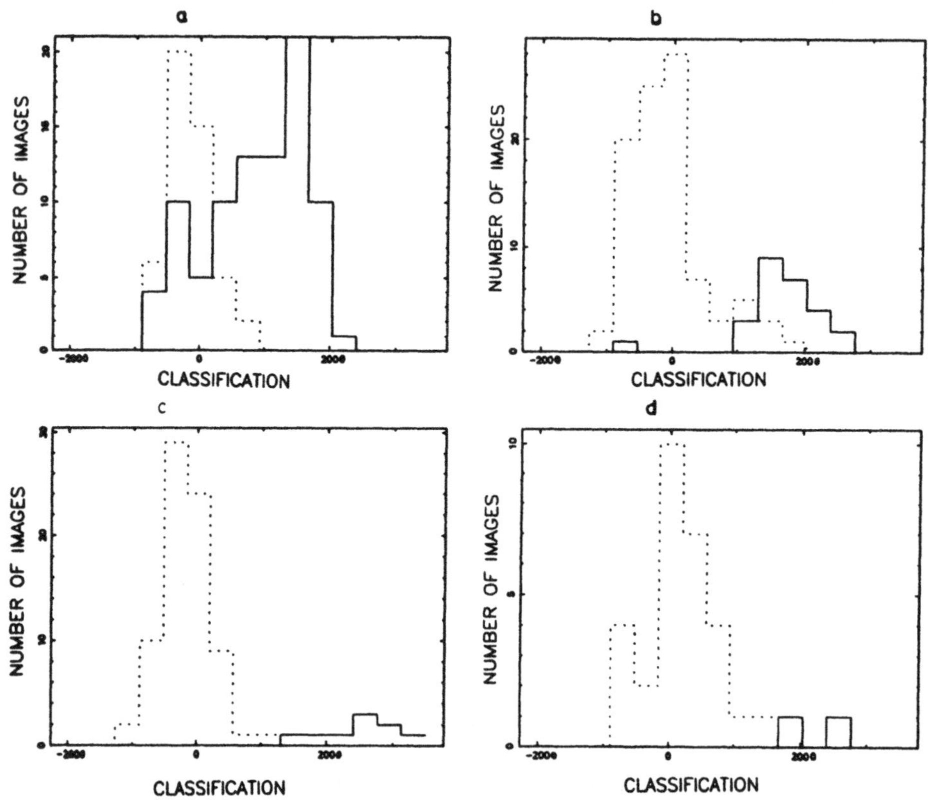

Fig. 8. Comparison of visual classification checks and ψ for 4 magnitude ranges as in *Fig. 6*. In each plot the dotted histogram shows the number of objects visually classified as stars as a function of ψ, and the solid histogram the number of objects visually classified as galaxies.

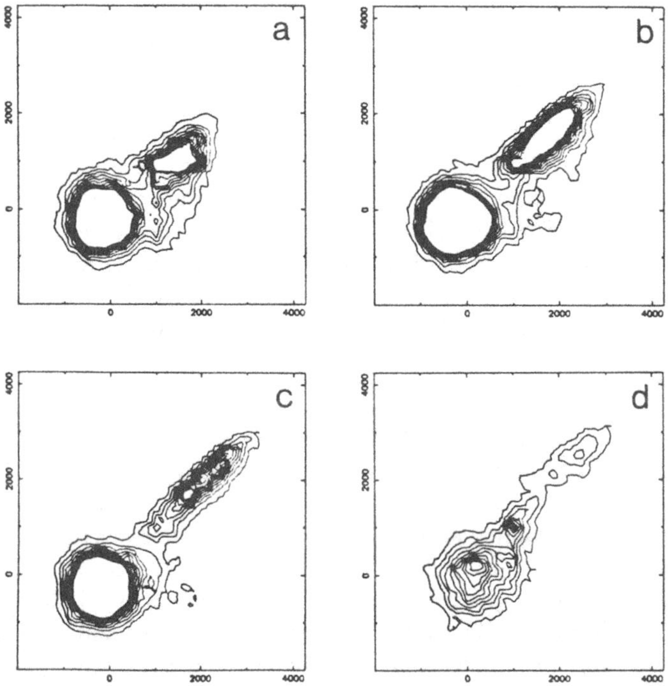

Fig. 9. Comparison of the ψ classification value parameter for images in the overlap between field 286 and field 287 for 4 magnitude ranges similar to those in *Fig. 6*. The axes of each plot are 1000 $\log_{10}(\psi_{286})$ and 1000 $\log_{10}(\psi_{287})$. The contours show the number of images in the cells of a 50×50 grid, smoothed by a Gaussian with $\sigma = 1$ cell. The lowest contour is for 4 images per cell and the interval between the levels is 2 images per cell. If the normalisation of the noise were exactly correct, the stars would be centred on (0,0) and the slope of the galaxy ridges would be 1.

The combined classification parameter ψ, measured from a field, can be compared with rescans of the same field and neighbouring fields to measure the repeatability of the classification. Comparison of the classifications of objects from repeat scans of the same plate shows that the ψ parameter is very repeatable. Similar comparisons of the classification parameter of objects in the overlaps show that the parameter is also repeatable for different plates and gives a reliable separation between stars and galaxies. A typical overlap is shown in *Fig. 9*, which is a series of contour maps of image number density in the ψ_1^2 *vs.* ψ_2^2 plane. The maps are for the same magnitude ranges as in *Fig. 8*. The stars form the large circular concentration of images at $(0,0)$ and the galaxies fall in the elongation towards $(2000, 2000)$. Over the magnitude range $9 < M < 13$ there is a clear separation between stars and galaxies that is the same for each plate.

The uniformity of galaxy selection from these parameters is very important when measuring the clustering properties of the distribution. If the selection varies from plate to plate, the fraction of galaxies that are included, and the fraction of contaminating stars will be different for each field within the survey. Such variations introduce spurious clustering in a similar way to changes in the magnitude limit. Therefore in our final catalogue we will match the classification parameters between plates in the same way as the magnitudes. The ψ limit used to define the galaxy sample can then be adjusted for each field so that the completeness and stellar contamination are kept constant over the survey. Even without the matching, the selection function is quite accurately uniform over large areas. *Fig. 10* shows the galaxies in the whole survey area selected using a fixed cut of the ψ classifier. The confidence level chosen is fairly high so that the variations in plate quality are not very significant and the boundaries between most fields are not visible (*cf. Fig. 2*). With this cut in ψ the completeness of the galaxy sample is about 70% at a limit of $B_J = 20.5$.

7 Summary

A preliminary reduction of the APM galaxy survey including relative and absolute magnitude calibration, and star galaxy separation has been completed. A more sophisticated and reliable star galaxy separation technique has been developed and applied to the survey. The technique allows full matching of the galaxy selection function on different fields. Also, a visually classified sample of 11 963 galaxies with $B_J < 16.3$ has been completed. This sample provides reliable classifications for the brighter galaxies which are difficult to classify from the image parameters, and so complements the deeper sample. Detailed correlation and cluster analyses of the galaxy distribution will be described elsewhere.

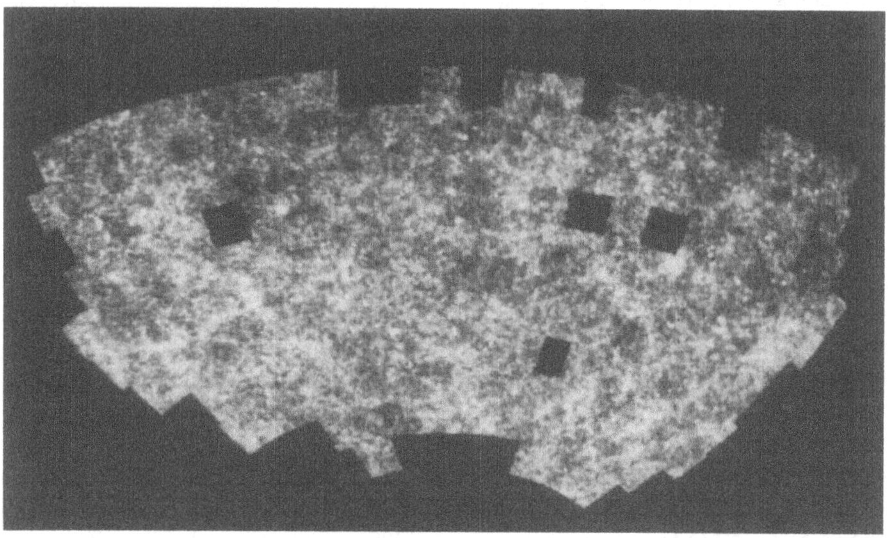

Fig. 10. Equal area projection of the surface density of 2.6×10^6 galaxies in the APM survey at a limit of $B_J = 20.5$ before matching the classification selection. A few field boundaries can be picked out, but most boundaries are not visible even though no classification adjustments have been made. The grey levels are set so that black = 0 image density, and white = 1.8×10^3 images per square degree.

Acknowledgements

We are indebted to the APM group, E.J. Kibblewhite, M.J. Irwin, P. Bunclark and M. Bridgeland, for their constant advice during this project. This work has been supported by the SERC and by the Nuffield Foundation. SJM, JL and WJS acknowledge the award of SERC research studentships.

References

Abell, G.O., 1958. *Astrophys. J. Suppl.*, **3**, 211.

Baker, A.E., 1925. *Proc. R. Soc. Edinburgh*, **45**, 166.

Blair, M., Gilmore, G., 1982. *Publ. astr. Soc. Pacific*, **94**, 742.

Godwin, J.G, Metcalfe, N., Peach, J.V., 1983. *Mon. Not. R. astr. Soc.*, **202**, 113.

Cawson, M.G.M., Kibblewhite, E.J., Disney, M.J., Phillipps, S., 1987. *Mon. Not. R. astr. Soc.*, **224**, 557.

Colless, M., Hewett, P., 1987. *Mon. Not. R. astr. Soc.*, **224**, 453.

Dawe, J.A., Metcalfe, N., 1982. *Proc. astr. Soc. Australia*, **4**, 466.

Groth, E.J., Peebles, P.J.E., 1986. *Astrophys. J.*, **310**, 507.

Høg, E., von der Heide, J., von Fischer-Trenenfeld, F., Holst, G., Loibl, B., Ziegler, U., Nikoloff, I., Helmer, L., Behr, A., 1970. *Abhandl. Sternw. Hamburg-Bergedorf*, Band 9.

Jarvis, J.F., Tyson, J.A., 1981. *Astr. J.*, **86**, 476.

Kibblewhite, E.J., Bridgeland, M.T., Bunclark, P., Irwin, M.J., 1984. In *Astronomical Microdensitometry Conference*, ed. Klinglesmith, D.A., NASA CP-2317, p. 277.

MacGillivray, H.T., Martin, R., Pratt, N.M., Reddish, V.C., Seddon, H., Alexander, L.W.G., Walker, G.S., Williams, P.R., 1976. *Mon. Not. R. astr. Soc.*, **176**, 265.

Maddox, S.J., Efstathiou, G., Loveday, J., 1988. In *IAU Symp. 130, The Structure of the Universe*, eds. Audouze, J., Szalay, A., Kluwer, Dordrecht (in press).

Murray, C.A., 1983. *Vectorial Astrometry*, Hilger, Bristol.

Peebles, P.J.E., 1980. *The Large-Scale Structure of the Universe*, Princeton Univ. Press, Princeton.

Sebok, W.L., 1979. *Astr. J.*, **84**, 1526.

Seldner, M., Siebers, B., Groth, E.J., Peebles, P.J.E., 1977. *Astr. J.*, **82**, 249.

Shane, C.D., Wirtanen, C.A., 1967. *Publ. Lick Obs.*, **22**, Part 1.

Walker, A.R., 1984. *Mon. Not. R. astr. Soc.*, **209**, 83.

West, R.M., 1978. In *Modern Techniques in Astronomical Photography*, eds. West, R.M., Heudier, J.L., ESO Publ., Geneva, p. 193.

Zwicky, F., Herzog, E., Wild, P., Karpowicz, M., Kowal, C.T., 1961-1968. *Catalogue of Galaxies and Clusters of Galaxies*, **I - VI**, California Institute of Technology, Pasadena.

Comparison of the Correlation Functions of Bright and Faint Galaxies

J. Hollósi
Space Telescope Science Institute
Baltimore, MD, USA, and
Roland Eötvös University
Budapest, Hungary

G. Efstathiou
Institute of Astronomy
Cambridge, England, UK

Abstract

It is a generally accepted assumption that correlation is independent of absolute magnitude. We choose two samples from the CfA galaxy catalogue: *bright galaxies* with magnitudes $M \leq -19.16$ and *faint galaxies* with $M \geq -17.16$. We have found that the bright galaxies are *significantly stronger* correlated than the faint galaxies. In the most characteristic range of separations the ratio of pair-frequencies is nearly constant: $(\xi_{bright}(r)+1)/(\xi_{faint}(r)+1) = 2.15 \pm 0.1$ for $2\ \mathrm{Mpc} \leq r \leq 8\ \mathrm{Mpc}$.

1 Motivation

The main question is the following:

Does the correlation of galaxies depend on absolute magnitude?

If we had a large volume-limited catalogue of *bright* and *faint* galaxies in the same region, we could answer this question easily. Unfortunately we do not. Thus we have to try to find the answer based on the known catalogues. The result should be important to check the theories of galaxy evolution.

2 Data

The CfA redshift catalogue was used for the investigation. This magnitude-limited catalogue ($m \leq 14.5$) within the geometrical boundaries $b \geq 40°$; $\delta \geq 0°$ or $b \leq -30°$; $\delta \geq -2.5°$ contains 2381 galaxies.

Throughout this article we will use $H_0 = 100\,\mathrm{km\,s^{-1}\,Mpc^{-1}}$.

3 The luminosity function

The luminosity function can be approximated by the Schechter form:

$$\Phi(M)\,dM = \frac{2}{5}\ln(10)\cdot\Phi^*\cdot\left[10^{0.4(M^*-M)}\right]^{\alpha+1}\cdot\exp\left[-10^{0.4(M^*-M)}\right]\,dM\;, \quad (1)$$

where $M^* = -19.16$ and $\alpha = -0.96$ for $v_r \geq 300\ \mathrm{km\,s^{-1}}$.

The numerical parameters of the luminosity function were determined by Efstathiou *et al.* (1988). The most important parameter for us is M^* because the following definitions are based on its value.

We call a galaxy *bright* if its absolute magnitude is less than M^*, and *faint* if its absolute magnitude is larger than $M^* + 2$.

Fig. 1 shows the best fitting Schechter function.

4 The selection function

The selection function of *bright* and *faint* galaxies can be derived from the luminosity function by the simple formula:

$$p(x) \propto \int_{M_{min}(x)}^{\infty} \Phi(M)\,dM\;, \quad (2)$$

where $M_{min}(x)$ is the minimal magnitude which can be observed from the distance x.

Figures 2 and *3* show that we cannot find any region where the selection functions are approximately the same, thus we have to estimate the correlation function of *bright* and *faint* galaxies separately.

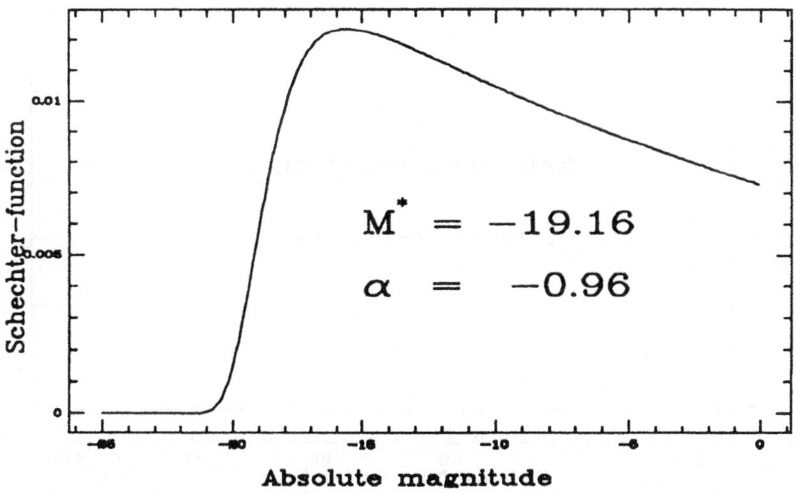

Fig. 1. The best fitting Schechter function.

5 Estimation of the correlation functions

Several random catalogues were generated using the luminosity function. The *bright* and *faint* galaxies were picked out separately and the following estimation was used:

$$\xi(r) \approx \frac{\langle DD \rangle_r}{\langle DR \rangle_r} - 1 \,. \tag{3}$$

The symbol $\langle \ \ \rangle_r$ means weighted number of *Data* – *Data* and *Data* – *Random* pairs whose distances are r. Every pair was weighted with $1/\big(p(x_1) \cdot p(x_2)\big)$, where x_1 and x_2 denote the distances from us. To avoid large errors coming from small values of $p(x)$ we cut the samples where $p(x)$ decreases to 0.1. This method simulates volume limited samples.

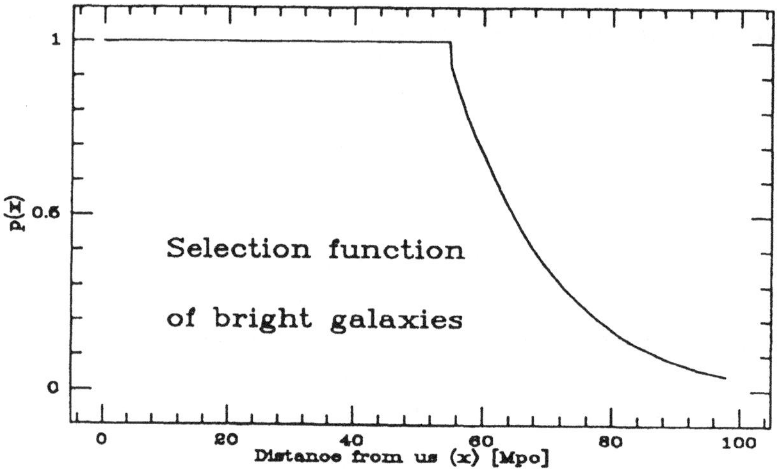

Fig. 2. The selection function of *bright* galaxies.

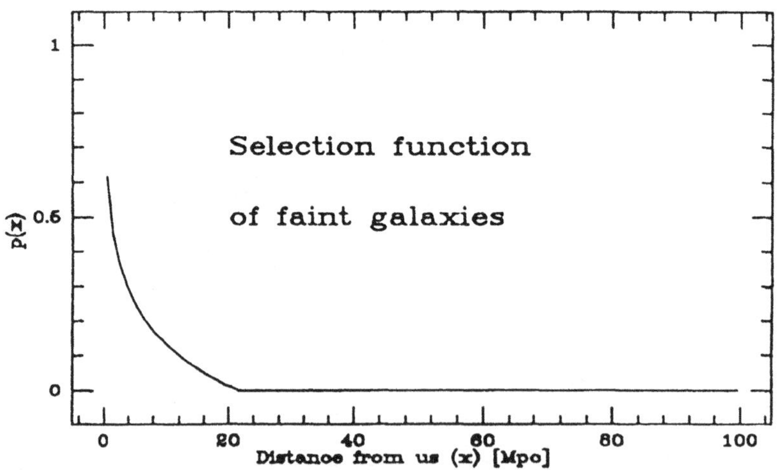

Fig. 3. The selection function of *faint* galaxies.

6 Result

The estimated correlation functions are plotted in *Fig. 4.* Note that the correlation function of *faint* galaxies breaks at about 8 Mpc because of its distance limit.

Let $\chi(r)$ be defined as follows:

$$\chi(r) = \frac{1 + \xi_{bright}(r)}{1 + \xi_{faint}(r)} . \tag{4}$$

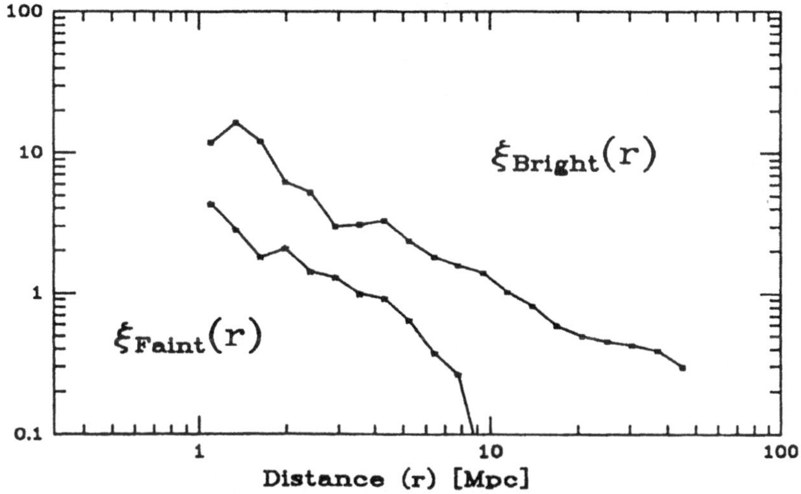

Fig. 4. The spatial correlation functions of *bright* and *faint* galaxies.

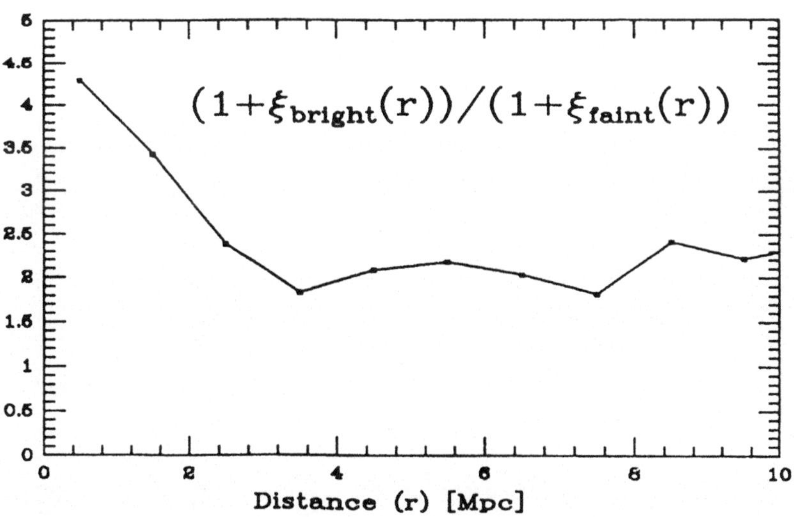

Fig. 5. Ratio of the normalized *bright* and *faint* pair frequencies.

Then we can establish (*Fig. 5*) that

$$2.5 < \chi(r) < 4.5 \qquad\qquad r < 2\,\mathrm{Mpc}\,,$$

and

$$\chi(r) = 2.15 \pm 0.1 \qquad\qquad 2\,\mathrm{Mpc} < r < 8\,\mathrm{Mpc}\,.$$

Above 10 Mpc ξ_{faint} is too noisy to estimate the ratio.

The constant ratio for distances 2 – 8 Mpc is interesting, but our most important result is that on small scales the *bright* galaxies are significantly stronger correlated than the *faint* galaxies. This contradicts the widespread and accepted assumption that correlation is independent of absolute magnitude.

Further studies are necessary to obtain a final answer.

References

Efstathiou, G., Ellis, R.S., Peterson, B.A., 1988. *Mon. Not. R. astr. Soc*, **232**, 431.

The Muenster Redshift Project. Automated Analysis of Galaxy Clustering on Schmidt Plates

H. Horstmann
Astronomisches Institut
Westfälische Wilhelms-Universität
Münster, F.R. Germany

Abstract

The Muenster Redshift Project (MRSP) collects information about the three-dimensional distribution of galaxies by processing pairs of direct and objective prism Schmidt plates. This contribution describes the role of direct plates in the survey. Methods of analysis and their errors are discussed. Maps of two-dimensional galaxy distributions are presented, including the distributions of different types of galaxies (using ellipticity as a coarse morphological criterion). We present preliminary results for a region near the South Galactic Pole, containing several clusters of galaxies, of which five form a physical group at redshift $z = 0.11$.

1 Introduction

The basic goal of the MRSP is to obtain information on the three-dimensional distribution of galaxies and the properties of matter and space which can be derived from it. The basic data are obtained by processing automatically pairs of direct and objective prism Schmidt plates. The material consists of ESO/SRC atlas plates (film copies of IIIa-J plates) and film copies of IIIa-J objective prism plates with a dispersion of 246 nm mm^{-1} at Hγ. Both kinds of plates are taken with the UK Schmidt telescope.

The faint limiting magnitudes ($21\overset{m}{.}5$ on the direct and $20\overset{m}{.}5$ on the prism plates) and the large area of the sky covered, even by a single plate, enables us to study extended and relatively distant structures in the distribution of galaxies. The large numbers of objects detected, typically 150 000 on the direct and 50 000 on the objective prism plates at high galactic latitudes require fully automatic reduction procedures on the level of single objects.

The processing of objective prism plates, in particular automatic redshift measurements, are described in a paper by Schuecker (1988). The present contribution describes the reduction of the direct plates and, in particular, the study of internal properties of clusters of galaxies based on these data.

2 The role of direct plates in the MRSP

2.1 Service functions of the direct plates

The primary function of the direct plates is to supply information which can be obtained more easily and/or with higher precision from direct images. For example, the **segmentation** (search for objects) can cause substantial problems when executed on the objective prism plate, especially in regions of high object density (clusters of galaxies), because the relatively long spectra often overlap. Even if these overlaps are very slight, it is difficult to recognize them properly on the objective prism plates. Other examples are the object positions in R.A. and Decl., star/galaxy separation, and measurement of apparent magnitude.

The transport of information from the direct to the objective prism plate is possible after coordinate transformations have been applied, using 12 to 15 reference stars. Object positions on the objective prism plate are predicted with an accuracy of $1''$ to $2''$ over the whole plate, which is sufficiently precise for object identification.

A more sophisticated use of the direct plate is made in the determination of wavelength zero points for the objective prism spectra. A widely used procedure takes the emulsion cutoff as wavelength reference. This, however, introduces large systematic errors due to the brightness and colour dependence of the cut-off position. In the MRSP a high accuracy plate transformation is employed, using approximately 1 000 reference stars and a higher order transformation model. This procedure is described in detail in the paper of Tucholke (1988).

2.2 Deeper surveys from direct plates

There is, of course, more information available on the direct plates than is needed for processing the objective prism plates: the direct plates reach fainter objects and supply (coarse) morphological parameters of the galaxies. When deep direct plate studies are applied to clusters of galaxies – whose distances are measured on the objective prism plates – shapes, orientations, density profiles, morphological contents and luminosity functions of the clusters can by derived, while the redshift data help to disentangle clusters superimposed on each other along the line of sight.

3 Plate digitization and segmentation

3.1 Digitization

The plates are scanned with the PDS 2020 GMplus microdensitometer at the Astronomical Institute Muenster (AIM). The central $300 \times 300\,\mathrm{mm}^2$ of the plates are digitized with a step size of $15\,\mu m$ or about $1''$. This corresponds to an area of $5.5 \times 5.5\,\mathrm{deg}^2$ on the sky. The scan time for a whole plate is presently about 20 hours, if the maximum mechanical speed of the machine is used. The density which is measured by the PDS ranges from density 0.0 to 4.5, with a resolution of 0.00125 over the whole measurable range.

The search for objects is performed online. An *object* is defined as a group of connected pixels (6 pixels in the case of the ESO/SRC atlas plates) which exceed a minimum density threshold above sky background. The latter is determined by applying a strong median filter to a scanline (or, in practice, to each 10th line, since the sky background is varies very slowly) the significance level is defined at 3σ above sky background. The **search** process leads to typically 150 000 objects in a high galactic latitude field, varying between 125 000 and 195 000 for the 9 plates processed so far. During the online search, only a few *object parameters* are computed: object position, obtained by density weighted object moments, area covered by the object (in pixels), extent of the object in x and y, local plate density and local standard deviation of the plate density.

3.2 Segmentation

Together with the object parameters, the images themselves are stored permanently in the form of quadratic scan segments, $21 \times 21\,\mathrm{arcsec}^2$, called **picture frames**. The objects found on one plate can be saved on one to several high density tapes. The permanent storage permits direct access to an individual star or galaxy image (not only to a list of image parameters). This makes subsequent processing possible without going back to the plate. All classification processes become very fast by simply using the digitized images.

The following image analysis, including the computation of effective radius, central intensity, ellipticity, orientation, apparent magnitude and other parameters, uses these picture frames. It is obvious that the limited size of the frames leads to systematic errors for the brighter objects whose outer parts are lost. For galaxies these errors arise at about $16\overset{\mathrm{m}}{.}5$ or brighter, depending also on morphological type. The most important error is the magnitude error. Because of this, bright objects must either be excluded from the survey or looked at with larger picture frames.

Generally, the restrictions can be tolerated, mainly for two reasons. The first is that galaxies brighter than about 17^{m} are rare objects compared with the total number of galaxies found in deep surveys. The second reason is, that we cannot measure redshifts for these extended objects with the slitless technique. The bright dominant galaxies in the cluster centers are treated interactively during investigation of the cluster morphology (see Sect. 6).

4 Object classification

4.1 Star/Galaxy separation process

In the following, a brief description of the **star/galaxy classification** algorithm is given. For information about complications due to instabilities of the plate parameters see also Horstmann (1988).

For star/galaxy separation the difference in the *peak density/effective radius* relations for stellar and non-stellar images are used. *Fig. 1* shows a diagram with several thousand objects detected on an ESO/SRC atlas plate. The classification *stellar/non-*

stellar image is performed by defining a discriminating function along the right hand side of the stellar image strip. This curve (spline interpolation) is established automatically by fitting Gauss curves to the stellar density distributions at several positions of the stellar curve (see *Fig. 2*).

4.1.1 Field dependent star/galaxy separation

The curve separating stars from galaxies cannot be assumed constant over the whole plate. The vertical as well as the horizontal part of the stellar curve in *Fig. 1* changes slightly with position on the plate. Reasons for this are primarily the variations of the plate background density which in turn is caused mainly by desensitization during plate exposure (Malin 1983, Dawe *et al.* 1984). Some maps of the background density are shown in Horstmann (1988). Other reasons may be coordinate-dependent

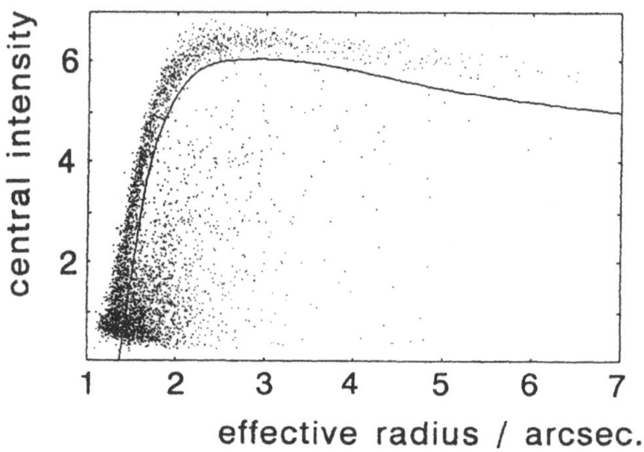

Fig. 1. Central intensity vs. effective radius for several thousand objects from the ESO/SRC field No. 411. The solid line separates stars from galaxies.

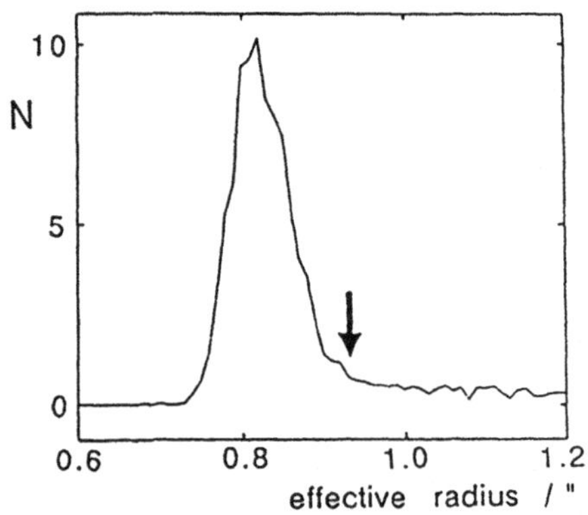

Fig. 2. Tracing through the stellar distribution in *Fig. 1* at about half the maximal intensity. The arrow indicates the 3σ point which defines the location of the star/galaxy separating function.

defocussing during the exposure of the plates and/or during scanning. We correct for this effect by dividing the plate into 25 fields. For each of these fields the discriminating function is determined separately with the methods described above. For each position between the centers of these reference fields, the functions are interpolated.

4.1.2 The discriminating curve

Figure 2 shows a tracing across the stellar curve at about half maximum intensity. The location for the fitting points is chosen at the $3\,\sigma$ level obtained from the Gauss fit and is indicated by an arrow in *Fig. 2*. If the distribution is Gaussian, this position leads to a misclassification rate for stars of about 0.1 %. As the plot shows, the tracing is not strictly Gaussian, in particular, the object density on the right side is significantly larger, making the stellar distribution asymmetric. Nevertheless, from inspection by eye we have obtained an error rate for objects brighter than about 20^{m} of no more than 2 %.

For fainter objects star/galaxy separation becomes more difficult. Stellar and galaxy regions overlap and it is hard to establish a discriminating function which optimally separates stars from galaxies. A reference sample cannot be obtained from classification by eye, because the personal errors become very large at faint magnitudes and strongly influence the result. The discriminating function is thus extrapolated into the region of the faintest objects. We expect that the classification error up to the completeness limit of a given plate, typically 21^{m}, is never larger than a few %. This is also supported by the two point correlation functions for the star and galaxy distributions (Sect. 4.2).

4.1.3 Recognition of double stars

After distinguishing between stellar and non-stellar images, all *non-galaxies* among the non-stellar images must be eliminated. They are blends of star and galaxy images and plate flaws, dust *etc.* For a non-disturbed galaxy we expect, in the majority of cases, that the position of the density center does not depend on the density- or intensity-threshold used in image analysis, and that the maximum density is reached at the position of the density center. By testing the rejection parameters interactively, we have obtained $0\overset{''}{.}75$ as a suitable maximum for the permitted deviation between object centers at various density levels. For a undisturbed galaxy image we expect that the maximum image intensity does not exceed the central intensity by more than 15 %.

The rejection rate as a function of apparent brightness is shown in *Fig. 3.* The increase of the rejection rate with decreasing object brightness follows from the fact that an object can be influenced only significantly by objects of comparable brightness, and that therefore the probability of being disturbed by a neighbour increases for faint images. For objects fainter than 22^{m} (most of which are nothing but plate noise), the rejection rate increases steeply to 100 %.

It is apparent from the description of the classification algorithm, that it cannot distinguish between stars and galaxies when the images are saturated. The apparent magnitude at which this occurs depends for galaxies on the morphology of the object; but normally only galaxies brighter than 15^m to 16^m will reach maximum plate density. For these objects we have developed a method which makes use of the fact that on Schmidt plates bright stellar images show spikes, while galaxy images do not. For the data described in the following, this classifier was not employed, since the images were already excluded because they exceed the limit of the picture frames.

4.2 Results from the star/galaxy separation

As one result of the star/galaxy separation process, *Fig. 4* shows the two-point angular correlation functions $w(\theta)$ for stars and galaxies up to the limiting magnitude 21^m. We have counted stars and galaxies in cells of $3'.3 \times 3'.3$ and used the estimator (Peebles 1975, Hewett 1982)

$$w(\theta) = \frac{\langle n_1 n_2 \rangle}{\langle n_1 n_2 \rangle} - 1 \,, \tag{1}$$

where the n_1, n_2 are cell counts and $\langle\ \rangle$ denotes averaging over all pairs $n_1\ n_2$ with cell separation θ. The functions indicate the expected null correlation for a purely stellar sample and the expected correlation on the scale of 1° for the galaxies. All data are taken from field No. 411.

5 Morphological classification of galaxies

Since the work of Dressler (1980), Tully and Shaya (1984) and others, the distribution of different morphological types of galaxies has become of great interest. The MRSP data seem well suited for such studies, because they are complete up to faint magnitudes (at least to $20^m.5$) and because large areas of the sky are covered. Additionally, the automated classification has the advantage of being free of any personal bias. On the other hand the problem arises of how to obtain morphological information at apparent magnitudes around 20^m. Classification using object colours is possible, but it requires the scanning of additional direct plates in different spectral regions. They

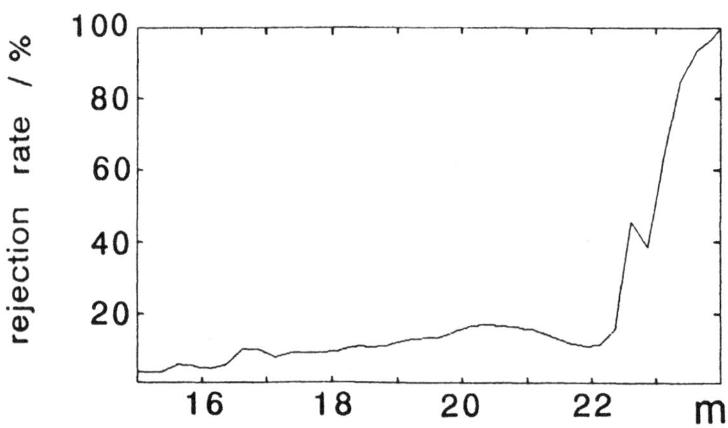

Fig. 3. Object rejection rate as a function of apparent magnitude. See text for details.

are not available at the moment (even the ESO red survey is still very incomplete). Colour measurements from the objective prism spectra cannot be used because the blue parts of the spectra are missing at these faint magnitudes. We have attempted to perform a classification using intensity profiles, but it appeared that these tracings, with a length of typically four or five pixels, do not contain enough information.

The quantity finally used for the classification is the *apparent ellipticity* of the galaxy images. The basic idea is that the images of elliptical galaxies seldom show extreme axial ratios. Ellipticities larger than about 0.5 occur for much less than 10 % of all E type galaxies (Sandage *et al.* 1970, Schechter 1987 and references therein). The images of spiral galaxies, on the other hand, show high axial ratios when seen more or less edge-on. Ellipticities larger than 0.5 should occur for about 30 % of all images of S type galaxies. Thus, the content of objects with large ellipticities should be an indicatior of the morphological content of a given cluster of galaxies or of a given region on the sky.

Because the classification is statistical, one cannot decide for a single object whether it is an E- or S-type galaxy. Additionally, some problems arise because fainter objects appear to have systematically lower ellipticities, as is shown in *Fig. 5.* This is due to seeing effects, which are more severe for faint objects where they tend to reduce the ellipticity. Also, experience shows that objects consisting of very few pixels tend to be 'rounder' than brighter and larger images, partially due to aperture effects from the scans.

Ellipticities are computed as

$$E = \frac{\sqrt{(M_{2,0} - M_{0,2})^2 + 4M_{1,1}^2}}{M_{0,2} + M_{2,0}} \tag{2}$$

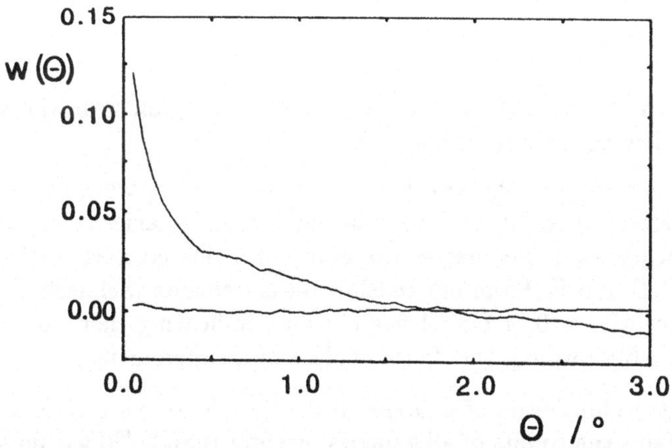

Fig. 4. Two-point angular correlation function for all galaxies (upper curve) and stars up to 21^{m} from field No. 411. Note that the stellar images do not show any significant clustering, as is expected in the direction of the galactic pole.

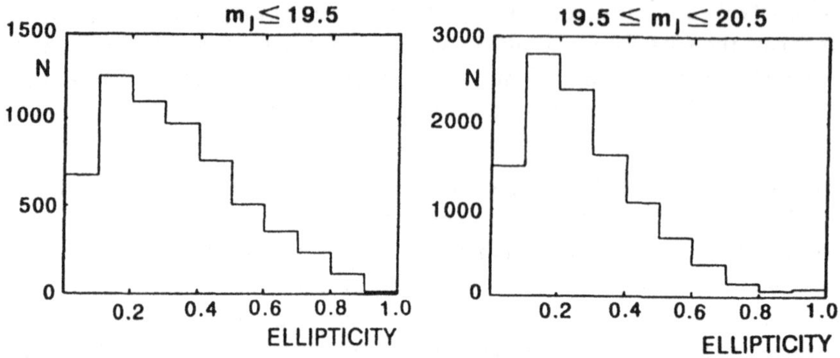

Fig. 5. Apparent ellipticity histograms for galaxies in field No. 411. Left: Galaxies brighter than $19^{\mathrm{m}}.5$. Right: Galaxies between $19^{\mathrm{m}}.5$ and $20^{\mathrm{m}}.5$. The 'rounder' images at fainter magnitudes are due to seeing and digitization effects.

with the moments $M_{i,j}$ defined as

$$M_{i,j} = \sum_{A} (x - x_c)^i \, (y - y_c)^j \, I(x, y) \tag{3}$$

where A is the summation over all pixels of a given object, $I(x, y)$ the intensity above the local sky background intensity, and x_c and y_c are the coordinates of the object center.

The equations from which the ellipticity is derived include contributions from all intensity levels; the axial ratios are not determined at a fixed intensity level only. There is a noticeable lack of objects in the interval 0.0 to 0.1 in *Fig.* 5, which may be an artefact, although this has also been found by other authors (*e.g.* Benacchio and Galetta 1980), and, recently, by Davies *et al.* (1988), who find a similar deficit of 'round' systems for dwarf ellipticals in the Fornax Cluster region.

6 Results

We have applied the methods described above to a region near the south galactic pole. In the following, first results are given.

Figure 6 shows a map of the galaxy distribution in three adjacent ESO/SRC fields for all galaxies brighter than $20^{\mathrm{m}}.5$. The total number of galaxies is about 60 000. The most striking feature on this map is the group of 5 rich clusters in the central field, marked A, B, C, D and E. From our redshift measurements (Schuecker 1988) we have found a distance of $z = 0.11$ for all five clusters, indicating that they are physically connected, probably forming the dense nucleus of a supercluster.

Histograms of the ellipticities of galaxies in the five clusters are given in *Fig. 7*. Each histogram contains the counts of all galaxies brighter than $20^{\mathrm{m}}.0$ within a radius of 15' around the center of the corresponding cluster. As the histograms show, the counts at higher ellipticity levels are significantly lower than one would expect from the histograms of all galaxies from the whole plate. With our interpretation of ellipticities

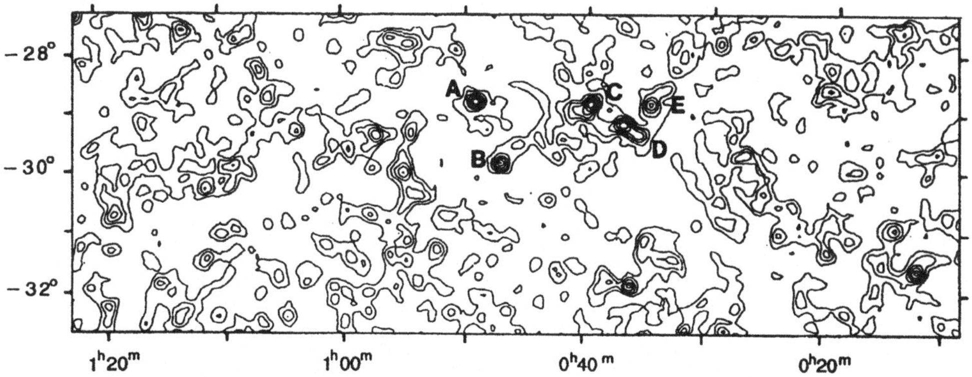

Fig. 6. Maps of the galaxy distribution obtained from three fields, Nos. 410, 411 and 412 (from right to left), near the South Galactic Pole. The isopleths correspond to 1.1, 1.4, 1.7... times the mean galaxy backround density. The five clusters A to E are at the same distance and probably form the nucleus of a galaxy supercluster. See the text for details.

this means that compared to the field *spiral galaxies are underrepresented in rich clusters*, in agreement with the results of Dressler (1980) and others.

Our result is also supported by the correlation functions for objects in different ellipticity intervals. *Fig. 8* shows plots of the two-point angular correlation functions for the three fields of *Fig. 6*, for galaxies with ellipticities below and above 0.4 separately. For fields 412 and 410, both showing few prominent clusters, the two curves are nearly identical. For the central field 411, containing the five large clusters, the low ellipticity galaxies have significantly stronger correlations than those of higher ellipticity.

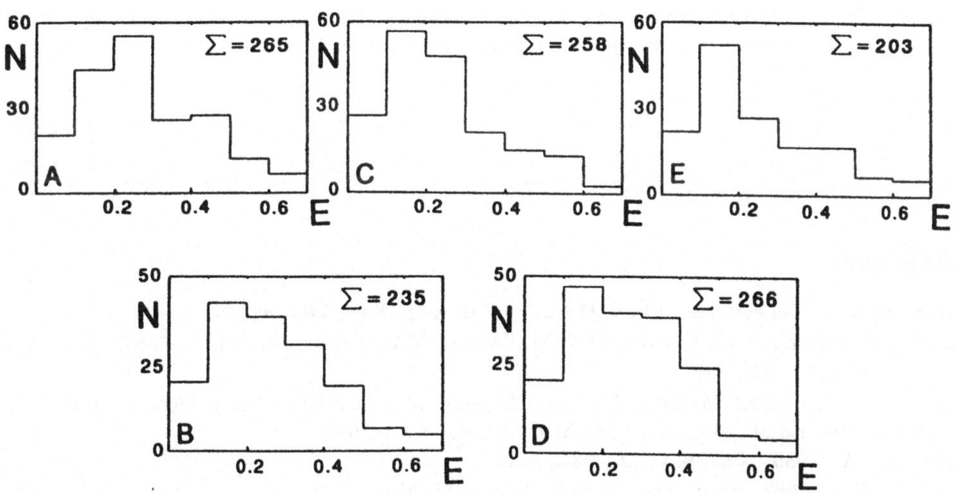

Fig. 7. Histograms of galaxy ellipticities for the five clusters marked in *Fig. 6*.

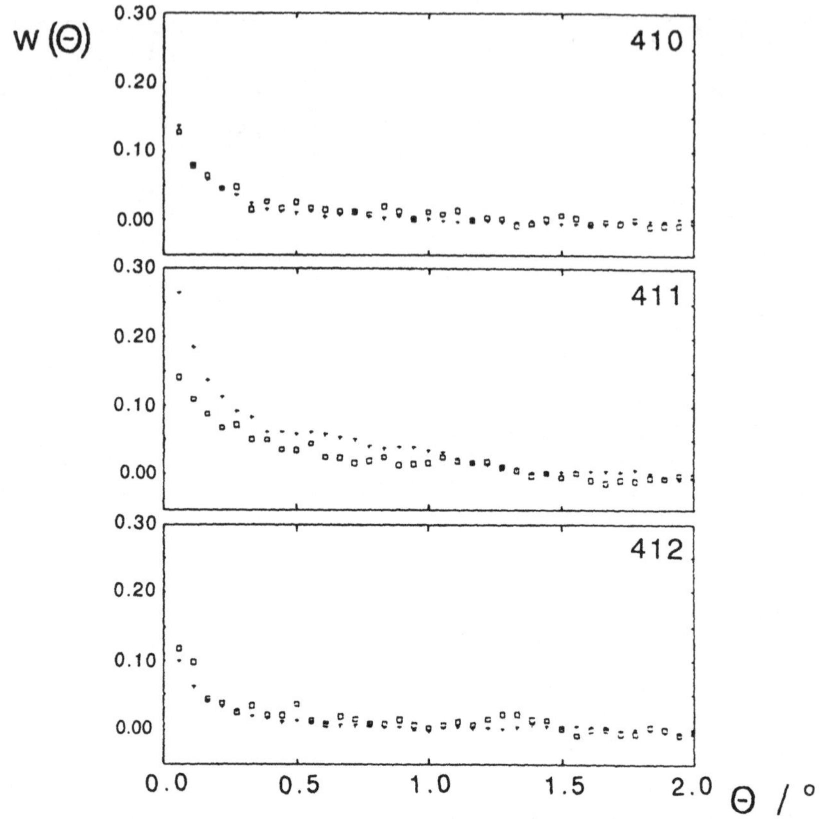

Fig. 8. Two-point angular correlation functions for galaxies brighter than $20^{m}.0$ from ESO/SRC fields 410, 411 and 412. In field 411, containing five prominent clusters, galaxies with apparent ellipticities less than 0.4 (crosses) show stronger clustering than those of higher ellipticity (rectangles). In the two other fields, containing few prominent clusters, the correlation functions for both populations show no significant differences.

References

Benacchio, L., Galetta, G., 1980. *Mon. Not. R. astr. Soc.*, **193**, 885.

Davies, J.I., Phillips, S., Cawson, M.G.M., Disney, M.J., Kibblewhite, E.J., 1988. *Mon. Not. R. astr. Soc.*, **232**, 239.

Dawe, J.A., Coyte, E., Metcalfe, N., 1984. In *Astronomical Photography*, Occasional Reports Royal Obs. Edinb., **14**, eds. Sim, M.E., Ishida, K., p. 59.

Dressler, A., 1980. *Astrophys. J.*, **236**, 351.

Hewett, P.C., 1982. *Mon. Not. R. astr. Soc.*, **201**, 867.

Horstmann, H., 1988. In *Proc. IAU Workshop Astrophotography*, ed. Marx, S., Springer, Berlin, p. 186.

Malin, D.F., 1983. In *IAU Coll. No. 78, Astronomy with Schmidt–Type Telescopes*, ed. Capaccioli, M., Reidel, Dordrecht, p. 57.
Peebles, P.J.E., 1975. *Astrophys. J.*, **196**, 647.
Sandage, A., Freeman, K.C., Stokes, N.R., 1970. *Astrophys. J.*, **160**, 831.
Schechter, P.L., 1987. In *IAU Symposium No. 127, Structure and Dynamics of Elliptical Galaxies*, ed. de Zeeuw, T., Reidel, Dordrecht, p. 217.
Schuecker, P., 1988. *These proceedings*, p. 142.
Tucholke, H.J., 1988. *These proceedings*, p. 136.

Part II

Three-Dimensional Distribution of Galaxies

Heinz–Albert Ott

Peter Schuecker

Hans–Joachim Tucholke

Waltraut Seitter

On the Use of Redshift Surveys in Observational Cosmology

H.-A. Ott
Astronomisches Institut
Westfälische Wilhelms-Universität
Münster, F.R. Germany

Abstract

Redshift surveys of galaxies are an important tool for studying the large-scale distribution and motion of luminous matter. The shape and size of the survey volume, the sampling procedure, the number density of objects, and the redshift error characterize the *power* of a survey.

1 Introduction

Our universe contains a population of discrete physical objects called *galaxies*. The observation of their spatial distribution and motions is one of the subjects of empirical cosmology.

Within standard Big Bang models aggregations of baryonic matter, like galaxies, are rather late events in the course of universal evolution. When using galaxies as tracers of matter distribution and motion, there are, however, good reasons to expect the relics of primordial phases "frozen" into the phenomena observable today.

One of such relics is the Hubble flow (HF). Its high isotropy with an intrinsic dispersion of less than 15 % is suggested empirically (Sandage 1987). HF isotropy is also expected theoretically, even when large-scale inhomogeneities occur throughout the distribution of matter (Silk 1974). The Hubble law, apparent in the HF, proves to be the most important tool for studying the large-scale spatial distribution of galaxies. It opens the opportunity to investigate relics of primordial density fluctuations and global properties of the matter dominated universe.

Among the three distance-dependent parameters of basic cosmological interest (redshift, apparent size and apparent luminosity of galaxies) redshift is the only one not affected by the physical properties of the objects, but directly related to their motions. On large scales, the HF dominates over the local "noise" components in the redshifts, defining a cosmological domain comparable to the continuous fluid of the world models.

Besides isotropic bulk motion, the standard world models demand also homogeneous distribution of matter. Empirically, we know that our deepest observations to date have not yet encountered a global scale limit, above which galaxies appear to be distributed homogeneously. This does not necessarily mean that we must look farther into space. It may be that our *surveying strategies* are insufficient for treating the problem in practice, although the observational range might already permit a solution.

Appropriate strategies have to include all relevant observational aspects like coverage, completeness, redshift errors *etc.*, in order to optimize the route to a pre-defined goal. It is beyond the scope of this contribution to present perfect strategies of this kind. Instead, we will test the *power* of different concepts pertaining to redshift surveys.

2 Problem solving through observational statistics

In contrast to the space-time continuum of theory observational cosmology deals with the discrete distribution of galaxies. Mapping individual galaxies in redshift space provides the data base for structural investigations, which necessarily require the use of statistical methods.

Because we can only take *samples* of the real distribution, the procedure how to do it is of major importance for a survey. In order to get a fair, *i.e.* representative sample, different observational circumstances have to be taken into account: sample size, volume fraction of the universe investigated, statistical access to the data, etc. The completeness, often quoted as a relevant factor, though important, depends on the chosen parameter limitations, it does not guarantee a fair sample a priori! For example: observing a small volume fraction of the accessible universe always means observing a small parameter subspace. Therefore, according to a basic statistical theorem, one must expect a poor approximation of the whole population characteristics, irrespective of the completeness level. The best available sampling procedure is that which provides a set of least biased data.

To analyse these data again statistical methods have to be used. They constitute the link to all physical interpretations of the galaxy distribution – on global scales as well as on smaller local scales. Redshift, as an indicator of radial position and of the age of galaxies, opens the door to several cosmological applications, not possible with two-dimensional statistics only. Some examples are given below:

GLOBAL STRUCTURE

Global investigations aim at properties of the universe which have been predicted from world models. The Cosmological Principle, as a basic ingredient of the relativistic Friedmann models, implies the idea of spatial homogeneity of the cosmic density field above some scale limit. With redshift surveys of galaxies, the homogeneity can be tested, at least in principle, by a simple spatial averaging procedure (Stoeger *et al.* 1987). The degree of homogeneity, found in the observed distribution, indicates the statistical reliability and the justification of global investigations.

The universal expansion parameters H_0, q_0 can be determined *e.g.* by using redshift-flux relations for objects of known luminosity and number-counts in different redshift bins. The latter is effectively a redshift-volume test, less affected by galaxy evolution than the flux-number test (*e.g.* Weinberg 1972) and, furthermore, sensitive for all kinds of gravitating matter, including non-baryonic matter (Loh 1988). With properly estimated values of H_0, q_0 deeper surveys open the chance of studying the evolution of galaxies and large-scale structures of galaxies. Combining the redshift-volume and redshift-flux tests and using assumptions concerning luminosity and/or density evolution, the present curvature of the universe may be estimated (Ehlers and Rindler 1987).

LOCAL STRUCTURES

Large-scale structures of different sizes and shapes can be investigated using a variety of statistical methods (for a review of statistical methods see Murtagh and Heck 1987).

Statistically, the simplest technique is multiple binning into different volumes of space, but (aside from technical questions) the interpretation of galaxy counts has to deal with a major problem: the scale mixing of density fluctuations within the bins.

Mixing effects also influence the results of covariance functions (or N-point correlation functions), which measure the average degree of irregularity on a characteristic scale length. The same applies to many other statistical measures (Peebles 1973).

The mixing problem does not occur in the power spectrum analysis of clustering (Peebles 1973, Webster 1976), a sensitive and flexible method in order to investigate intrinsic clustering on different scales. The multiplicity function of galaxy clusters (Bhavsar *et al.* 1981) and percolation statistics (Dekel and West 1985) represent alternative, scale-sensitive techniques. All methods mentioned so far are hardly able to discriminate between different types of clustering.

In order to detect the shape of structures (*e.g.* spherical clusters and voids, filaments, sheets) several additional methods have been proposed, *e.g.* the nearest neighbor statistics (Kuhn and Uson 1982), the minimal spanning tree formalism from graph theory (*e.g.* Tucker 1980), different tests for elongation and alignment of positions (*e.g.* Fry 1986). These and similar techniques are more complicated and less meaningful in two-dimensional than in three-dimensional application.

With fully three-dimensional information about the distribution of galaxies, the Gaussian curvature per unit volume of boundary surfaces between high- and low-density regions may represent the most versatile measure of the large-scale structures (Gott III et al. 1986). This topological method is sensitive to all types of structures (isolated or connected) as a function of threshold density.

In the linear regime of Gaussian density fluctuations with random phases and a given initial power spectrum the present curvature can be estimated (Hamilton *et al.* 1986). By this direct connection to (assumed) physical conditions in the early universe, the curvature method seems to be useful for evolutionary interpretations of the observed large-scale structures.

The general problem of local structures is much more complex than that of global structure, where the standard Friedmann models are commonly accepted reference frames for comparison. The topology of the density field, the shape and size of isolated structures, their distribution and evolution require sophisticated statistical methods. To the degree that theoreticians succeed in unifying conceptually the different statistical approaches, observational cosmologists can follow in unifying the different empirical pictures of local structures in the universe. A considerable driving force is provided by the extensive redshift surveys which are now becoming available.

3 Criteria determining the power of a redshift survey

Galaxies shall be treated as identical statistical objects, distinguishable from each other only by their position in redshift space. One would prefer to study all objects within reach with the highest available accuracy of redshift measurements. The *economy* of observational/reductional techniques, however, does not allow this ideal procedure. Instead one has to optimize the methods in order to achieve the best *solving power* for a given task in observational cosmology.

The practical economy in redshift surveys results in several observational "philosophies" concerning the appropriate optimization by assigning different weights to the two complementary parameters number of observed objects N_o and mean error of redshift measurement e_z. The much smaller errors in spherical positions can be neglected. Three main types of *optimizing strategies* can be distinguished:

I *maximize* N_o, *e.g.* by measuring redshifts for all accessible objects from a wide-angle plate or a deep CCD frame, a procedure giving only moderate or low precision for the resulting redshifts;

II *minimize* e_z, *e.g.* by using high quality slit spectra of individual objects, obtained with large telescopes, a procedure only practicable for a small number fraction of accessible objects in the field of a typical Schmidt plate;

III *select* a number of interesting objects or fields employing models for the definition of selection criteria, *e.g.* positions leading to "clusters" of objects, or physical properties leading to specified classes like radiogalaxies or "quasars".

Several existing, generally magnitude-limited redshift surveys can be assigned to these classes, *e.g.*:

class I – photometric redshifts from a deep CCD of small field size: Loh 1988; spectroscopic redshifts from objective prism Schmidt plates: MRSP (Horstmann 1988, Schuecker 1988)

class II – accurate spectroscopic redshifts of medium depth: CfA (Geller *et al.* 1987), its southern extension (da Costa et al. 1988)

class III – statistically selected fields, several projects: Kirshner *et al.* 1978, 1981; special fields, *e.g.* Coma/A 1307: Gregory *et al.* 1978; Hercules/A 2199: Chincarini *et al.* 1981

The first two strategies apply the statistical concept of *random access*, (RA), assuming that all objects in the field are equally accessible and therefore will be picked up randomly like identical balls from a box. In principle, this procedure leads to unbiased samples. The third strategy, mainly used in conjunction with high-accuracy redshifts, represents a conceptually different approach, because it is *model-dependent*, and gives results only for predefined (entities of) objects. The interpretation of the results depends crucially on the assumed role of the selected objects within the cosmological context, a role which can be elucidated only through observations using RA strategies. We conclude that *selection* is an important empirical tool for *follow-up* observations, based on the results of prior RA surveys.

The large variety of existing selection models shall not be discussed here, instead we will concentrate on the model-independent RA strategies and their different uses.

3.1 Spatial resolution

For the description of spatial structures it is necessary to estimate *densities* of objects within various space volumes. We consider only objects in redshift space, neglecting the problem of superpositions of cosmological and non-cosmological redshifts.

Estimating the densities means *counting* the numbers of objects in their respective volume elements. A volume element V is given by a solid angle ω, extending from the position of the observer, the distance z of its centre, and its radial range dz. dz be approximated by the difference between the two extreme measured values z_{min} and z_{max}. Let $e(V)$, $e(n/V)$ be the mean errors of V and of the number density n/V. Since we neglect all error contributions except the count noise (assumed to be Poisson noise) and the redshift error, the relative error of the number density is calculated by error propagation to

$$\frac{e(n/V)}{(n/V)} = \sqrt{\frac{1}{n} + \left(\frac{e(V)}{V}\right)^2} = 1/(\mathrm{S/N}) . \tag{1}$$

S/N is the signal/noise ratio of the estimated density.

The number of objects on a typical Schmidt plate be N_P, its redshift range z_P, the total solid angle covered A_P, the total volume V_P; n is approximated by the mean value

$$n = N_P \frac{V}{V_P} . \tag{2}$$

The survey is assumed to be distance-limited. *The lower limit of the spatial resolution* given through the parameters ω and dz is also a function of S/N.

The coordinates in the two following figures are the two observational scales dz parallel to the line of sight, and ω perpendicular to the line of sight, here given by

$$\omega = \frac{n\, V_P}{N_P} f(z, dz) , \tag{3}$$

where f is a known function.

Figure 1 shows two variants of random access strategies with parameters chosen in such a way that the observational (reductional) amount of work be roughly equal. The depth of the survey $z_{max} = 0.3$, the central redshift of the volume investigated, $z = 0.1$, and the spherical field size, $A_P = 0.01$ sterad, are kept constant. The mean redshift error e_z and the number of objects N_P are varied. The first case is: $e_z = 0.01$, $N_P = 10^4$, the index "n", indicating a number–optimizing method. The second case is $e_z = 10^{-4}$, $N_P = 100$, the index "z" indicating a redshift-optimizing method,

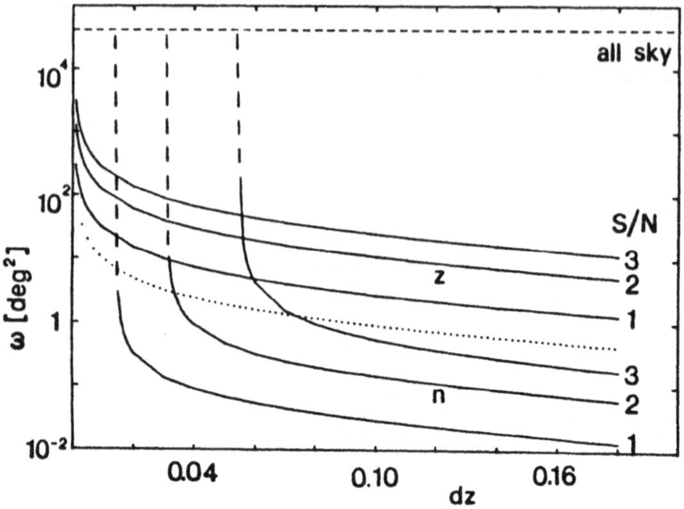

Fig. 1. Resolution in solid angle (ω) and redshift (dz) as a function of the signal-to-noise ratio (S/N) for two hypothetical surveys using random access strategies. Details and assumed parameters see text. The dotted line shows the loci of constant volume resolution.

characteristic of a survey using slit spectra. The spatial resolutions are shown for the S/N values indicated at the right end of each curve.

The diagram exhibits some general features of the random access methods.

1. For fixed S/N, the z-method gives the highest redshift resolution attainable, the n-method the highest angular resolution, as was expected. The limit of redshift resolution is determined by the redshift errors.

2. For fixed S/N, the n-method gives the highest volume resolution attainable above a fixed dz. The dotted line corresponds to a given constant volume resolution (*i.e.* $V(\omega, dz, z) = \text{const}$). All volume resolution curves have the same shape and lie successively lower in the diagram, the better (smaller) the volume V.

3. An empty region in the lower left part of the diagram indicates resolution values not covered by either of the chosen methods. Only the use of larger numbers of objects and/or improved redshifts, *i.e.* merging of the two approaches, permits to fill this region.

At present, only selection strategies combining precise redshifts and a large number-density of objects over a small solid angle are able to cover the empty region. Within a selected or randomly chosen small volume of space the RA–z–method can be applied again for obtaining an unbiased density estimate with high S/N.

For large volumes, there seem to be no realistic cases where RA–z–strategies should be preferred, the high redshift accuracies do not balance the effects of small number

Table 1. Adopted survey parameters

survey	limiting magnitude $m(B)$	depth redshift z_{max}	mean error e_z	total solid angle A (sr)	total object number/A N	most freq. redshift z_c
CfA	15.5	0.06	10^{-4}	0.42	5500	0.02
MRSP	20	0.3	0.01	0.01/plate	6000/plate	0.13
Loh	∼25(J=22)	1.0	0.05	$2.4 \cdot 10^{-5}$	1000	0.5

CfA: Geller *et al.* 1987; MRSP: Horstmann 1988, Schuecker 1988; Loh: Loh 1988.

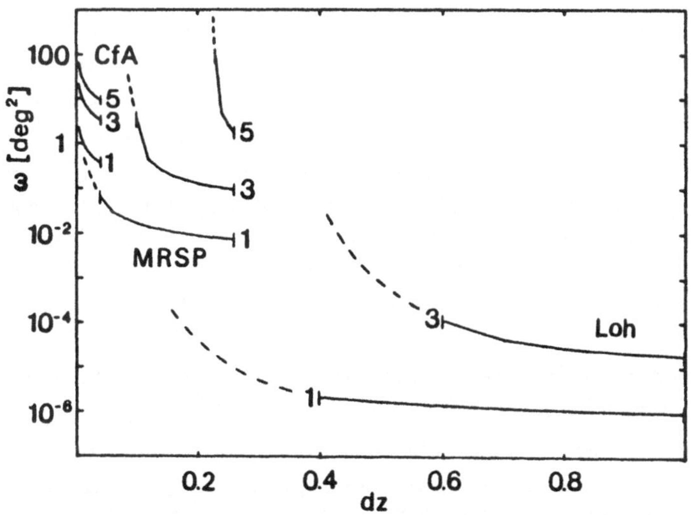

Fig. 2. Resolution (as in Fig. 1) for three existing surveys with the parameters given in Table 1.

statistics, except in the case of special shapes of structures, *e.g.* radially thin shells over the whole sky. Because the S/N increases with increasing numbers, the RA–n–method is particularly well suited for investigations of large voids and scarcely populated regions between conspicuous clusters.

Figure 2 shows the spatial resolution of three existing surveys. The adopted parameters are listed in Table 1. Because these surveys are magnitude-limited, we used the number of objects expected within the redshift interval dz, centered on the most frequent redshift z_c, as calculated from the respective number/redshift distributions. The curves in *Fig. 2* are limited on the right hand side by the depth of the survey. The left hand limits, which are determined by the redshift errors, lie at the all-sky limit not shown here. The left break-off bars are entered in the diagram at somewhat larger values of dz. The S/N ratios are indicated.

Each survey covers a distinct regime with characteristic depth and resolution. The CfA project with its high redshift resolution is strongly limited by its small number density over large solid angles. Loh's survey combines very low redshift resolution (which does not allow $S/N \geq 5$) with high number density over a small solid angle. The MRSP appears to be the "missing link" between both, with medium redshift resolution and medium number density.

3.2 Spatial inclusion

As a matter of fact we suppose the existence of *systematic*, coherent density fluctuations within the universal galaxy population. The necessary separation of structures built up by these fluctuations requires methods defining the boundaries of adjacent regions (*e.g.* by the density gradient, its sign or derivative).

Linear scales of structures, estimated with the separation criteria, are related to the spatial directions investigated and also characterize the whole structures if two relevant shapes are considered: an elongated prolate one and a spherical one.

The *inclusion power* of a survey, its ability of including a structure of given shape and maximum size, characterizes the survey as a tool for studying large-scale structures. In the following we consider briefly the case of linear structures, but concentrate on the more general case of spherical ones, because the spherical inclusion is of specific relevance: it does not favour any spatial direction, so it represents a mean, or averaged, ability of the survey to catch a structure, even if it is arbitrarily shaped, but randomly oriented.

The upper limit for the inclusion of linear structures is given by the maximum linear extension of the volume in redshift space.

If this volume is of complex shape, the exact computation of this scale might be difficult. We get reliable estimations by assuming typical survey volumes with a compact, circular or roughly quadratic, "window" (= total solid angle) with "diameter" D (in radians) and a radial range given by the redshift difference $dz := z_{\max} - z_{\min}$. We simply assume the typical *linear inclusion scale* to be: $S_L = \max(dz, Dz_{\max})$. S_L represents the maximum length of a linear structure to be included in the survey. For most of the modern redshift surveys having $z_{min} = 0$, S_L is given by z_{max}, if $D \leq 1$ (*i.e.* $< 57°$), a condition nearly always fulfilled.

In the case of spherical structures we get a first approximation by computing $d := \dfrac{2z_{\max} \sin(D/2)}{1 + \sin(D/2)}$ and assume the typical *spherical inclusion scale* to be: $S_S = \min(dz, d)$. S_S represents the maximum linear diameter of a spherical structure to be included in the survey. For many existing surveys $z_{\min} = 0$, so that S_S is given by d.

In principle, the inclusion power of a survey should be limited by its parameter *range* (*e.g.* $z_{\min}, z_{\max}, D$), not by its observational *errors* (*e.g.* e_z). Furthermore, a really range-dependent inclusion may neglect the influence of observational errors and of small-number statistics. A look at existing surveys, however, reveals an important fact:

Several survey projects investigate small windows of the sky combined with large depth, the so-called "pencil beams" (PB). Their inclusion power is strongly limited by the redshift errors which, in some cases (*e.g.* Loh 1988), are much larger than the maximum inclusion scale. Such PB surveys are not suited for the investigation of spherical structures at all, they only provide "cross sections" through large structures, which eventually serve as helpful hints for detecting the *existence* of structures.

The *shape* of structures, however, can be determined only by a range-dependent survey where the linear scale (L) investigated at a distance z has to be much larger than the redshift error e_z, for instance: $L > ke_z$ with $k = 5 \ldots 10$.

Using two assumptions about the redshift errors we are able to derive simple window conditions related to the inclusion power of a survey. If we choose L to be the diameter of a spherical structure, z_c the redshift of its center, then a necessary condition for the angular "diameter" of a survey window can be estimated easily:

$$D > 2 \arcsin \left(\frac{k}{2} \frac{e_z}{z_c} \right)$$

or, for small arguments (small e_z):

$$D > k \frac{e_z}{z_c} \, .$$

We notice that two special cases lead to particularly simple and intuitive conditions: $D >$ const. for constant *relative* redshift errors and k, and $Dz_c >$ const. for constant and small *absolute* redshift errors and k.

To fulfill the first condition, the only possibility to increase the inclusion power is to enlarge the window of the survey. Here, we introduce a new kind of empirical strategy, the "large area" (LA) survey, where large spherical structures can be detected despite the observational errors mentioned above.

To fulfill the second condition, either the window or the depth of the survey must be enlarged, the latter one leading to a *deep* pencil beam (PB) survey. These complementary strategies open the chance of investigating the same types of spherical structures at different evolutionary times, for the present with LA and for the earlier epochs with PB. Though smaller windows with more accurate redshifts also fulfill the above condition, this combination does not increase the inclusion power of a survey, as long as one stays in the range-dependent domain.

For the depths and mean redshift errors of the three surveys mentioned, *Fig. 3* shows the spherical inclusion scale in units of redshift error as function of corresponding window *diameter D*. The diameter is estimated from the total solid angle. The present status of each project is indicated.

With $k = 1$, Loh's PB survey does not fulfill the window condition even for the largest possible scales, and is thus unable to include spherical structures (it was not designed to do this!). The CfA project represents a typical range-dependent LA survey. The MRSP in its present status begins to turn from the error-dependent regime to the LA domain.

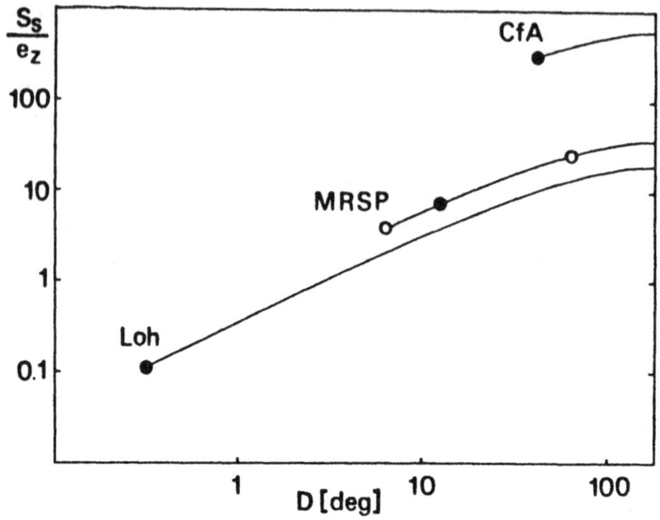

Fig. 3. Maximum linear diameter of a spherical structure in redshift space (normalized to redshift error) as function of window "diameter" D for the surveys of Table 1. Filled circles represent the 1987 status of the projects. For the MRSP it corresponds to 4 plates; the open circles indicate 1 and 100 plates respectively. A normalized diameter smaller than 1 means that an inclusion of spherical structures is not possible at all, due to the limiting redshift error.

Table 2. Survey properties and applications.

Cosmic features to be investigated	Relevant survey property ⟶ relevant information (goals)
spatial structures of small scale	*size* of volume element ⟶ spatial resolution *shape* of volume element ⟶ resolution for different spatial directions
spatial structures of large scale	*size* of volume ⟶ spatial range *shape* of volume ⟶ shapes or sections of structures
global structure	*size* of volume ⟶ homogeneity, isotropy
all	*depth* ⟶ evolution *statistical access* ⟶ minimal bias

The different depths of the two LA projects result in different scale limits, 0.03 (CfA), 0.03 and 0.20, respectively, (MRSP: 1 plate, 100 plates). The MRSP represents a continuous extension of the CfA survey, which is close to its depth-dependent limit for one hemisphere.

A qualitative summary of the most relevant properties of redshift surveys is given in Table 2.

With regard to a redshift of 0.25, it was still possible in 1985 to make the following statement: "...a magnitude-limited galaxy survey out to that sort of distance is totally unthinkable with current technology." (Batuski *et al.* 1985).

Three years ago, this was already wrong, and even more so today there is no excuse for carrying out *only* selection strategies and/or PB observations up to medium redshifts (< 0.3), in the near future even to larger scales. The observational material for medium-depth LA surveys, Schmidt plates covering appreciable parts of the sky, exists and awaits extensive measurements and reductions, which are particularly time-saving in comparison to other techniques.

References

Batuski, D.J., Burns, J.O., 1985. *Astr.J.*, **98**, 1413.

Bhavsar, S.P., Gott III, J.R., Aarseth, S.J., 1981. *Astrophys. J.*, **246**, 656.

Chincarini, G., Rood, H.J., Thompson, L.A., 1981. *Astrophys. J.*, **249**, L47.

Da Costa, L.N., Pellegrini, P.S., Sargent, W.L.W., Tonry, J., Davis, M., Meiksin, A., Latham, D.W., Menzies, J.W., Coulson, I.A., 1988. *Astrophys. J.*, **327**, 544.

Dekel, A., West, M.J., 1985. *Astrophys. J.*, **288**, 411.

Ehlers, J., Rindler, W., 1987. *Astr. Astrophys.*, **174**, 1.

Fry, J.N., 1986. *Astrophys. J.*, **306**, 366.

Geller, M.J., Huchra, J.P., de Lapparent, V., 1987. In *IAU Symp. 124, Observational Cosmology*, eds. Hewitt, A., Burbidge, G.R., Li, Z.F., Reidel, Dordrecht, p. 301.

Gott III, J.R., Melott, A.L., Dickinson, M., 1986. *Astrophys.J.*, **306**, 341.

Gregory, S.A., Thompson, L.A., 1978. *Astrophys. J.*, **222**, 784.

Hamilton, A.J.S., Gott III, J.R., Weinberg, D., 1986. *Astrophys. J.*, **309**, 1.

Horstmann, H., 1988b. *These proceedings*, p. 111.

Kirshner, R.P., Oemler, A., Schechter, P.L., 1978. *Astr. J.*, **83**, 1549.

Kirshner, R.P., Oemler, A., Schechter, P.L., Shectman, S.A., 1981. *Astrophys. J.*, **248**, L57.

Kuhn, J.R., Uson, J.M., 1982. *Astrophys. J.*, **263**, L47.

Loh, E.D., 1988. *Astrophys. J.*, submitted; see also *These proceedings*, p. 267.

Murtagh, F., Heck, A., 1987. *Multivariate Data Analysis*, Reidel, Dordrecht.

Peebles, P.J.E., 1973. *Astrophys. J.*, **185**, 413.

Sandage, A., 1987. In *IAU Symp. 124, Observational Cosmology*, eds. Hewitt, A., Burbidge, G.R., Li, Z.F., Reidel, Dordrecht, p. 1.

Schuecker, P., 1988. *These proceedings*, p. 142.

Silk, J., 1974. *Astrophys. J.*, **193**, 525.

Stoeger, W.R., Ellis, G.F.R., Hellaby, C., 1987. *Mon. Not. R. astr. Soc.*, **226**, 373.

Tucker, A., 1980. *Applied Combinatorics*, Wiley, New York.

Weinberg, S., 1972. *Gravitation and Cosmology*, Wiley, New York.

Wavelength Calibration of Objective Prism Plates by Transformation from Direct Plates

Hans-Joachim Tucholke
Astronomisches Institut
Westfälische Wilhelms-Universität
Münster, F.R. Germany

Abstract

The zero points for galaxy redshift measurements from objective prism plates (dispersion 246 nm mm^{-1} at Hγ) are obtained through transformation of object positions from the corresponding direct plates. Approximately 1000 G-stars per plate, classified automatically, are used.

On the direct plate, positions in x and y are computed from intensity-weighted first moments. On the objective prism plate object positions are given in x through the Ca II-break at 400 nm and in y through marginal fits to the unwidened spectra. Redshifts are obtained from the difference between the expected and the measured Ca II-break positions in the galaxy spectra. The transformation equations include quadratic terms in the direction of dispersion. The inclusion of third-order and colour terms is discussed.

We present the transformation characteristics for three adjacent fields near the South Galactic Pole. The mean residuals are approximately 5 μm, corresponding to a redshift error of about 700 km s^{-1} at $z = 0$.

This is an extended version of the contribution presented at the IAU Colloquium No. 100 at Belgrade (Tucholke *et al.* 1988).

1 Introduction

The Muenster Redshift Project (MRSP), which is outlined in Horstmann (1988) and Schuecker (1988), investigates three-dimensional structures in the Universe using galaxy redshifts from low-dispersion objective prism plates. Presently, redshifts up to $z = 0.3$ with an accuracy of $dz = 0.01$ are reached for galaxies with m_J magnitudes $16^{\mathrm{m}}.0$ to $20^{\mathrm{m}}.5$ on plates taken with the UK Schmidt Telescope. As described by Gericke (1988), methods for the automatic detection of QSO's and the investigation of their distribution are developed.

2 Wavelength calibrations of objective prism plates

For the determination of redshifts free from systematic and large random errors, a proper choice of the wavelength reference point is of primary importance. The position of the IIIa-J emulsion cutoff, which is frequently used as reference point, depends on magnitude and colour and leads to systematic errors of 50 μm (Beard *et al.* 1986). This corresponds to a redshift error of $\leq 6500\,\mathrm{km\,s^{-1}}$ near the Ca II-break at 400 nm and of $\leq 11\,500\,\mathrm{km\,s^{-1}}$ at a highly redshifted Ca II-break above 500 nm, using the dispersion curve by Nandy *et al.* (1977). After correcting for the systematic effect there are still random errors of $\leq 20\,\mu$m, corresponding to $\leq 2600\,\mathrm{km\,s^{-1}}$ at 400 nm and $\leq 4600\,\mathrm{km\,s^{-1}}$ at 500 nm.

In order to avoid these difficulties, the MRSP uses the transformation from the direct plate positions of G-type stars to the positions of the Ca II-break in their spectra on the objective prism plate. The spectral appearance of the G-stars is similar to that of the majority of nearby normal galaxies. The stellar velocity dispersion $\leq 100\,\mathrm{km\,s^{-1}}$ is small compared to the redshift accuracy attainable for galaxies from objective prism plates.

3 The transformation model

A quadratic model is used to transform the direct plate positions (x, y) (x is measured in the direction of dispersion) to positions (x', y') on the objective prism plate:

$$\begin{aligned} x' &= a_0 + a_1 x + a_2 y + a_3 x^2 + a_4 xy + a_5 y^2 \\ y' &= b_0 + b_1 x + b_2 y + b_3 x^2 + b_4 xy + b_5 y^2 \,. \end{aligned} \tag{1}$$

The linear terms allow for zero point difference of the plate scans, relative rotation between the plates and differential refraction. The field distortions caused by the objective prism introduce quadratic terms with $a_3 \approx a_5$, the other quadratic terms are negligible. This transformation model has been applied successfully to astrometry and radial velocity measurements (Stock and Osborn 1980, Weis *et al.* 1981, Stock 1984, Stock 1986) and spectrophotometry (Clowes *et al.* 1980) from objective prism plates. A third order model, as discussed by Stock and Osborn (1980), was tested, but did not improve the data: The residuals remained the same, the additional terms were barely significant and the error of the other constants increased.

A colour term in x, however, might be important, because the spectral features on the objective prism plate are monochromatic, while the positions of the objects on the direct plate are affected by atmospheric dispersion. We estimate the size of this effect for the UK Schmidt Telescope using the colour–refraction curves given by Murray and Corben (1979) and Murray (1984). For objects of extreme colours, the displacement relative to the G-type stars used in the transformation amounts to $0''.25$ at a zenith distance of 60°. This leads to a maximum error of the chosen reference feature of 3.7 μm. The actual error is negligible for most galaxies at low redshifts. For a redshift $z = 0.3$ the maximum error is $900\,\mathrm{km\,s^{-1}}$ and has to be corrected. As soon as our programme of colour measurement and spectral classification for all types of stars is finished, we shall compute the transformation equations for galaxies of high redshifts and for very blue objects from a wider range of stellar colours and with a colour term included in the transformation model.

4 Measurements and reductions

The observational material, on which the results of the MRSP are based so far, consists of film copies of direct and objective prism plates, taken with the UK Schmidt Telescope (IIIa-J emulsion, dispersion 246 nm mm^{-1} at Hγ). The direct plates are taken from the ESO/SERC-Atlas. The film copies were scanned with the PDS 2020 GM plus microdensitometer at the Astronomical Institute Muenster (mean positional accuracy for stellar positions 0.7 μm, Tucholke 1983). On each plate an area of 300×300 mm^2 ($5.5° \times 5.5°$) is measured.

The object positions on the direct plate are computed by first-order moments. They are intensity-weighted positions corresponding to those parts of the object which also contribute most of the light to the spectra. This leads to reliable measurements even for galaxies with complex light distribution, and with unusual colours of the brightest regions, when a colour term is taken into account. When the star positions are computed with the more time-consuming method of Gauss-fits to the density profiles, neither the constants nor the residuals of the transformation are significantly altered, because the accuracy is limited by the measurement of the Ca II-break.

The transformation stars are classified automatically through cross correlations of the rectified spectra and the application of a fuzzy set classifier (Schuecker 1988). To each candidate a normalized quality factor Q is assigned, whose value is a measure for the fuzzy set possibility of the object being a G-star. The use of objects with high Q values only excludes over- and underexposed spectra, which would lead to small magnitude-dependent effects presently neglected.

The half intensity point at the Ca II-break defines the reference point in the stellar spectra. The x'-position of this point is obtained by differential filtering and subsequent Gauss-fit to a small spectral range centered on the break. The $y'-$position is derived from a Gauss-fit to the marginal sum of the spectrum taken along the direction of dispersion. Errors in y' enter into the x'-position with less than 1 %.

5 Tests and results

The wavelength calibration method described above was applied to the ESO-SRC Atlas fields Nos. 351, 411, and 474 near the SGP. Initially $\approx 10^3$ G-star candidates were selected for each field. *Fig. 1* shows the distributions of these objects in the three fields. As required for a reliable transformation, the distributions are homogeneous. Small gaps in the distributions result from crowding with clustered objects (Sculptor dwarf galaxy in field 351, globular cluster NGC 288 in field 474; *Figs. 1a, c*, respectively), but these minor inhomegeneities do not significantly affect the quality of the transformation.

The G-star sample was cleaned iteratively from stars showing large residuals until the improvement of the mean residual fell below the preset limit of 1 %. As a test for the possible dependence of the transformation constants on plate position, the transformation was computed separately for the four plate quadrants and for three sections each in the directions of right ascension and declination.

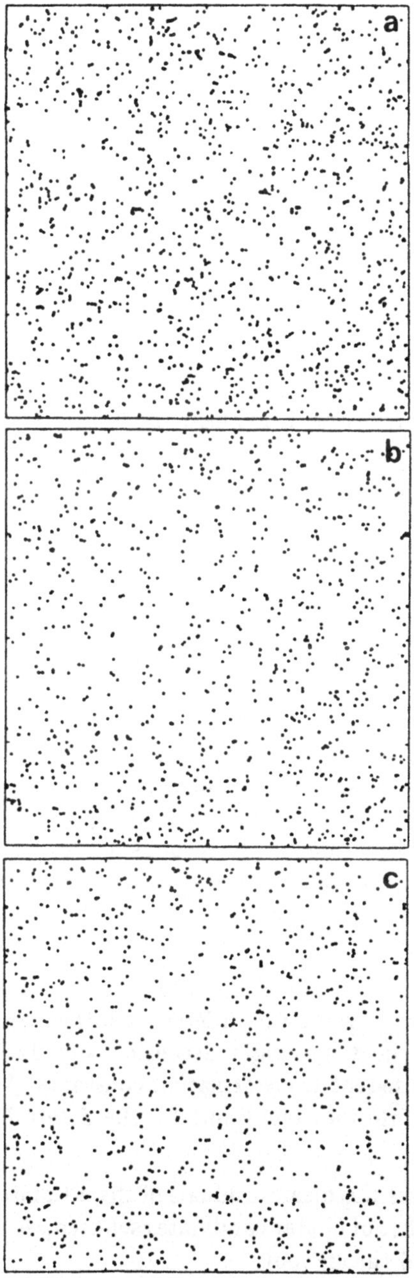

Fig. 1. (a) Distribution of G-star candidates selected for the ESO-SRC field No. 351. Note the overall homogeneous distribution. The gap in the NE quadrant is caused by overlap with the Sculptor dwarf galaxy. 10 mm correspond to 1°. North is up and east to the left.
(b) Same as above for ESO-SRC field No. 411.
(c) Same as above for ESO-SRC field No. 474. A gap in the distribution near the SE corner is caused by overlap with the globular cluster NGC 288.

Table 1: Transformation characteristics for field No. 474

Quadrant		Q	σ_x (arcsec)	σ_y (arcsec)	N_+	N_-
I	NE	0.75	0.42	0.37	237	32
II	SE	0.75	0.41	0.35	209	43
III	NW	0.75	0.34	0.32	238	36
IV	SW	0.75	0.38	0.32	224	31
All		0.80	0.37	0.36	575	70

Table 2: Mean residuals for three fields near the SGP

Field	N	N_+	σ_x (arcsec)	σ_y (arcsec)
351	1297	659	0.41	0.37
411	983	232	0.38	0.34
474	1050	575	0.37	0.36

As predicted by theory, the only significant quadratic terms in the transformation are a_3 and a_5, which are approximately equal. Table 1 shows the results for field 474, where the initial number of G-star candidates with $Q \geq 0.75$ was 1050. The residuals σ_x and σ_y of the transformation in arc seconds are listed along with the numbers N_+ and N_- of the stars kept and removed from the transformation for the four plate quadrants and for the whole plate. For the all-plate solution a higher quality limit Q was applied.

No significant differences in the transformations for the different plate quadrants are found. In the overlapping regions the different solutions agree within $0''.10$ to $0''.20$. The subdivisions of the field into right ascension and declination sections showed similar residuals and consistent transformation constants. We now routinely use the subdivision of the plate into four quadrants for the determination of the wavelength reference point.

Table 2 briefly summarizes the mean residuals of the all-plate solution for three fields. N denotes the total number of G-star candidates selected. A lower limit for the quality factor Q of 0.80 was used throughout.

The residuals in the dispersion direction x slightly exceed those in y. The accuracy in the determination of the wavelength reference point is limited by the measurement of the Ca II-break in the stellar spectrum. An error of $0''.39$ (mean σ_x of Table 2) corresponds to 5.9 μm on the plate and to a redshift error of $dz = 0.0024$ or $v = 820\,\mathrm{km\,s^{-1}}$ at $z = 0.0$ and $dz = 0.0047$ or $v = 1400\,\mathrm{km\,s^{-1}}$ at $z = 0.3$. This limits the accuracy attainable for the galaxy redshifts.

In the near future we will test the relative astrometric accuracy of original plates, glass copies and film copies (used here) from the UKST. It is possible, that the use of original plates not only reveals fainter objects, but also gives higher accuracy of the wavelength reference points.

Acknowledgements

The MRSP is the cooperative work of all members of the Astronomical Institute of Muenster. My thanks go to all of them, but especially to P. Schuecker and H. Horstmann. – This work is based on film copies of objective prism plates kindly provided by the UKSTU.

References

Beard, S.M., Cooke, J.A., Emerson, D., MacGillivray, H.T., Kelly, B.D., 1986. *Mon. Not. R. astr. Soc.*, **219**, 251.

Clowes, R.G., Emerson, D., Smith, M.G., Wallace, P.T., Cannon, R.D., Savage, A., Boksenberg, A., 1980. *Mon. Not. R. astr. Soc.*, **193**, 415.

Gericke, V., 1988. *These Proceedings*, p. 235.

Horstmann, H., 1988. *These Proceedings*, p. 111.

Murray, C.A., 1984. In *IAU Coll. No. 78, Astronomy with Schmidt-Type Telescopes*, ed. Capaccioli, M., Reidel, Dordrecht, p. 217.

Murray, C.A., Corben, P.M., 1979. *Mon. Not. R. astr. Soc.*, **187**, 723.

Nandy, K., Reddish, V.C., Tritton, K.P., Cooke, J.A., Emerson, D., 1977. *Mon. Not. R. astr. Soc.*, **187**, 63p.

Schuecker, P., 1988. *These Proceedings*, p. 142.

Stock, J., 1984. *Rev. Mexicana Astr. Astrofis.*, **9**, 77.

Stock, J., 1986. In *IAU Symposium No. 109, Astrometric Techniques*, eds. Eichhorn, H.K., Leacock, R.J., Reidel, Dordrecht, p. 253.

Stock, J., Osborn, W., 1980. *Astr. J.*, **85**, 1366.

Tucholke, H.-J., 1983. *Diploma Thesis*, Astr. Inst. Univ. Münster.

Tucholke, H.-J., Schuecker, P., Horstmann, H., Seitter, W., 1988. In *IAU Coll. No. 100, Fundamentals of Astrometry*, eds. Pakvor, I., Eichhorn, H.K., in press.

Weis, E.W., Upgren, A.R., Dawson, D.W., 1981. *Astr. J.*, **86**, 246.

The Muenster Redshift Project. Automated Redshift Measurements from Low-Dispersion Objective Prism Schmidt Plates

P. Schuecker
Astronomisches Institut
Westfälische Wilhelms-Universität
Münster, F.R. Germany

Abstract

A three-dimensional galaxy survey at faint magnitudes and over large volumes of space is currently carried out at Muenster. In order to enhance the reliability of the redshifts measured from objective prism plates, three different methods are used: the *correlation method*, the *least-squares method* and the *break method* where continuum breaks are identified directly. The redshift errors of the individual methods are $dz =$ 0.007 (correlation), 0.011 (direct identification) and 0.016 (least-squares). In the majority of cases two *independent* measurements are possible leading to mean redshift values with errors of 0.008. To date redshifts of 24 000 galaxies with $17^{\mathrm{m}} < m_J < 20^{\mathrm{m}}$ have been obtained. The current rate is an increase of 6 000 galaxy redshifts per measuring week.

1 Introduction

In 1986 the Astronomical Institute Muenster (AIM) has started a three-dimensional redshift survey, the Muenster Redshift Project (MRSP). A major goal of MRSP is to study the complex structures of clusters and superclusters of galaxies on large scales, *i.e.* in large solid angles and up to at least $0.3\,c$. The observational data are obtained from direct and objective prism Schmidt plates. The direct plates are used for the identification of galaxies and clusters of galaxies, the very low dispersion objective prism plates for measuring galaxy positions in redshift space. From these positions the locations of galaxies in the depth of real space and information about their membership in clusters and superclusters are obtained.

All plates are digitized with the PDS 2020 GMplus and processed fully automatically with the software system ADAS (Astronomical Data Analysing System, ADAS 1987) of AIM (Teuber 1988). MRSP provides galaxy positions with accuracies of about $0\farcs4$ (Tucholke 1988), galaxy magnitudes $16\fm5 \leq m_J \leq 21\fm5$ with accuracies $0\fm1$ (Horstmann 1988) and redshifts $0 \leq z \leq 0.3$ with an average accuracy ≤ 0.008 in the range $17^{\mathrm{m}} < m_J < 20^{\mathrm{m}}$. One plate covers 30 square degrees and includes about 40 000 galaxies in fields near the South Galactic Pole. Typical subsamples of galaxies contain 25 000 galaxies with rough *morphological types* from the direct Schmidt plate and 6 000 galaxies with *redshifts* from the objective prism plate. For one objective prism plate the total reduction time is one week.

In the present paper three methods for automated redshift measurements are discussed. In Sect. 2 the reduction of the spectra prior to redshift determination is introduced. Sect. 3 describes the three methods of automated redshift determination. Sect. 4 contains the error estimates and Sect. 5 the conclusions and future plans.

2 Reduction of the objective prism spectra

2.1 Spectral features at low dispersions

Studies concerning the determination of radial velocities of faint galaxies from Curtis and UK Schmidt telescope plates by Cooke *et al.* (1977, 1981) show the possibility of determining galaxy redshifts using very low dispersions (140 and 246 nm mm^{-1} at Hγ). In these cases, however, radial velocities cannot be determined by measuring line positions: the size of the seeing disc during exposure and the effective angular size of the galaxy limit the spectral resolution and smear out all smaller spectral features. The only features seen in the galaxy spectra up to redshifts 0.3 on IIIa-J plates with a dispersion 246 nm mm^{-1} at Hγ are

- a strong absorption feature at 400 nm (400 nm break), mainly caused by the blend of Ca II H, K with contributions of Hϵ and metallic blends. This is the most prominent spectral feature.

- a feature at 430 nm (430 nm break) corresponding to the G-band with Hγ and metallic blends, enhanced by an emulsion dip

- a weak absorption feature at 365 nm caused by Fe I blends.

Simulations with model spectra of Kurucz (1979) show that weak absorption systems of Fe II 233–263 nm and Mg II 280 nm can be used in galaxy spectra with $0.3 < z < 1.0$ (Schuecker 1986a).

Redshifts from objective prism spectra are obtained by measuring the distances between the breaks and a suitably chosen reference point. Beard *et al.* (1986) and Parker *et al.* (1987) used the position of the emulsion cutoff of the J-emulsion near 538 nm. In the present investigation a zero point transformation from the direct plate is employed (see Sect. 2.4).

So far, one major problem in obtaining large numbers of galaxy redshifts is the time consuming *interactive* measuring process. Some investigations toward automatization are given in Cooke *et al.* (1983, 1984) and Bunclark (1984). In the present communication various methods of *automated* redshift measurements from low-dispersion spectra are introduced. The observational material consists of film copies of unfiltered IIIa-J objective prism plates with a dispersion of 246 nm mm^{-1} at Hγ taken with the UK Schmidt telescope and prism 1. Each plate covers 41 square degrees of which 23 square degrees are nominally unvignetted. The limiting magnitude is $m_J = 20\overset{\text{m}}{.}5$ (Tritton 1983).

2.2 Digitization and automated detection of objective prism spectra

Film copies from direct UK Schmidt plates as distributed in the ESO/SRC Atlas and copies of the corresponding UK objective prism plates are digitized with the microdensitometer PDS 2020 GMplus of AIM with step size 15 μm, aperture $20 \times 20\,\mu m^2$, and density range 0.0 – 4.5. After *segmentation* of the objective prism plate (Horstmann 1988) the spectra of all objects brighter than $20^{\mathrm{m}}.5$ on the direct plate, e.g. 50 000 spectra in high galactic latitude fields, 35 000 of starlike objects and 15 000 of galaxies, are stored for further processing. Each spectrum is represented by an arrray of 101×15 pixels. In addition, information is given about the position of the object, the positions of its nearest neighbours and their apparent magnitudes, object type (star or galaxy as classified on the direct plate), object size and shape, and the local sky background and background noise. The information about the nearest neighbours is used to identify overlapping spectra.

2.3 Preprocessing of the spectra

The density of each pixel in the stored spectrum array is transformed into a relative intensity I_{ij} using the step wedges given on the UKST plates. The spectral profile I_i (*Fig. 1*) is obtained from marginal sums perpendicular to the direction of dispersion, where each pixel I_{ij} is weighted according to a mean intensity distribution function DF_j. Unlike the point spread function, the DF_j for extended objects comprises both the distortions along the light path and the effects of light distribution over the extended image. It is thus essential that the DF_j is determined individually for each object. The DF_j is estimated from the marginal sums in the direction of dispersion, and normalized to values between 0 and 1 (i: in the direction of dispersion, j: perpendicular to i):

$$I_i = \frac{\sum_j I_{ij} DF_j}{\sum_j DF_j} \qquad (1)$$

with

$$DF_j = \frac{\Phi_j - \Phi_{min}}{\Phi_{max} - \Phi_{min}}, \quad \Phi_j = \sum_i I_{ij} \, .$$

Rebinning of the pixel coordinates x, represented in Eqn. 1 by the discrete coordinates i, into other coordinates λ, *e.g.* wavelengths, leads to corrections of the intensities which are large in the case of large nonlinear terms in the transformation curve $\lambda(x)$:

$$I(\lambda) = I(x) \left[\frac{\partial \lambda(x)}{\partial x} \right]^{-1} . \qquad (2)$$

The transformation of the plate scale x into a wavelength $\lambda(x)$ or a logarithmic wavelength scale uses the dispersion curve of the prism. For prism 1 of the UKST the dispersion curve was determined by Nandy *et al.* (1977). The final rebinning is computed with cubic spline interpolations. The nonlinear transformation (Eqn. 2) leads to a wavelength dependence of the noise and a (partial) restauration of the spectral features on the linear wavelength scale.

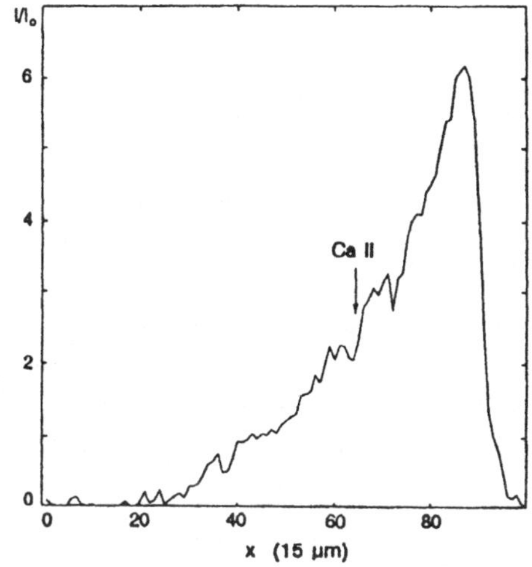

Fig. 1. Intensity profile of a low-dispersion objective prism galaxy spectrum.

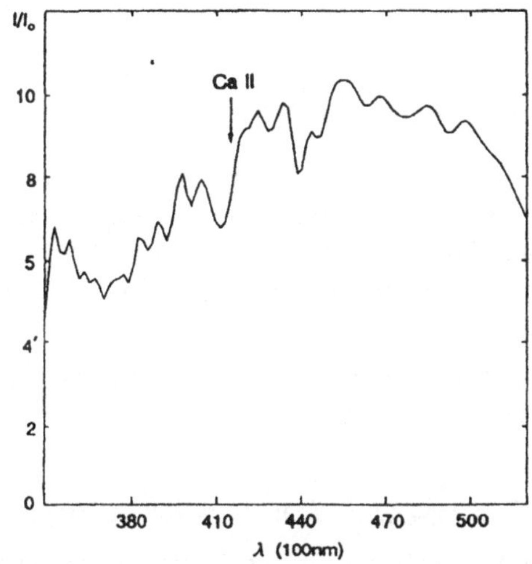

Fig. 2. Intensity profile of *Fig. 1* corrected for atmospheric extinction, transmission of the prism and the achromatic correcting plate and the sensitivity of the emulsion.

Intensity profiles corrected for atmospheric extinction $E(\lambda)$, transmission of the prism $P(\lambda)$ and the achromatic correcting plate $A(\lambda)$, and the sensitivity of the emulsion $S(\lambda)$ are needed for some purposes, *e.g.* for the calculation of equivalent widths in emission line objects and for galaxy redshift measurements by least-squares comparisons of spectral profiles. Corrected profiles (*Fig. 2*) are obtained from Eqn. 3, the filter functions are taken from Clowes *et al.* (1980):

$$I(\lambda)_{corr} = I(\lambda)\,[E(\lambda)\,P(\lambda)\,A(\lambda)\,S(\lambda)]^{-1} \ . \tag{3}$$

2.4 The wavelength reference point

A critical process for reliable wavelength measurements is the determination of the wavelength reference point. Emerson (1981, 1983) and Beard *et al.* (1986) showed that the frequently used emulsion cutoff introduces a systematic error $\lesssim 50\,\mu$m depending on magnitude and colour of the object. This leads to radial velocity errors from $\lesssim 6\,600$ km s^{-1} at 400 nm to $\lesssim 11\,500$ km s^{-1} above 500 nm.

The MRSP uses the transformation from direct plate positions of automatically selected G-type stars to the positions of their 400 nm breaks on the objective prism plate. The error obtained in x is $0\farcs33$, corresponding to 700 km s^{-1} at $z = 0.0$ and 1 200 km s^{-1} at $z = 0.3$. This is the limit of the redshift accuracy attainable. More information about the wavelength calibration is given by Tucholke *et al.* (1988) and Tucholke (1988).

Details about the automated classification, especially for M- and G-type stars, are given by Schuecker (1986b) and Schuecker *et al.* (1986, 1988a). The classification will be extended to all MK-spectral classes using low-dispersion classification criteria given by Seitter (1988).

2.5 Rectification

2.5.1 About the use of fuzzy algebra

A basic problem in automated redshift measurement is the correct identification of the chosen spectral feature. In order to make such a feature visible at low dispersions, it is essential that a suitable reference continuum is defined. Usual methods, like polynomial fits *etc.*, fix the continuum with regard to *one* criterion only. In order to get solutions which optimize *several* criteria, *fuzzy algebra* is applied.

In a fuzzy set C the transition from membership to nonmembership is continuous and for any point characterized by a declining sequence of numbers μ_C (Zadeh 1965). C is defined as a set of n pairs

$$C = \{(\mu_C(L_i),\, L_i); \quad i = 1, 2, \ldots, n\} \ . \tag{4}$$

For each point L_i, $\mu_C(L_i)$ gives the degree of membership in C. Therefore, C includes the concept of ordinary sets if $\mu_C = 0$ or 1. A standard expression of the membership function $\mu_C(L)$ is the S-function (Zadeh 1975):

$$S(L; a, b, c) = \begin{cases} 0 & a > L \\ 2(\frac{L-a}{c-a})^2 & a \leq L < b \\ 1 - 2(\frac{L-c}{c-a})^2 & b \leq L \leq c \\ 1 & c < L \qquad c \neq a \neq 0 \end{cases} \tag{5}$$

with

$$b = \frac{a+c}{2} \ .$$

It is an S-shaped function with the parameters a, b, c where b determines the cross-over point. S- or other membership functions are similar to probability distribution functions, but conceptually different (Pal and Dutta Majumder 1986): while the probability distribution function describes how *frequently* values $\leq L$ occur, the membership function indicates how *closely* L resembles an ideal element. The choice of the function for the solution of a special problem is *ad hoc* and only based on the fact that it gives useful results. The unique evaluation of membership functions is a ongoing major controversy in fuzzy set theory.

The intersection between different fuzzy sets C_i is calculated from the values of the contributing membership functions (Zadeh 1965, 1968, 1973):

$$\mu_{intersection}(L) = \min_L\{\mu_{C_1}(L), \mu_{C_2}(L), \ldots\} . \tag{6}$$

Weighting of the different sets is attainable *e.g.* with exponential fuzzifiers e:

$$\mu_{weighted}(L) = [\mu(L)]^e . \tag{7}$$

They also offer the possibility to modify the slope of the membership function.

2.5.2 Rectification with fuzzy algebra

In terms of fuzzy sets, the problem of rectification can be formulated as follows. Given is a set of different reference continua $\{R_i^L\}$ with parameter L. L, *e.g.* the filter width, is chosen for the calculation of the different continua. The problem is to find the continuum (*i.e.* the value of L) which optimizes the criteria and thus yields the properties of the wanted continuum. Each criterion must be represented by a fuzzy set. The intersection of the different fuzzy sets (Eqn. 6) represents the combination of the different criteria. The value L with the highest membership gives the continuum which satisfies all criteria *best*. The criteria chosen here are:

1 : clear protrusion of breaks above the reference continuum

2 : smoothness of the continuum (short length)

3 : good approximation of the spectrum (low *rms*).

The different reference continua R_i^L are computed by cubic spline interpolations between k knots selected from the intensity profile I_i. The knots are defined by the minima found in k wavelength intervals of I_i. The parameter L is the length of the interval. For low values of L the approximation of I_i increases, whereas the smoothness of the continuum and the protrusion of the breaks decrease. For large L the breaks are more clearly visible but the approximation of I_i decreases.

The three criteria constitute three fuzzy sets. To evaluate the corresponding membership functions, *e.g.* the parameters a, b, c of the S-function (Eqn. 5), the criteria are given in quantitative forms:

$$\begin{aligned} &\text{Criterion 1:}\ \max\delta_L = \textstyle\sum_i (I_i - R_i^L) \\ &\text{Criterion 2:}\ \min\theta_L = \textstyle\sum_i \sqrt{1 + \left(\frac{\partial R_i^L(\lambda)}{\partial \log\lambda}\right)^2} \\ &\text{Criterion 3:}\ \min\epsilon_L = \textstyle\sum_i (I_i - R_i^L)^2 \end{aligned} \tag{8}$$

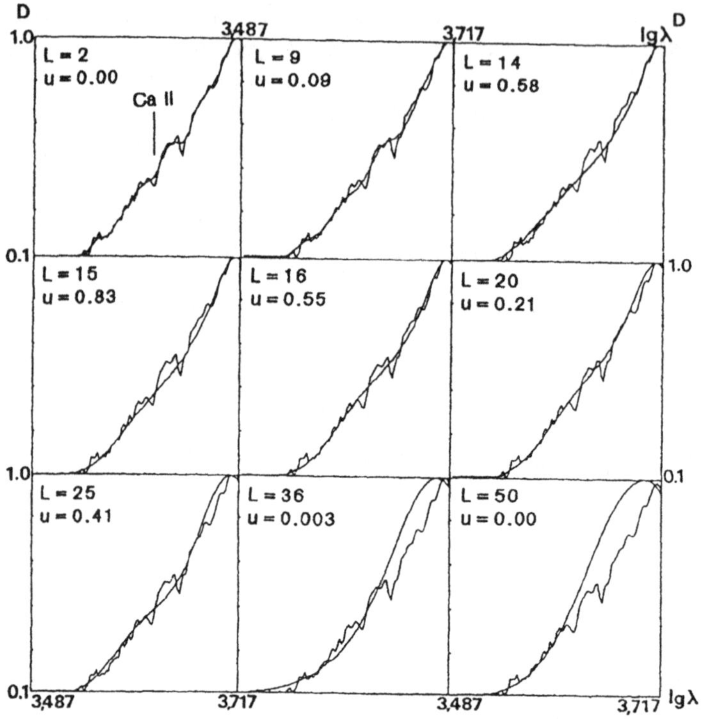

Fig. 3. Reference continua superimposed on a galaxy spectrum profile. L is the parameter of the reference continuum, μ is the membership value of the intersecting fuzzy set. The chosen reference continuum has the hightest membership value $\mu = 0.83$ and $L = 15$.

The quantities δ_L, θ_L and ϵ_L are calculated for each reference continuum R_i^L. Physically meaningful values of L lead to sets $\{\delta_L\}$, $\{\theta_L\}$ and $\{\epsilon_L\}$, from which convenient membership functions can be obtained. If the S-function is used, the extrema of δ_L, θ_L and ϵ_L define the parameters a, b, c:

$$\begin{array}{lll} a_\delta = \min_L\{\delta_L\} & c_\delta = \max_L\{\delta_L\} & b_\delta = \frac{a_\delta + c_\delta}{2} \\ a_\theta = \min_L\{\theta_L\} & c_\theta = \max_L\{\theta_L\} & \ldots \\ a_\epsilon = \min_L\{\epsilon_L\} & c_\epsilon = \max_L\{\epsilon_L\} & \ldots \end{array} \tag{9}$$

The corresponding membership functions are:

$$\begin{array}{l} \mu_\delta(L) = \{\quad S(L; a_\delta, b_\delta, c_\delta)\ \}^{e_1} \\ \mu_\theta(L) = \{1 - S(L; a_\theta, b_\theta, c_\theta)\ \}^{e_2} \\ \mu_\epsilon(L) = \{1 - S(L; a_\epsilon, b_\epsilon, c_\epsilon)\ \}^{e_3} \end{array} \tag{10}$$

with the exponential fuzzifiers e_1, e_2 and e_3. $\mu_\delta = 1$ corresponds to a continuum showing the breaks in the spectrum I_i best. $\mu_\delta = 0$ is the membership value for the continuum with the worst representation of the breaks. Corresponding relations hold for μ_θ and μ_ϵ. Eqn. 6 gives the intersection of the fuzzy sets. The value L with the highest membership of the resulting fuzzy set yields the wanted reference continuum.

Figure 3 shows reference continua with different search intervals L for a galaxy spectrum (type Sc, $m_J = 16\overset{m}{.}3$). They are chosen from a total of 49 different continua. L varies between $L = 2$ ($d\log\lambda = 0.0046$) and $L = 50$ (0.115). The adjusted values for the exponential fuzzifiers are $e_1 = 10.0$, $e_2 = 1.0$ and $e_3 = 0.1$. Membership values from Eqn. 6 lie between $\mu = 0.00$ and $\mu = 0.83$. Curves with $\mu = 0.00$ represent continua where one or more criteria completely fail: for $L = 2$ and 50 no clear protrusion of the 400 nm break is achieved. In addition, the continuum for $LL = 2$ is rather long. The *best* reference continuum has $L = 15$.

The different S/N ratios of the galaxies lead to different values of L for the optimal reference continua. Therefore, the procedure described above must be applied individually to each spectrum.

3 Methods for redshift determination

3.1 The correlation method

After rebinning the x-scales of the spectra into logarithmic wavelength scales (Eqn. 2), redshifts correspond to uniform linear shifts of the spectral features. In order to find the redshift z of a galaxy spectrum I_i, one has to find the template spectrum T_i^z most similar to I_i from a set of templates with known redshifts $\{T_i^z\}$. This is performed with the generalized correlation function (*e.g.* Bracewell 1978):

$$c_z = \frac{\sum_i I_i T_i^z}{\sqrt{\sum_i (I_i)^2 \sum_i (T_i^z)^2}} . \tag{11}$$

The normalization factor leads to correlations c_z between -1 (anticorrelation) and $+1$ (ideal correlation, autocorrelation). The template T_i^z which maximizes c_z is the chosen template. The position of the correlation peak gives the redshift of the programme galaxy.

Figure 4 presents correlation functions between a G-type star template and three galaxy spectra. The correct redshifts of the galaxy spectra, taken from Parker *et al.* (1986) and West and Frandsen (1981), are indicated by arrows. A G-type star template was chosen because most of the integral galaxy spectra are of this spectral type (Humason *et al.* 1956). The set of templates $\{T_i^z\}$ is created by artificially shifting the G-type spectrum relative to the galaxy spectrum.

The correlation is more sensitive to redshift when rectified spectra with their continua subtracted are matched. This mode was used here. For rectification the procedure outlined in Sect. 2.5 is applied. In order to measure redshifts up to twice as large as those obtained by using periodic boundary conditions (*e.g.* FFT, Simkin 1974) zeros are appended to both ends of the spectra. Larger shifts between T_i^z and I_i lead, however, to lower numbers of spectrum pixels available for the calculation of the cross term in Eqn. 11. Therefore, correlation peaks at large shifts have lower weight than peaks at small shifts.

In *Fig. 4a* the position of the highest correlation peak gives the correct redshift. The neighbouring peaks correspond to the 430 nm breaks. *Figs. 4b* and *c* illustrate two

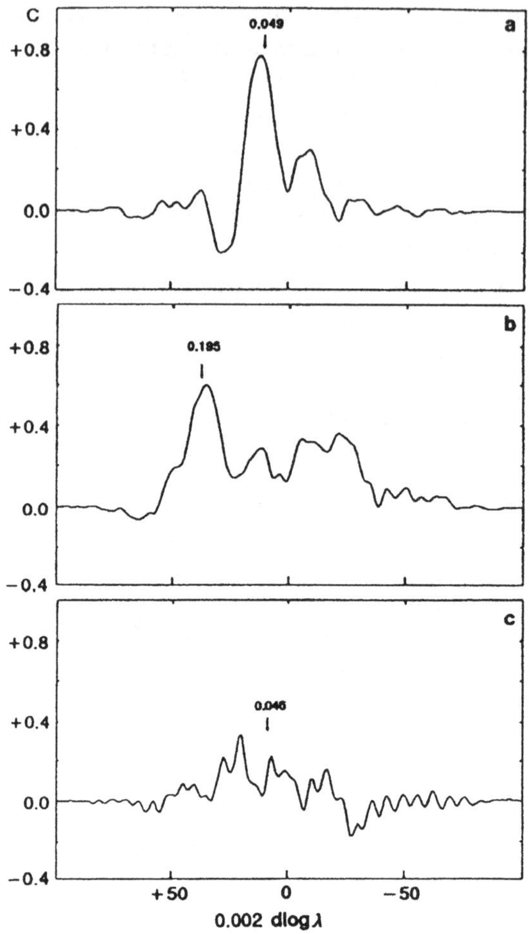

Fig. 4. Correlation functions c between a G-type star template and three galaxy spectra. The correct redshifts are indicated by arrows.
4a: The position of the highest correlation peak gives the correct redshift.
4b: Correlation function with local registration error.
4c: Correlation function with false acquisition error.

important correlation errors (Ryan and Hunt 1981):

- *Local registration errors* occur when the breaks are smeared out, resulting in broader correlation peaks with shifted maximum positions (*Fig. 4b*). The distortions of the breaks are caused by seeing effects and grain noise of the emulsion. Because of the highly nonlinear dispersion curve of the prism, local registration errors are larger in the red spectral range. Estimates of the correlation errors, *e.g.* from the asymmetric component of c_z (Tonry and Davis 1979), are thus dependent on wavelength, *i.e.* redshift, and cannot be computed with simple analytic expressions.

- *False acquisition errors* occur when different breaks or large noise peaks in the spectra lead to more than one prominent correlation peak. The position of the highest peak may not give the correct redshift (*Fig. 4c*). In contrast to local registration errors, which are more or less random, false acquisition errors can lead to large systematic redshift errors. To circumvent this, additional methods for redshift measurements must be applied. They determine the *redshift range* in the correlation function where the correct correlation peak is expected. Two such methods are given in the following sections.

Details about the redshift accuracy attainable with the correlation method are given in Sect. 4.

3.2 The least-squares method

The method of measuring redshifts by comparing intensity distributions of galaxies of unknown redshifts with those of known redshifts was originally developed by Baum (1962). In order to measure the redshifts of clusters of galaxies, he averaged multicolour photometric intensities of galaxies in each cluster, thus increasing the S/N ratios of the distributions, and compared the intensity distributions to those of galaxies with known redshifts. Oke (1971) improved this observational procedure and was able to compare the photometric data of *individual* galaxies with suitable templates. More recent work using the photometric redshift method is reported by Loh (1988).

A corresponding method can also be used with spectra. It is to find the *best* template T_i^z which leads to a minimum sum of (unweighted) squared deviations between the actual programme spectrum I_i and the different templates T_i^z:

$$s_z^2 = \sum_i (I_i - T_i^z)^2 \ . \tag{12}$$

In most cases the S/N ratios of individual programme galaxy spectra are low. If the S/N ratios of the templates are also low, s_z^2 is dominated by noise and therefore not sensitive to redshift. To reduce this problem, templates with high S/N ratios should be used.

For measuring redshifts from objective prism spectra with this method, intensity profiles corrected with Eqn. 3 are compared. The profiles are normalized to equal intensities at 520 nm. The templates $\{T_i^z\}$ are created by shifting one template along the logarithmic wavelength axis to artificial redshifts. As in the case of template matching via correlations, the programme spectrum has to be extended on both sides, here using the mean intensities at the beginning and at the end of the spectra. The template which minimizes Eqn. 12 determines the redshift.

Figure 5 gives three cases of s^2-values vs. differences in logarithmic wavelengths between the programme galaxies and a G-type star. *Fig. 5a* illustrates a correct least-squares mininum. *Fig. 5b* and *c* show errors analogous to the local registration and false acquisition errors (Sect. 3.1). The fact that the same types of errors occur in both methods relies on similarities between *least-squares minimization* and *correlation maximization* as shown by Ryan and Hunt (1981).

In contrast to the correlation method where rectified spectra are compared, the least-squares method is perfomed with intensity profiles, including both the line spectra and continua. While the correlation method is more sensitive to redshift, resulting in sharp peaks of the correlation function, the advantage of the smoother least-squares curves is that the effects of noise or other spurious features are suppressed. Here the possible *systematic errors* mentioned above are more clearly visible.

Another important redshift error of both methods is caused by different intrinsic colours of galaxies. To reduce this error, template galaxies (or artificially redshifted stars) of different spectral types must be used.

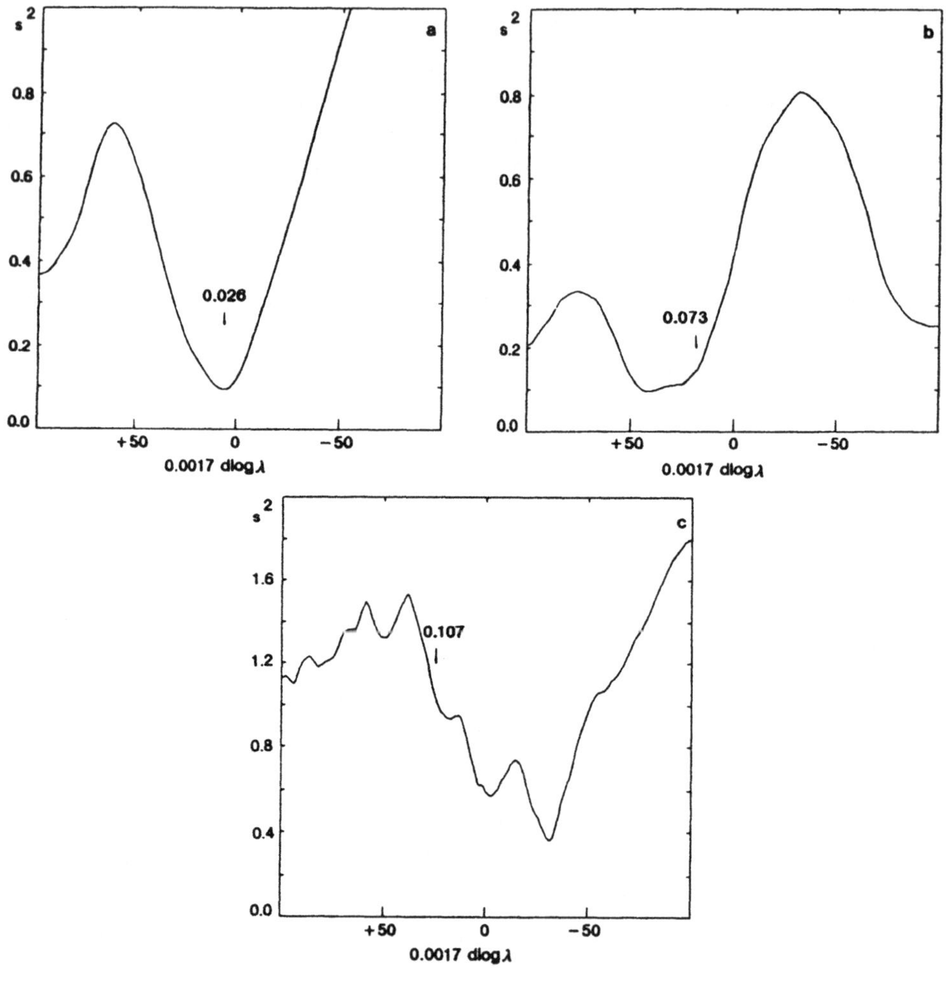

Fig. 5. Least-squares functions s^2 between a G-type star template and three galaxy spectra. The correct redshifts are indicated by arrows.
5a: The position of the s^2-minimum gives the correct redshift.
5b: s^2-function with local registration error.
5c: s^2-function with false acquisition error.

3.3 The break measuring method

The two methods described in Sects. 3.1 and 3.2 give integral measures of the galaxy redshifts. They use all spectral features without identifying them individually. In this section a method for the automated identification of breaks is described.

3.3.1 Search for individual breaks

From Eqn. 2 it is apparent that the nonlinear dispersion curve of the prism affects the intensity profiles of all spectra in a characteristic way: the spectral intensities increase with increasing wavelengths. Thus, strong absorption systems have depressed intensities at their short wavelength end, leading to a break-like appearence. This suggests the following break detection method:

At a certain number of pre-determined intensity levels of an intensity profile $I(\lambda)$ the corresponding wavelengths are determined. Several neighbouring intensity values found within a small wavelength interval indicate the presence of a break. The combination of numbers of intensity levels and width of the wavelength interval must satisfy preset criteria. These given, the wavelength groups are found by *cluster analysis.*

3.3.2 Identification of breaks

Spectrum profiles with different numbers of breaks are given in *Fig. 6.* For a one break spectrum it is always assumed that the detected feature is the strongest one, *i.e.* the 400 nm break. A two break spectrum is most likely to show the 400 nm and the 430 nm break, the two most prominent features in a galaxy spectrum. In order to ensure correct identifications, the two detected breaks must give similar redshifts. Spectra of this kind are found for galaxies with redshifts $z < 0.25$; at higher redshifts the 430 nm break is not observable on the IIIaJ-emulsion used.

Spectra with no break are frequently overexposed and thus rejected. Peculiar galaxy profiles may have more than two prominent breaks. For faint spectra it is expected that some of them show the Mg II or Fe II absorptions, indicating large redshifts and extremly high absolute luminosities. A special study of those objects is in preparation.

After identification of the breaks the final redshift is obtained by measuring the position of half maximum intensity of the 400 nm break using polynomial fits. Additional fits to the 430 nm break, when present, are not necessary because the position of this break is not as well determined as that of the 400 nm break, leading to higher redshift errors.

4 Estimates of redshift accuracies

The redshift accuracies for the three methods described in Sect. 3 are discussed here. Because of the small number of galaxy redshifts obtained from slit spectra in the relevant magnitude and redshift ranges ($17^{\mathrm{m}}.0 < m_J < 20^{\mathrm{m}}$; $z \leq 0.3$) the redshift accuracies are estimated by comparing the redshifts measured with the individual methods. The redshifts are obtained for SRC field No. 411. In the following the sub-

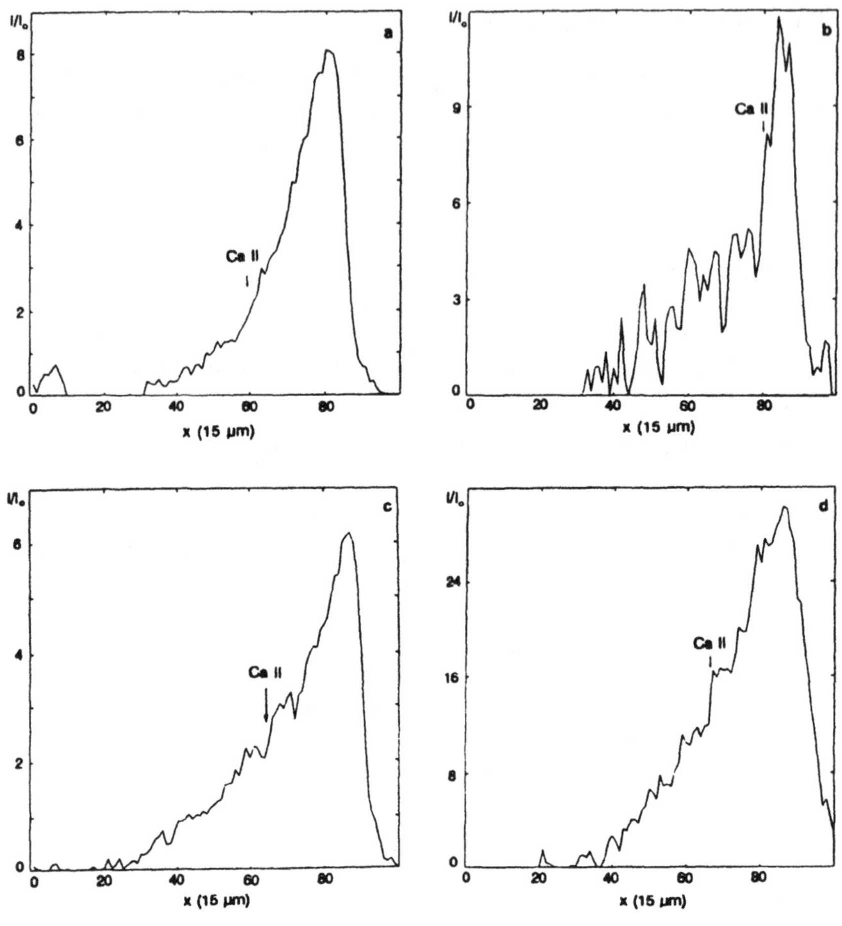

Fig. 6. Galaxy spectrum profiles with
6a: no break, 6b: one break, 6c: two breaks, 6d: three breaks.

script 'C' denotes *correlation method* (Sect. 3.1), 'L' *least-squares method* (Sect. 3.2) and 'B' *break measuring method* (Sect. 3.3)

4.1 Systematic errors

In the first step *systematic errors* are detected and corrected. For this the redshifts z_L and z_B are compared. One break galaxies are plotted in *Fig. 7a*, two break galaxies in *Fig. 7b*. In contrast to one break galaxies, two break galaxies might have z_L redshifts with false acquisition errors, the minima of the s^2-relation are caused by the 430 nm break and not by the 400 nm break. Galaxy redshifts with this error are located in a region nearly parallel to the diagonal in *Fig. 7b*. *Fig. 7c* shows the relation between redshifts *after correcting* z_L for the false acquisition error. Galaxies are rejected when large redshift differences $z_L - z_B$ are found which cannot be explained by systematic confusions of spectral features.

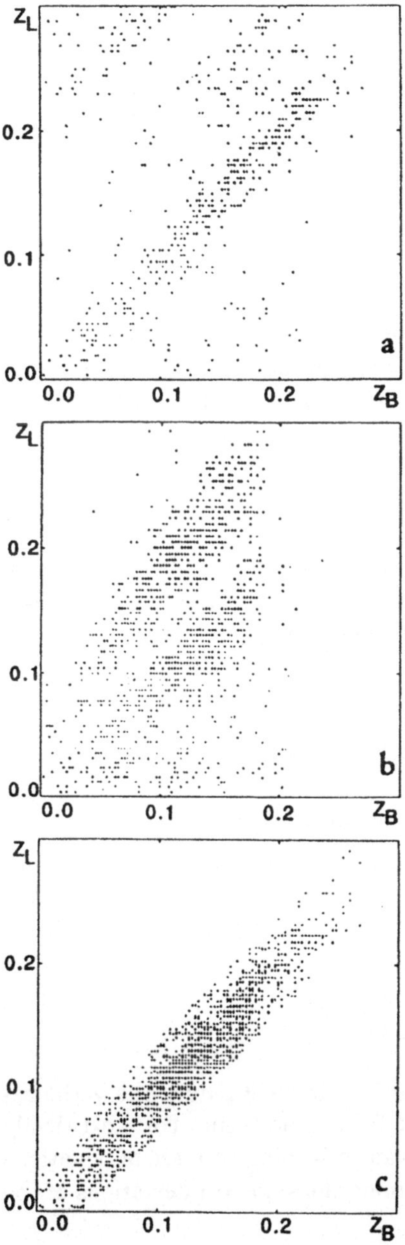

Fig. 7. Redshifts z_L obtained from the least-squares method vs. redshifts z_B obtained from direct break identifications.
7a: One break galaxies
7b: Two break galaxies
7c: z_L-redshifts corrected for false acquisition errors.

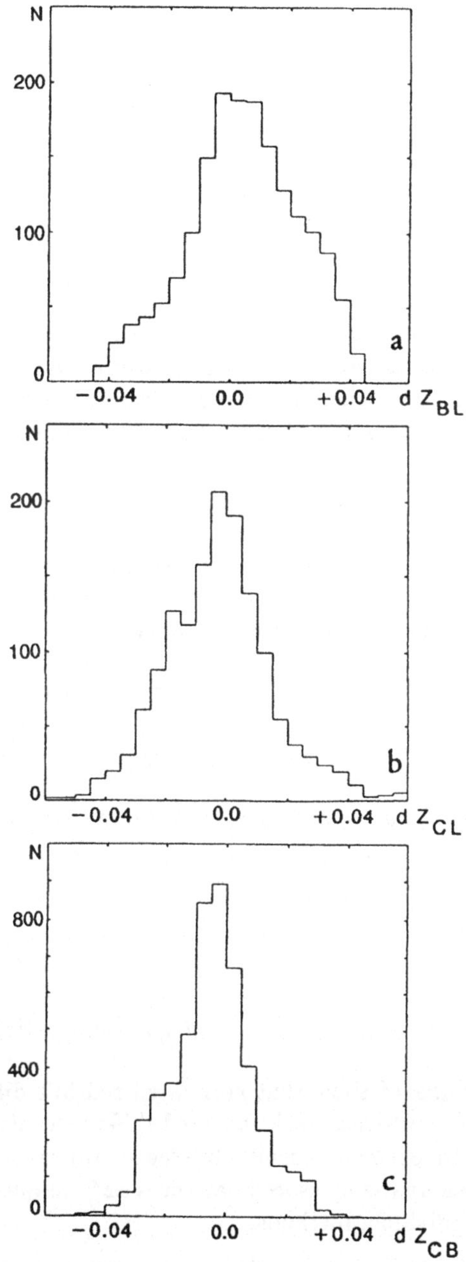

Fig. 8. Histograms of the redshift differences.
8a: dz_{BL}
8b: dz_{CL}
8c: dz_{CB}
The z_L-redshifts are corrected for false acquisition errors.

Table 1. Errors of the redshift measurement methods

relative errors	mean errors	systematic errors	mean sample redshifts	number of galaxies
$dz_{BL} = 0.0196$	$dz_L = 0.0160$	$\delta z_{BL} = +0.0058$	$\langle z_B \rangle = 0.107$	$N = 1\,766$
$dz_{CL} = 0.0177$	$dz_B = 0.0110$	$\delta z_{CL} = -0.0025$	$\langle z_C \rangle = 0.125$	$N = 1\,452$
$dz_{CB} = 0.0135$	$dz_C = 0.0074$	$\delta z_{CB} = -0.0046$	$\langle z_L \rangle = 0.112$	$N = 4\,945$

For each galaxy the averaged redshift $z_0 = \frac{z_L + z_B}{2}$, or $z_0 = z_B$ or z_L when only one reliable redshift could be measured, determines the central value for the redshift range $z_0 \pm \delta z$ where the redshift from the correlation method z_C should be looked for. δz is chosen according to empirical values.

4.2 Statistical errors

In the next step the distributions of redshift differences are used to compute the *statistical* errors of the individual methods dz_B, dz_L, dz_C. If z_T is the true (unknown) galaxy redshift, error differences are given by the equations:

$$\begin{aligned} dz_{BL} &= z_B - z_L = (z_T + dz_B) - (z_T + dz_L) = dz_B - dz_L \\ dz_{CL} &= z_C - z_L = (z_T + dz_C) - (z_T + dz_L) = dz_C - dz_L \\ dz_{CB} &= z_C - z_B = (z_T + dz_C) - (z_T + dz_B) = dz_C - dz_B\,. \end{aligned} \tag{13}$$

dz_{BL}, dz_{CL} and dz_{CB} are obtained from the redshift measurements.

In order to get reliable estimates of dz_L, dz_B and dz_C, variances are calculated. They contain cross terms $\sum dz_i \cdot dz_j$ corresponding to correlated redshift differences dz_i, dz_j:

$$\begin{aligned} dz_{BL}^2 &= dz_B^2 + dz_L^2 - 2\sum dz_B \cdot dz_L \\ dz_{CL}^2 &= dz_C^2 + dz_C^2 - 2\sum dz_C \cdot dz_L \\ dz_{CB}^2 &= dz_C^2 + dz_B^2 - 2\sum dz_C \cdot dz_B\,. \end{aligned} \tag{14}$$

Eqns. 14 show that correlated redshift differences $dz_i \cdot dz_j > 0$ reduce the variances dz_{ij}^2 whereas anticorrelated differences $dz_i \cdot dz_j < 0$ increase them. For statistically independent redshift differences the cross terms vanish leading to a set of equations where the measured variances dz_{ij}^2 uniquely determine the standard deviations of the individual methods.

Table 1 contains the redshift errors of the individual methods obtained from the relative errors. The mean redshift differences δz_{BL}, δz_{CL}, δz_{CB}, the mean redshifts $\langle z_L \rangle$, $\langle z_B \rangle$, $\langle z_C \rangle$ and the numbers N of the galaxies used for the computations are also given. The distributions of dz_{BL}, dz_{CL} and dz_{CB} are shown in *Fig. 8*.

The computations of the individual errors are in general agreement with values derived by Cooke *et al.* (1977, 1981) and Beard *et al.* (1986) for interactive redshift measurements using visual break identifications.

The correlation method is the most accurate redshift measurement technique whereas the least-squares method has the lowest accuracy. This is partially caused by the fact that the correlation method matches *spectral features* while the least-squares method uses the *whole* of the spectrum profiles.

Another reason is that for the correlation method a preset redshift interval $z_0 \pm 0.03$ ties the results to one or both of the redshifts determined by the other methods. With the redshifts not being independent, the corresponding *relative errors* are smaller and so are the resulting *individual errors.* The break measurement technique gives intermediate accuracies: the break measurements are more affected by noise than the correlation measurements.

The systematic differences between the redshifts (δz_{ij}) are small relative to the standard deviations dz_i so that they are presently neglected.

The number of galaxies used for error calculations are different for the individual methods. Most galaxy redshifts are obtained with the break measurement and correlation method. The least-squares method needs the spectral continuum and, therefore, gives reliable redshifts only for the brighter galaxies. This is supported by the computed mean redshifts $\langle z \rangle$ which are lower for the least-squares method.

5 Conclusions and future plans

In Sects. 2 and 3 the automated preprocessing and redshift measurement algorithms used in MRSP were described. Traditional reduction techniques like spectrophotometry, wavelength calibrations, correlation and least-squares methods and new algorithms of pattern recognition, like fuzzy algebra, were applied. The calculated redshift errors of the individual methods are about $dz = 0.01$, corresponding to 3 000 km s^{-1} for single redshift measurements and 0.008, coresponding to 2 300 km s^{-1} for the mean from the two independent redshifts, *e.g.* 25 % of all quoted measurements.

With the method described here it is possible to obtain about 6 000 galaxy redshifts from one objective prism plate at high galactic latitudes for objects with $m_J < 20$. So far, four plates (ESO/SRC fields Nos. 351, 411, 412 and 474) covering more than 100 square degrees, were analyzed. A total number of 200 000 spectra, stellar and non-stellar, were processed. The number of redshifts obtained is 24 000. The accuracies of redshift determinations have since been improved so that now more than 75 % of them are found from two independent methods. By the end of 1988 twelve or more objective prism plates will have been measured with more than 600 000 spectra and more than 70 000 redshifts. With the reduction time of one week per spectral plate (one template), progress is mainly limited by the supply of objective prism plates. Astronomical and cosmological interpretations of the data from field No. 411 are given by Schuecker *et al.* (1988b, c), Schuecker (1988), and Ott (1988).

Acknowledgements

This work is supported by the Deutsche Forschungsgemeinschaft. I especially thank the staff of the UKSTU Edinburgh for the film copies of the objective prism plates.

References

ADAS, 1987. *ADAS Manual*, ed. Teuber, D., Astr. Inst. Univ. Muenster.

Baum, W.A., 1962. In *IAU Symposium No. 15, Problems of Extra-Galactic Research*, ed. McVittie, G.C., Macmillan, New York, p. 390.

Beard, S.M., Cooke, J.A., Emerson, D., MacGillivray, H.T., Kelly, B.D., 1986, *Mon. Not. R. astr. Soc.*, **219**, 251.

Bracewell, R.N., 1978. *The Fourier Transform and its Applications*, McGraw-Hill, Tokyo, p. 46.

Bunclark, P., 1984. *PhD thesis*, Univ. Cambridge, UK.

Clowes, R.G., Emerson, D., Smith, M.G., Wallace, P.T., Cannon, R.D., Savage, A., Boksenberg, A., 1980. *Mon. Not. R. astr. Soc.*, **193**, 415.

Cooke, J.A., Beard, S.M., Emerson, D., 1984. In *IAU Coll. 78, Astronomy with Schmidt-type Telescopes*, ed. Capaccioli, M., Reidel, Dordrecht, p. 401.

Cooke, J.A., Emerson, D., Nandy, K., Reddish, V.C., Smith, M.G., 1977. *Mon. Not. R. astr. Soc.*, **178**, 687.

Cooke, J.A., Emerson, D., Kelly, B.D., MacGillivray, H.T., Dood, R.J., 1981. *Mon. Not. R. astr. Soc.*, **196**, 397.

Cooke, J.A., Emerson, D., Kelly, B.D., 1983. *Workshop on Astronomical Measuring Machines 1982*, eds. Stobie, R.S., MacInnes, B., Royal Observatory Edinburgh, p. 209.

Emerson, D., 1981. *Simulations of UK Schmidt objective prism spectra II: Stars*, Internal Report, Univ. Edinburgh.

Emerson, D., 1983. *Simulations of UK Schmidt objective prism spectra III : Elliptical Galaxies*, Internal Report, Univ. Edinburgh.

Horstmann, H., 1988. *These proceedings*, p. 111.

Humason, M.L., Mayall, N.U., Sandage, A.R., 1956. *Astr. J.*, **61**, 97.

Kurucz, R.L., 1979. *Astrophys. J. Suppl.*, 40, 1.

Loh, E.D., 1988. *These proceedings*, p. 267.

Nandy, K., Reddish, V.C., Tritton, K.P., Cooke, J.A., Emerson, D., 1977. *Mon. Not. R. astr. Soc.*, **178**, 63p.

Oke, J.B., 1971. *Astrophys. J.*, **170**, 193.

Ott., H.-A., 1988. *These proceedings*, p. 274.

Pal, S.K., Dutta Majumder, D.K., 1986. *Fuzzy Mathematical Approach to Pattern Recognition*, Wiley Eastern Limited, New Delhi.

Parker, Q.A., Beard, S.M., MacGillivray, H.T., 1987. *Astr. Astrophys.*, **173**, L5.

Parker, Q.A., MacGillivray, H.T., Hill, P.W., Dodd, R.J., 1986. *Mon. Not. R. astr. Soc.*, **220**, 901.

Ryan, T.W., Hunt, B.R., 1981. In *Progress in Pattern Recognition*, **1**, eds. Kanal, L.N., Rosenfeld, A., North-Holland, Amsterdam, p. 265.

Schuecker, P., 1986a. *Mitt. astr. Ges.*, **65**, 251.

Schuecker, P., 1986b. *Mitt. astr. Ges.*, **65**, 192.

Schuecker, P., 1988. *These proceedings*, p. 160.

Schuecker, P., Horstmann, H., Teuber, D., 1988a. *Proc. Klassifikation und Ordnung*, in press.

Schuecker, P., Horstmann, H., Seitter, W., 1988b. *IAU Symp. No. 130, The Structure of the Universe*, eds. Audouze, J., Szalay, A., Kluwer, Dordrecht, in press.

Schuecker, P., Ott, H.-A., Horstmann, H., Gericke, V., Seitter, W., 1988c. *3rd ESO-CERN Conf. on Cosmology*, Kluwer, Dordrecht, in press.

Schuecker, P., Horstmann, H., Volkmer, C.C., 1986. *Proc. 2nd Workshop on Data Analysis in Astronomy*, eds. Di Gesu, V., Scarsi, L., Crane, P., Friedman, J.H., Levialdi, S., Plenum

Press, New York and London, p. 109.
Seitter, W.C., 1988. In *IAU Workshop Astrophotography*, ed. Marx, S., Springer-Verlag, Berlin, p. 169.
Simkin, S.M., 1974. *Astr. Astrophys.*, **31**, 129.
Teuber, D., 1988. *These proceedings*, p. 323.
Tonry, J., Davis, M., 1979. *Astr. J.*, **84**, 1511.
Tritton, S., 1983. *UKSTU Handbook*, Royal Observatory Edinburgh.
Tucholke, H.-J., 1988. *These proceedings*, p. 136.
Tucholke, H.-J., Schuecker, P., Horstmann, H., Seitter, W.C., 1988. In *IAU Coll. 100, Fundamentals of Astrometry, Celest. Mech.*, eds. Eichhorn, H., Pakvor, I., in press.
West, R.M., Frandsen, S., 1981. *Astr. Astrophys. Suppl.*, **44**, 329.
Zadeh, L.A., 1965. *Inform. and Control*, **8**, 338.
Zadeh, L.A., 1968. *Inform. and Control*, **12**, 94.
Zadeh, L.A., 1973. *IEEE Trans.*, **SMC-3**, 28.
Zadeh, L.A., 1975. In *Fuzzy Sets and Their Applications to Cognitive and Decision Processes*, eds. Zadeh, L.A., Fu, K.S., Tanaka, K., Shimura, M., Academic Press, London, p. 1.

A Study of Galaxies with $z \leq 0.3$ in the ESO/SRC Atlas Field No. 411: Galaxy Distribution and Luminosity Functions

P. Schuecker
Astronomisches Institut
Westfälische Wilhelms-Universität
Münster, F.R. Germany

Abstract

The analysis of 6 300 galaxy redshifts obtained automatically from a single low-dispersion objective prism plate shows that the ESO/SRC Atlas field No. 411 is dominated by a galaxy supercluster at $z = 0.11 \pm 0.01$, ($330 \pm 30\, h^{-1} Mpc$). The supercluster consists of at least five rich clusters, connected by bridges of galaxies, constituting a filament of length $> 20\, h^{-1}$Mpc. Luminosity functions are calculated for the high and low density regions. The fraction of bright galaxies is higher for the high density region.

1 Introduction

A number of authors have drawn attention to the observation and confirmation of super large-scale structures with dimensions of several tens to several hundreds of megaparsecs. Among them are Bahcall and Soneira (1983, 1984), Schmidt (1984), Batuski and Burns (1985) and Kopylov *et al.* (1987). In these investigations the assumption is made that rich clusters of galaxies trace superclusters, that their absence constitutes voids, and that the underlying distribution of galaxies follows the distribution of the clusters (Abell 1961). Because of the limited observing times at large telescopes most cluster distances are obtained from a few galaxy redshifts only. The interpretation of cluster catalogues (*e.g.* Schmidt 1986, Struble and Rood 1987a, b) with respect to super large-scale topology is thus complicated and not unique.

The present paper is the first of a series of communications about results from the Muenster Redshift Project (MRSP). The large-scale distribution of galaxies will be investigated in individual and in adjoining wide angle fields, with about 6 000 measured galaxy redshifts per field up to $z \leq 0.3$, and accuracies $dz = 0.008$ corresponding to 2 400 km s^{-1}. Special emphasis is given to the identification of clusters and superclusters, but also to the distribution of field galaxies. Horstmann (1988a, hereafter HH) in collaboration with the present author studies the distribution of galaxies of different morphological types. Luminosity functions for low and high density regions, *i.e.* field and cluster galaxies, are determined. All redshifts are obtained automatically from UK objective prism Schmidt plates (Schuecker 1988, hereafter PS).

2 General properties of the redshift sample

2.1 The observational data

16 800 galaxy spectra with a preset magnitude limit $m_J \leq 20^{\rm m}\!\!.5$ were segmented from the objective prism plate of field No. 411 (HH). 3 000 of the spectra overlap and were automatically rejected. From the remainder, 6 300 galaxy redshifts were determined with the methods described in PS. The galaxies have redshifts $0.0 \leq z \leq 0.3$ and magnitudes $16^{\rm m}\!\!.5 \leq m_J \leq 20^{\rm m}\!\!.5$.

The magnitudes were obtained by transforming m_J to m'_B using the mean colour index of the objects $B - V = 0.8$ and calibrating with the photometric sequence of Hawkins (1981). m'_B magnitudes are corrected for the K-effect using the analytic expression given by Shanks *et al.* (1984) for E and S0 galaxies:

$$m_B = m'_B - (4.14z - 0.44z^2) \tag{1}$$

The choice of correction factors is determined by the fact that early-type galaxies constitute about $\frac{2}{3}$ of the galaxies as found in the morphological studies of HH.

Corrections for biased redshift and magnitude measurements are necessary before luminosity functions (Sect. 4) can be derived.

2.2 Statistical corrections

In *Fig. 1a* differential galaxy number counts $N_d(m_B)$ in intervals of 0.1 magnitudes obtained from the direct plate and differential number counts for galaxies with mea-

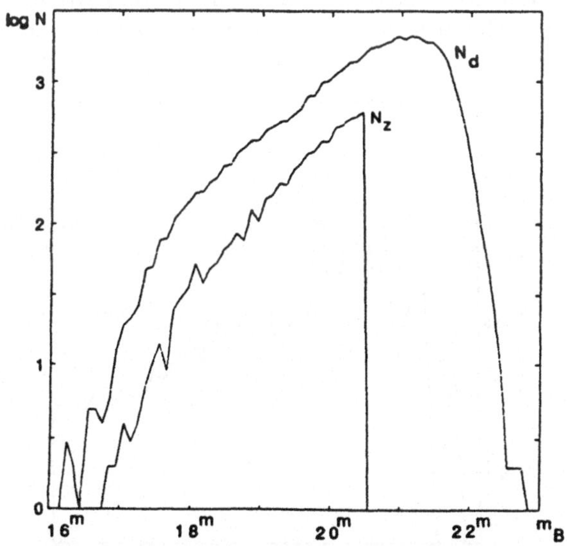

Fig. 1a. Differential galaxy number counts obtained from the direct plate $N_d(m_B)$ and for galaxies with measured redshifts $N_z(m_B)$ from the objective prism plate.

sured redshifts from the objective prism plate are shown. *Fig. 1b* gives the difference $s(m_B) = \log N_d(m_B) - \log N_z(m_B)$ for different magnitudes. The values of the $s(m_B)$-function measure the *incompleteness* of the redshift sample. The function is also sensitive to *biases* of the redshift sample relative to the sample from the direct plate.

Significant deviations from zero and from an unbiased flat distribution are found. Since higher values of $s(m_B)$ indicate a lower fraction of objects with measured redshifts it appears that increasingly more galaxies are missed going from faint to bright magnitudes down to about 17^m. This can be attributed to the fact that the galaxy images tend to be larger at bright magnitudes giving rise to a smeared out appearance and thus difficulties in measuring characteristic features. The small values of $s(m_B)$ for the brightest galaxies ($< 17^m$) are an artefact. This is supported by the fact that the noise in this region is also high, due to the low number of objects (small volume covered in near space and scarcity of absolutely bright galaxies at larger distances).

Equation 2 is a formal description of the above mentioned selection effects:

$$s(m_B) = \begin{cases} 0.6500\, m_B - 10.2883, & m_B < 17\overset{m}{.}0 \\ 0.1086\, m_B + 2.6079, & 17^m \leq m_B \leq 20^m \end{cases} \tag{2}$$

Eqn. 3 gives $N_z^c(m_B)$, the number of galaxies corrected for the selection effects, *i.e.* the expected unbiased number on the objective prism plate:

$$log N_z^c(m_B) = s(m_B) + \log N_z(m_B) \tag{3}$$

Because the correction factors are global they do not correct for differences in the redshift measurements of different galaxy types. The corrected sample is, of course, only complete and unbiased *relative* to the sample from the direct plate. Recent improvements of the techniques of redshift measurements show that our new data do not require a correction for bias. (Sect. 5 and Schuecker *et al.* 1988a).

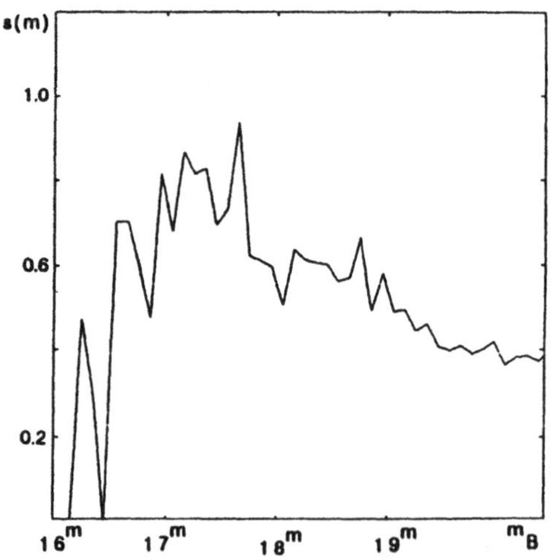

Fig. 1b. Logarithmic differences $s(m_B)$ for different magnitudes.

2.3 The $M(z)$-diagram

In *Fig. 1* of Ott (1988) absolute magnitudes M_B vs. redshifts z are plotted. The absolute magnitudes were calculated using the luminosity distance given by Mattig (1958), with the parameters $H_0 = 100$ km s^{-1} Mpc^{-1}, $q_0 = 0.5$ and $\Lambda = 0$:

$$\begin{aligned} M_B = \; & m_B + 5\log H_0 - 25 - \\ & 5\log\left\{\tfrac{c}{q_0^2}[q_0 z + (q_0 - 1)(\sqrt{1+2q_0 z} - 1)]\right\} \end{aligned} \tag{4}$$

In the $M_B(z)$ diagram the cutoff line is fixed by the limiting magnitude $m_B = 20\overset{m}{.}5$. The lack of apparently bright galaxies at low redshifts ($z < 0.1$) is explained in Sect. 2.2.

3 Morphological properties of the redshift sample

3.1 The large-scale distribution of galaxies

Results concerning the two-dimensional distribution of galaxies for field No. 411 are given in Dodd and MacGillivray (1986) and in Horstmann (1988 a, b). Information about the individual clusters, *i.e.* distribution of galaxies in two and three dimensions, cluster luminosity functions etc., can be found in Schuecker *et al.* (1988b). *Fig. 2* shows the isopleths of galaxies in this field as presented by HH. The field is divided into five sections, a) to d) at different declinations and fixed right ascensions, and e) over a narrow right ascension strip covering all declinations included in a) to d).

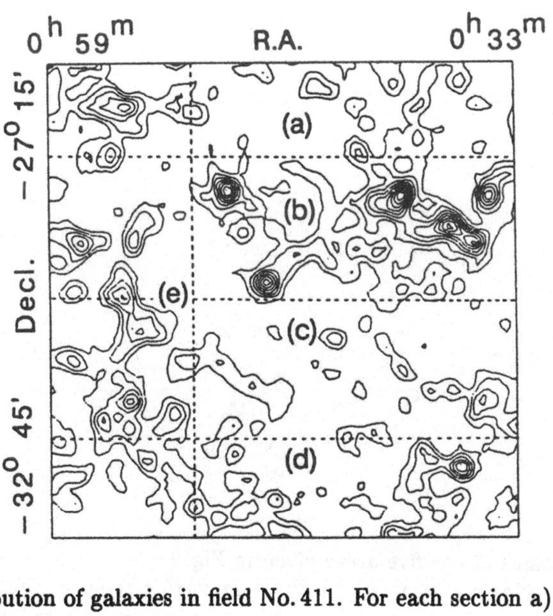

Fig. 2. The distribution of galaxies in field No. 411. For each section a) to e) corresponding wedge diagrams are shown in Fig. 3.

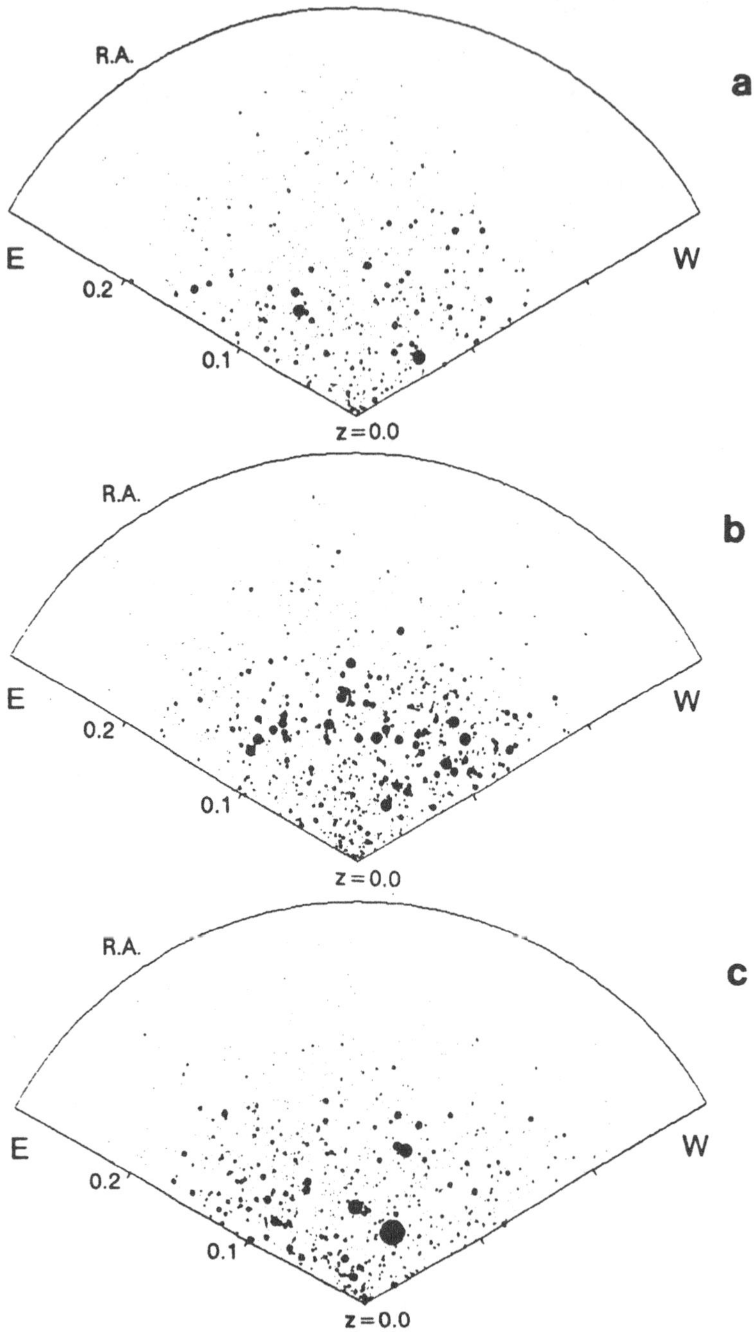

Fig. 3. Wedge diagrams of the five areas given in Fig. 2.

Fig. 3a: Section a) $0^h33^m \le$ R.A. $\le 0^h59^m$, $-28°21' \le$ Decl. $\le -27°15'$

Section b) $0^h33^m \le$ R.A. $\le 0^h59^m$, $-30°00' \le$ Decl. $\le -28°21'$

Section c) $0^h33^m \le$ R.A. $\le 0^h59^m$, $-31°39' \le$ Decl. $\le -30°00'$

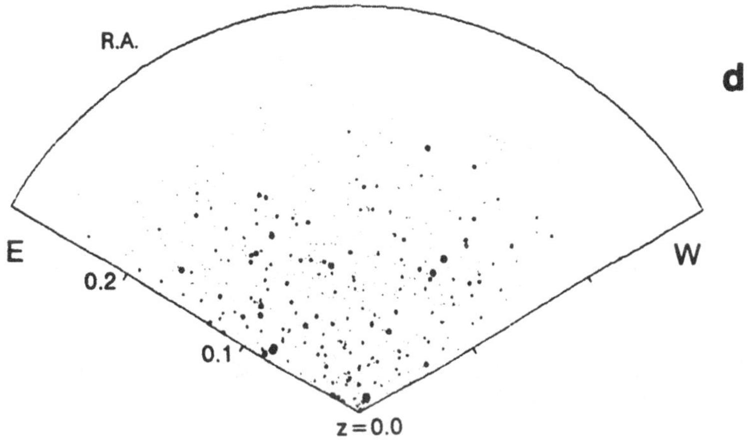

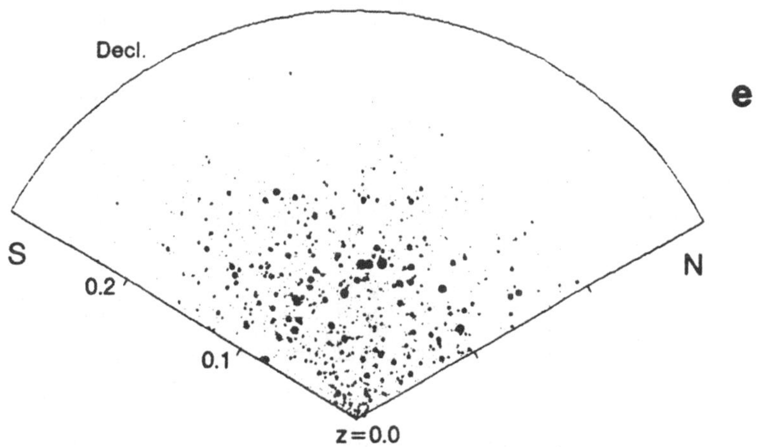

Fig. 3. cont.
Fig. 3b: Section d) $0^h33^m \leq$ R.A. $\leq 0^h59^m$, $-32°45' \leq$ Decl. $\leq -31°39'$
Section e) $0^h51^m \leq$ R.A. $\leq 0^h59^m$, $-32°45' \leq$ Decl. $\leq -27°15'$

Wedge diagrams of the five different areas are given in *Fig. 3.* The sizes of the symbols are proportional to the apparent brightness of the galaxies. *Fig. 3b* and e correspond to cluster-rich regions on the direct plate. The diagrams show that most of the luminous galaxies are concentrated in clusters near redshifts $z = 0.11$. Bright galaxies are also located at $z = 0.07$ (R.A. = 0^h46^m, Decl. = $-29°$) and $z = 0.12$ (R.A. = 0^h55^m, Decl. = $-30°$). In the other wedge diagrams no significant concentrations are found.

3.2 The redshift histograms

One hundred redshift histograms covering the whole field are presented in *Fig. 4.* Each histogram is calculated for an area of $33' \times 33'$. The redshift range is $0.0 \leq z \leq 0.3$. The galaxies are counted in bins $dz = 0.01$ leading to numbers of galaxies < 15 per bin. The average number per histogram is about 60.

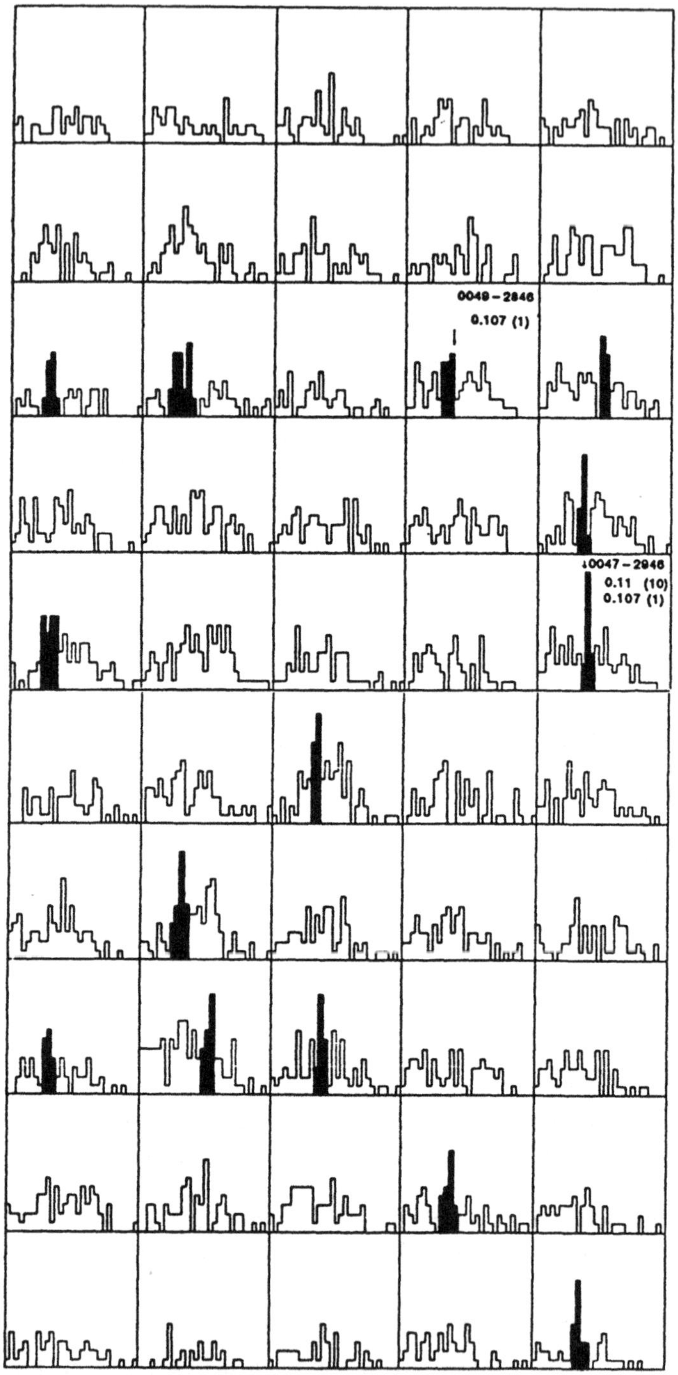

Fig. 4. Redshift histograms for field No. 411.
Fig. 4a: Section: $0^h46^m \leq$ R.A. $\leq 0^h59^m$, $-32°45' \leq$ Decl. $\leq -27°15'$

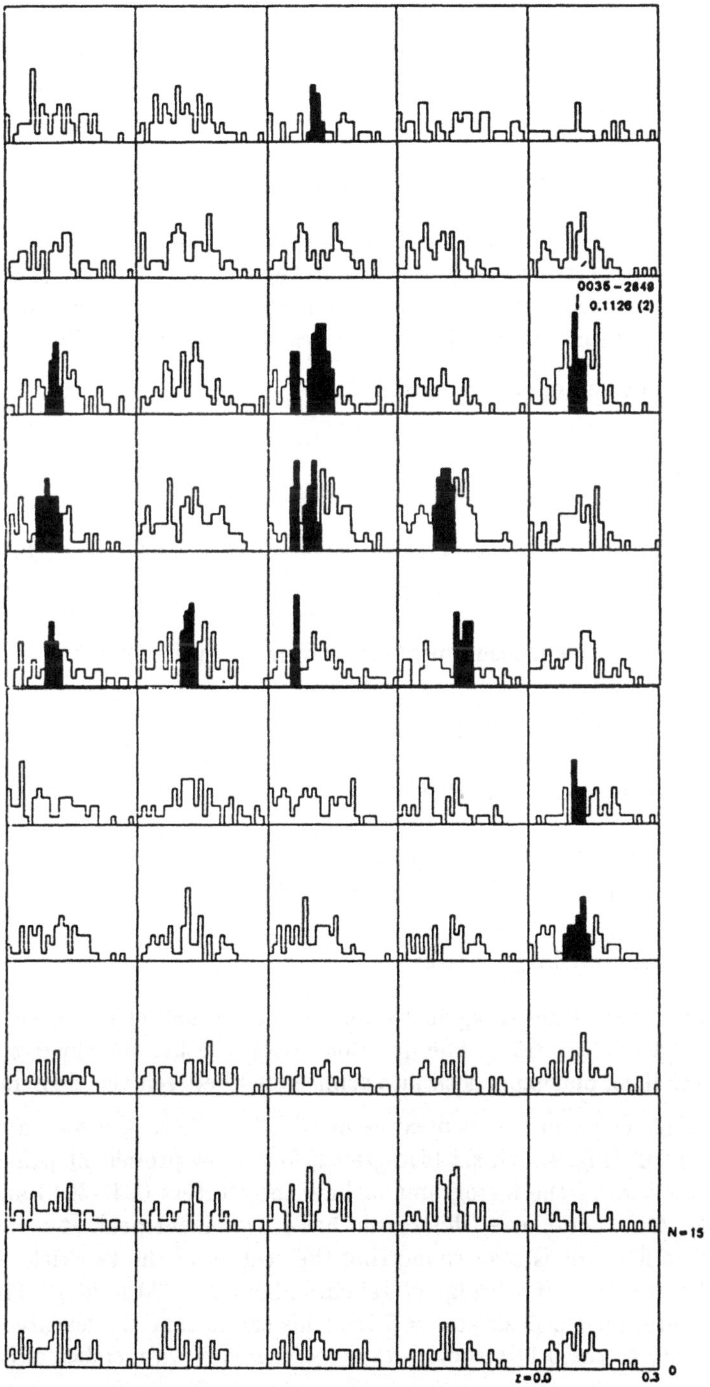

Fig. 4. cont.
Fig. 4b: Section: $0^h33^m \leq$ R.A. $\leq 0^h46^m$, $-32°45' \leq$ Decl. $\leq -27°15'$

As expected, the number of galaxies is larger in histograms from areas in the direction of the clusters than in the others. In these histograms the most significant columns are shaded. Most of them lie at redshifts $0.10 \le z \le 0.12$, again illustrating that the prominent clusters found on the direct plate are concentrated in this redshift range.

3.3 Comparison with previous measurements

Three of the five clusters have published redshifts obtained with slit spectra and objective prism spectra measured interactively. Table 1 lists the clusters, number of redshifts previously obtained, corresponding z-values, authors, number of galaxy redshifts measured in the MRSP and z-values obtained in the MRSP.

Table 1: Comparison of cluster redshifts

Cluster	N_{lit}	z	Authors	N_{MRSP}	z_{MRSP}
0035-2849	2	0.1126	1	20	0.105
0047-2946	1 (10*)	0.107 (0.11*)	2 (3*)	20	0.115
0049-2846	1	0.107	2	11	0.100

* interactive measurements from objective prism spectra

(1) West and Frandsen (1981)
(2) Ellis and Allen (1983)
(3) MacGillivray and Dodd (1979)

The external error of the MRSP cluster redshifts relative to those from other authors are not larger than the mean internal error of 0.008.

3.4 Clustering on large scales

The redshifts from Table 1 again support the existence of a concentration of the clusters at redshift $z = 0.11$. The question arises whether the clusters are part of a supercluster. If so, one expects to find connections between the clusters.

An interesting region in this context is at $0^h47^m \le$ R.A. $\le 0^h49^m$ and $-29°50' \le$ Decl. $\le -28°30'$ (*Fig. 4*). Of six histograms, four show prominent peaks at $z = 0.11$ or $z = 0.15$. Two of the histograms include the clusters 0047-2946 and 0049-2846, respectively. One 'non-cluster' histogram has a prominent peak at $z = 0.11$. On the direct plate a filament is seen connecting the centers of the two rich clusters. This suggests the existence of a bridge of galaxies about $6\,h^{-1}$Mpc long. The other non-cluster histogram has a peak at $z = 0.15$. This shows that the feature in spite of its close proximity to the 0.11-structure in projection is not a member of it.

Connections between the cluster pair 0047-2946, 0049-2846 in the eastern part and the cluster 0035-2849 in the western part of the 0.11-structure are apparent in the histograms for $0^h35^m \le$ R.A. $\le 0^h46^m$, $-28°15' \le$ Decl. $\le -30°00'$. Therefore, the galaxies at $0.11\,c$ form a supercluster. The minimum size of the supercluster is $20\,h^{-1}$ Mpc in the direction of R.A. and $6\,h^{-1}$Mpc in the direction of Decl.

Measurements in neighbouring fields suggest a much larger size of the total supercluster ($> 40\,h^{-1}$ Mpc) in which the 0.11-structure constitutes the dominant region of very high density (see *Fig. 6* of HH).

The histograms for $0^h36^m \leq$ R.A. $\leq 0^h43^m$, $-30°15' \leq$ Decl. $\leq -28°15'$ have significant maxima at $z = 0.07$. The structure is also marked by some bright galaxies in the wedge diagrams (*Fig. 3*). This foreground cluster seems to be elongated with a major axis lying nearly perpendicular to the 0.11-structure. The smaller distance and the orientation of the cluster implies non-membership in the 0.11-structure.

4 Physical properties of the redshift sample

4.1 Parametric representation of luminosity functions

In order to give a physical description of the MRSP redshift sample luminosity functions (LF) were calculated.

Luminosity functions should be measured in equal redshift intervals Δz at different redshifts z. This guarantees good coverage of both the faint and bright ends of the LF. The number of galaxies at redshifts $z \pm \Delta z$ with magnitudes between M_B and $M_B + \Delta M_B$ is according to Weinberg (1972):

$$\begin{aligned} N(z \pm \Delta z, M_B)\,\Delta M_B &= \\ \left(\frac{c}{H_0}\right)^3 \frac{1}{q_0^4} & \frac{[q_0 z + (q_0 - 1)(\sqrt{1+2q_0 z} - 1)]^2}{(1+z)^6\,\sqrt{1+2q_0 z}}\, n(z, M_B)\, 2\Delta z\, \Delta\theta\, \Delta M_B\,, \end{aligned} \quad (5)$$

$n(z, M_B)\,\Delta M_B$ is the number of galaxies per unit volume at redshift z with magnitudes between M and $M + \Delta M$.

If there is neither creation nor destruction and evolution of galaxies in the redshift range considered, the galaxy densities at redshifts z are related to the galaxy densities at $z = 0$ by

$$n(z, M_B) = (1+z)^3\, n(0, M_B)\,. \quad (6)$$

$n(0, M_B)$ is the general LF, *e.g.* the number of galaxies per proper unit volume.

According to the Schechter formalism (Schechter 1976) the LF is characterized by the parameters M_B^* (bright end magnitude), α (faint end slope) and n^* (normalization factor for galaxy densities). Whereas Schechter's analytic expression of the LF was originally formulated for luminosities, we use the magnitude dependent form:

$$\begin{aligned} \log[n(0, M_B)\,\Delta M_B] &= \log[0.4\ln 10] + \log(n^*) - 0.4\,(\alpha+1)\,(M_B - M_B^*) \\ &\quad - \log(e)\,\mathrm{dex}[-0.4\,(M_B - M_B^*)] + \log(\Delta M_B) \end{aligned} \quad (7)$$

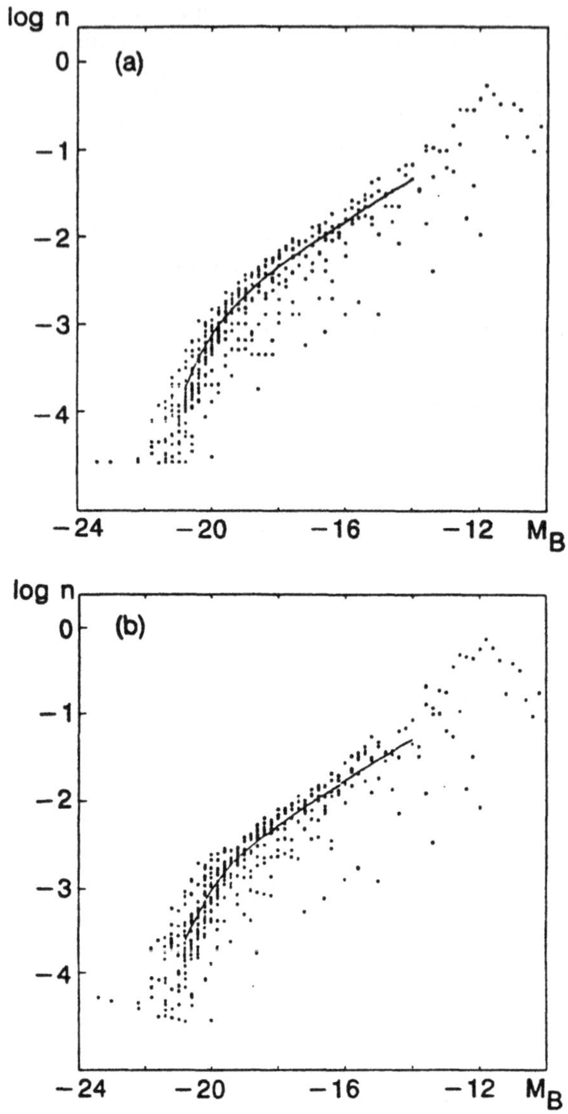

Fig. 5. Luminosity function of samples A (a) and B (b) with superimposed Schechter functions.

4.2 Luminosity functions of MRSP samples

In order to find possible differences between the LFs of the general field (low density) and the cluster regions (high density) five subsamples of the MRSP data were investigated:

Sample A: all galaxies with measured redshifts $0.0 < z \leq 0.3$
Sample B: sample A corrected for incompleteness and bias (Sect. 2.2)
Sample C: sample B for low density regions with $0.00 \leq z \leq 0.20$

Sample D: sample B for high density regions with $0.09 \leq z \leq 0.13$
Sample E: sample B for low density regions with $0.09 \leq z \leq 0.13$.

$n(0, M_B)$ is determined for the redshift intervals $2\Delta z = 0.01$ and the absolute magnitude intervals $\Delta M_B = 0.2$.

The comparison of samples A and B is used to show the influence of selection effects (Sect. 2.2) on the shape of the LF. Sample C is compared with the field LFs presented by Felten (1977). Samples D and E illustrate the differences between the shapes of the LFs for high and low density regions.

Figure 5 presents the LFs of samples A and B in the magnitude range $M_B = [-24, -10]$. The correction factors were computed separately for each $[z, z \pm \Delta z]$ and $[M_B, M_B + \Delta M_B]$ interval using Eqns. 1 through 4. Superimposed on *Fig. 5a* and *b* are the Schechter functions fitted to the data. The best-fit parameters and their formal errors are:

Sample A (uncorrected):
$M_B^* = -20\overset{\mathrm{m}}{.}2 \pm 0.2, \quad \alpha = -1.58 \pm 0.07, \quad n^* = (3.7 \pm 0.5)10^{-3}$
Sample B (corrected):
$M_B^* = -20\overset{\mathrm{m}}{.}2 \pm 0.2, \quad \alpha = -1.56 \pm 0.05, \quad n^* = (4.7 \pm 0.7)10^{-3}$

The logarithmic mean scatter of ± 0.2, corresponds to ratios of the $N(z \pm \Delta z, M_B)\,\Delta M_B$ of ± 1.6. The variances are caused by the clusters of galaxies found in the field and by the errors of the redshift and magnitude measurements. Measuring points with large deviations towards smaller n are systematically biased by the incompleteness of the sample; points with values $N(z \pm \Delta z, M_B)\,\Delta M_B < 5$ are not used for the fits with the Schechter function.

No significant differences between the LFs of the uncorrected and the corrected redshift samples are found. The use of the corrected galaxy number counts, however, increases n^*. The M_B^*-values are in general agreement with the values obtained *e.g.* by Sandage *et al.* (1985). The faint end slopes are smaller than the frequently quoted value $\alpha = -1.25$, but in good agreement with the data points in *Fig. 5*.

4.3 The field luminosity function

In *Fig. 6* the LF of sample C (field galaxies) is superimposed on the LFs assembled by Felten (1977). The mean number densities of field galaxies with equal absolute magnitudes in the present sample are determined using the same Hubble parameter $H_0 = 50\,\mathrm{km\,s^{-1}\,Mpc^{-1}}$ and the same magnitude bin size $\Delta m_B = 1$ mag as Felten. Because field No. 411 is near the SGP no corrections for galactic absorption are applied. The MRSP data follow the general shape of the LFs with slightly lower densities in the range $-21^{\mathrm{m}} \leq M_B \leq -18^{\mathrm{m}}$. The densities at high luminosities are obtained from a few galaxies only. In order to get more reliable densities in this magnitude range, additional measurements on other plates are in preparation.

Table 2: Comparison of luminosity functions

	M_B^* (mag)	α	n^* (10^{-3})
High density region	-20.4 ± 0.4	-1.4 ± 0.1	14.4 ± 4.3
Low density region	-22.5 ± 0.4	-1.7 ± 0.1	1.0 ± 0.2

4.4 Comparison of luminosity functions in high and low density regions

The LFs of samples D and E, fitted with Schechter functions, are shown in *Fig. 7*. The best-fit parameters are given in *Table 2*.

The LFs and the best-fit parameters suggest that the fraction of bright (giant) galaxies is higher in clusters than in the general field. At present it is not clear whether this effect is caused by the properties of field and cluster galaxies, selection effects depending on morphology (*e.g.* the fraction of spiral galaxies is higher for the general field) and/or instrumental effects (*e.g.* variation of the sensitivity of the emulsion across the plate).

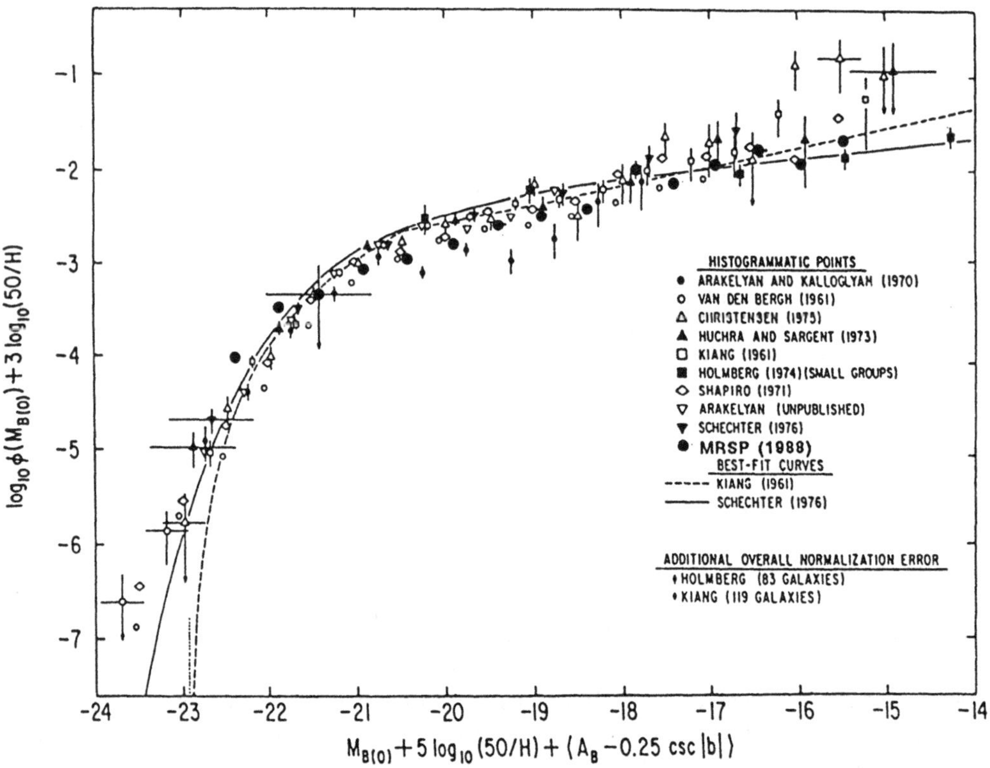

Fig. 6. The luminosity function of sample C superimposed on the luminosity functions assembled by Felten (1977).

5 Conclusions

A sample of 6 300 galaxies with redshifts $z \leq 0.3$ was analyzed. The results are:

- Five rich clusters are found at $z = 0.11$. For three of the clusters redshifts are available from slit spectra, supporting the present results.
- The rich clusters at $z = 0.11$ are connected and members of the **Sculptor supercluster** with a confirmed size of about $20\,h^{-1}$ Mpc and a possible extend $> 40\,h^{-1}$ Mpc. In the latter case the 0.11-structure constitutes a dominant (very high density) nucleus of the supercluster.
- The often used hypothesis that galaxies which are not members of rich clusters do follow the distribution of the clusters (Abell 1961, Einasto *et al.* 1980) is supported by the presence of extended bridges between the clusters.
- Possible clusters are found at $z = 0.07$ and $z = 0.15$, respectively.
- After correcting for the incompleteness of the redshift sample using corresponding galaxy number counts on the direct plate, luminosity functions were calculated. For the field galaxies the luminosity function is in general agreement with Felten (1977).
- Significant differences between the luminosity functions for high and low density regions are observed. It is not yet clear whether these differences are physically real.

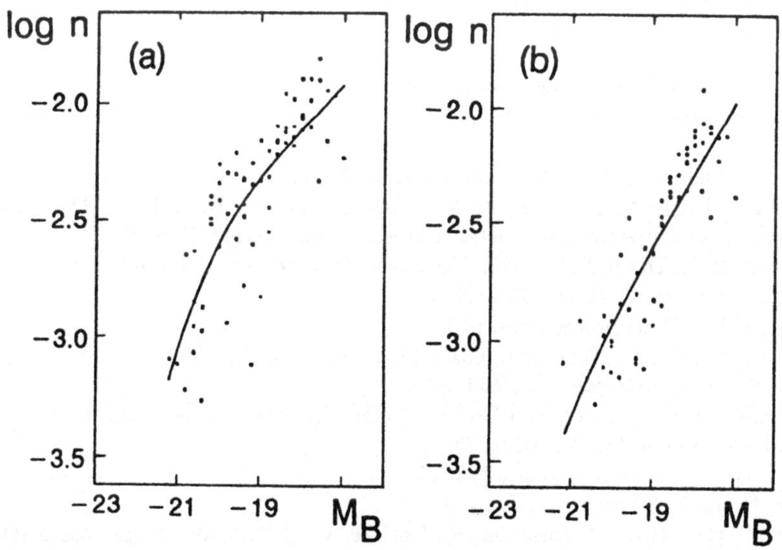

Fig. 7. Luminosity function of sample D (a) and E(b) with superimposed Schechter functions.

6 Prospects

Recent improvements of the redshift measurement methods have increased the completeness of the redshift sample, giving 7 200 galaxy redshifts with $m \leq 20^{\mathrm{m}}$ instead of 6 300 redshifts with $m_B \leq 20\overset{\mathrm{m}}{.}5$ for the sample discussed in this paper.

Redshifts were also obtained in fields adjacent to field No. 411. They yield 4 000 redshifts (No. 351), 4 300 redshifts (No. 474), 8 200 redshifts (No. 412). The large variations in the number of redshifts are mainly caused by differences in the quality of the objective prism plates. The redshift histograms of this enlarged sample of 23 700 galaxies support the existence of the 0.11 supercluster with possible extensions in field No. 412 and No. 474. Other concentrations of clusters of galaxies are found in field No. 351 suggesting the presence of another supercluster at $z = 0.14$.

Acknowledgements

This work is supported by the Deutsche Forschungsgemeinschaft. I thank H.-A. Ott for useful discussions about the luminosity functions.

References

Abell, G.O., 1961. *Astr. J.*, **66**, 607.
Bahcall, N.A., Soneira, R.M., 1983. *Astrophys. J.*, **270**, 20.
Bahcall, N.A., Soneira, R.M., 1984. *Astrophys. J.*, **277**, 27.
Batuski, D.J., Burns, J.O., 1985. *Astrophys. J.*, **299**, 5.
Dodd, R.J., MacGillivray, H.T., 1986. *Astr. J.*, **92**, 706. Einasto, J., Joeveer, M., Saar, E., 1980. *Mon. Not. R. astr. Soc.*, **193**, 353.
Ellis, R.S., Allen, D.A., 1983. *Mon. Not. R. astr. Soc.*, **203**, 685.
Felten, J.E., 1977. *Astr. J.*, **82**, 861.
Hawkins, M.R.S., 1981. *Mon. Not. R. astr. Soc.*, **194**, 1013.
Horstmann, H., 1988a. *These proceedings*, p. 111.
Horstmann, H., 1988b. In *IAU Workshop Astrophotography*, ed. Marx, S., Springer, Berlin, p. 186.
Hubble, E., Humason, M.L., 1931. *Astrophys. J.*, **74**, 43.
Kopylov, A.I., Kuznetsov, Yu., Fetisova, T.S., Shvartsman, V.F., 1987, *The Large-Scale Structure of the Universe*, Izd. Spets. Astrofiz. Obs., Stavropol', p. 39.
MacGillivray, H.T., Dodd, R.J., 1979. *Mon. Not. R. astr. Soc.*, **186**, 743.
Mattig, W., 1958. *Astr. Nachr.*, **284**, 109.
Ott, H.-A., 1988. *These proceedings*, p. 274.
Sandage, A., Bingelli, B., Tammann, G.A., 1985. *Astr. J.*, 90, 1759.
Schechter, P., 1976. *Astrophys. J.*, **203**, 297.
Schmidt, K.-H., 1984. *Clusters and Groups of Galaxies*, eds. Mardirossian, F., Giuricin, G., Mezzetti, M., Reidel, Dordrecht, p. 73.
Schmidt, K.-H., 1986. *Astr. Nachr.*, **307**, 69.
Schuecker, P., 1988. *These proceedings*, p. 142.
Schuecker, P., Ott., H.-A., Horstmann, H., Gericke, V., Seitter, W., 1988a. *3rd ESO-CERN Conf. on Cosmology*, Kluwer, Dordrecht, in press.
Schuecker, P., Horstmann, H., Seitter, W., 1988b. *IAU Symp. 130, The Structure of the Universe*, eds. Audouze, J., Szalay, A., Kluwer, Dordrecht, in press.

Shanks, T., Stevenson, P.R.F., Fong, R., MacGillivray, H.T., 1984. *Mon. Not. R. astr. Soc.*, **206**, 767.
Struble, M.F., Rood, H.J., 1987a, *Astr. J.*, **93**, 1035.
Struble, M.F., Rood, H.J., 1987b, *Astrophys. J. Suppl.*, **63**, 543.
Weinberg, S., 1972. *Gravitation and Cosmology*, Wiley, New York.
West, R.M., Frandsen, S., 1981. *Astr. Astrophys. Suppl.*, **44**, 329.

Part III

Clusters of Galaxies

Piotr Flin

Gabor Tóth

Guido Chincarini

Konrad Rudnicki

A Study of Nearby Clusters of Galaxies

Piotr Flin
Jagiellonian University Observatory
Kraków, Poland

Paul Hickson
Department of Geophysics and Astronomy
University of British Columbia
Vancouver, B.C., Canada

Giancarlo Pittella
Osservatorio Astronomico di Roma, sede Monte Porzio
Roma, Italy

Abstract

The paper describes a study, presently underway, of a sample of 150 nearby clusters of galaxies. We discuss the selection criteria, observational data, and methods of analysis, and present some illustrative results.

1 Introduction

In recent years much progress hass been made toward the understanding of nearby rich clusters of galaxies (Dressler 1984). In addition, studies of more distant clusters (Butcher and Oemler 1985) have revealed important differences between high-redshift clusters and nearby rich clusters. Despite this activity, we still know little about the nature of typical nearby clusters. As yet, there is no large well studied sample of typical low-redshift clusters, although some progress is being made toward this goal (Oegerle *et al.* 1986).

The lack of good published data for nearby clusters prompted PH to obtain photographic plates of a large sample of nearby clusters in the course of a study of structural properties of clusters (Hickson 1977a, b). These plates form the basis for the present study. Our primary objectives are to extend this previous work by including photometric and morphological properties of cluster galaxies, obtained by microdensitometer scanning and digitization of the plates. The resulting large homogeneous data set will allow us to examine statistical properties of clusters with accuracy. Such data are important, not only for the understanding of nearby clusters and the galaxies that they contain, but also to provide a firm basis for comparison with distant clusters.

In this paper, we discuss the observational data and selection criteria for the sample, the methods of analysis, and the present status of the project.

2 Observational data and sample selection

The observational material consists of 10-inch plates taken by PH with the 48 inch Palomar Schmidt telescope. Two emulsions were employed: Kodak 127-02 (a fine grain red-sensitive emulsion which preceded the more recent IIIa-F) was used for more distant clusters, and Kodak 098-04 (a more sensitive coarse grain emulsion) was used for nearby clusters. Both emulsions were used with 2 mm of Schott RG-1 glass, corresponding to the red photographic F-band of Oemler (1974). This band is centered at 6500Åand is almost identical to the r-band of Thuan and Gunn (1976). Plates were developed for eight minutes in MWP-2. Each plate was calibrated with the Palomar spot sensitometer, which exposes a corner of the plate to spots of increasing (by $\sqrt{2}$) intensities.

These plates are centred on clusters in Hickson's "nearby" sample (1977a), but their large format results in other clusters being serendipitously included. In addition to the central cluster, all Abell clusters also on the plate were included, if their galaxies were clearly visible. This resulted in a sample of 150 clusters, which forms the basis for our study.

3 Image analysis

Each selected field containing clusters as well as the sensitometer spots, was scanned at the Rome Observatory at Monte Porzio with a PDS 1010 G microdensitometer. The measurements were made in the transparency mode.

The plate scanning procedure produced a huge set of data recorded on magnetic tapes and to these data image processing software developed at the Rome Observatory was applied (Nanni *et al.* 1980, Pittella 1987). This software includes automatic object detection and identification, plate calibration and photometry, star-galaxy separation, and the determination of object shapes.

3.1 Object detection procedure

Object detection is the most essential procedure in the automatic image analysis. Its purpose is to build up a catalogue containing data on each of the objects detected. Objects were identified using the algorithm of Pittella and Vignato (1979), which operates in a line-by-line single pass mode (the image is read sequentially from mass memory only once with one line in the memory at a time). The algorithm is based on an image segmentation criterion which detects all sets of connected pixels above a suitable threshold level. To account for background variation, the local threshold level is defined using a smoothed surface derived from the original image by local averaging. A pixel is selected as part of an object if its transparency is lower than x % of the threshold value. In order to be accepted, an object must contain at least n pixels, where n is large enough to reject most random noise. The best values of x and n were determined by running the program interactively and displaying graphically the output from a region of the image.

This procedure produces the following data for each object: coordinates of the centroid, extent of the object (above threshold) in the coordinate directions, number of pixels n, maximum intensity, and threshold intensity.

3.2 Plate calibration and photometry

The calibration is obtained by fitting measured spot densities with a Fermi function of the form:

$$D = DF + \frac{DS - DF}{1 + Ae^{-\log \frac{I}{P}}}, \tag{1}$$

where DF, DS, A and P are free parameters. A typical calibration curve is shown in *Fig. 1.* The calibration curve serves to construct the look-up table.

Photometry is carried out as follows. First, a series of integrated flux measurements is made. These measurements integrate over pixels through diaphragms of increasing diameter to give the integrated object intensity profile. Then the local background intensity I_b is estimated by averaging five values nearest to the profile minimum. The intensity profile is converted to magnitudes according to the relation $m = -2.5\log(I/I_b)$. By using local background measurements we avoid errors associated with plate nonuniformity. The total magnitude of an object is computed as that corresponding to the largest diaphragm which does not include the background level.

3.3 Star-galaxy separation

Stars are distinguished from galaxies by comparing the magnitudes of all objects in diaphragms of two different sizes (Di Chio *et al.* 1983). In the present study, the diaphragm radii were 2.5 and 5 arcsec. Denoting the corresponding magnitudes by m_3 and m_5, respectively, the discriminating parameter between stars and galaxies

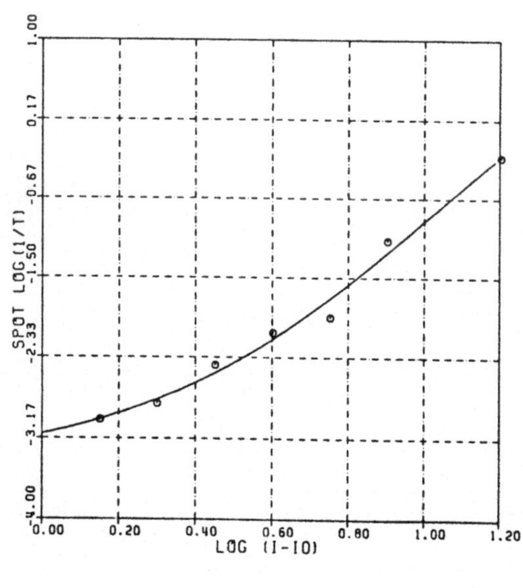

Fig. 1. Calibration curve (cluster A1661).

is: $\Delta m = m_3 - m_5$. Calculating the magnitude of a star with the intensity profile $I = I_0 f(r)$ as:

$$m = -2.5 \log \int_0^R I_0 f(r)\, r\, dr \ , \tag{2}$$

the discriminating parameter is:

$$\Delta m = 2.5 \log \frac{\int_0^5 f(r)\, r\, dr}{\int_0^3 f(r)\, r\, dr} \ . \tag{3}$$

Thus Δm is independent of magnitude, and assumes a minimum value for stars (*Fig. 2*).

3.4 Object description

The objects detected are structureless and small. Their shapes can be described by the values of their major and minor axes, their orientation is given through the position angle. These parameters are readily determined (Stobie 1980, Pittella 1987) from the computed second-order moments of an image.

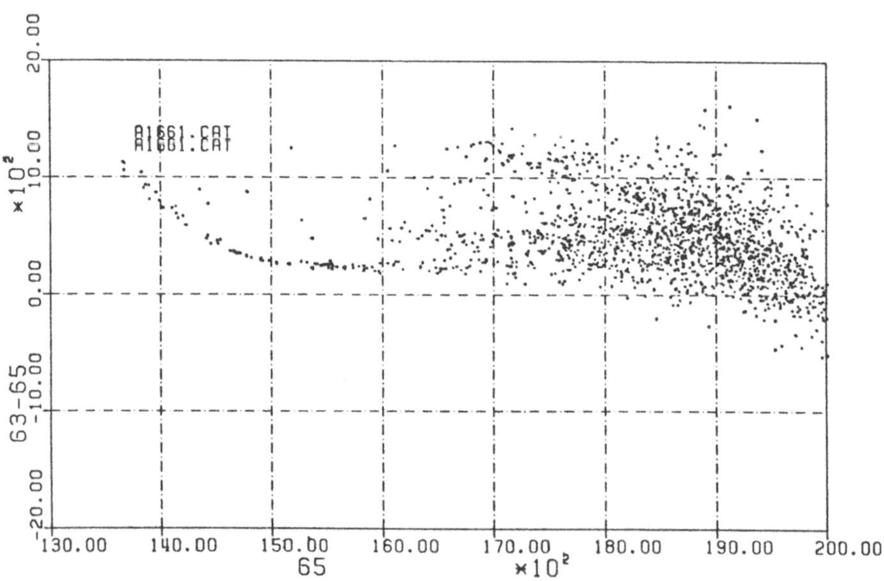

Fig. 2. Star-galaxy separation.

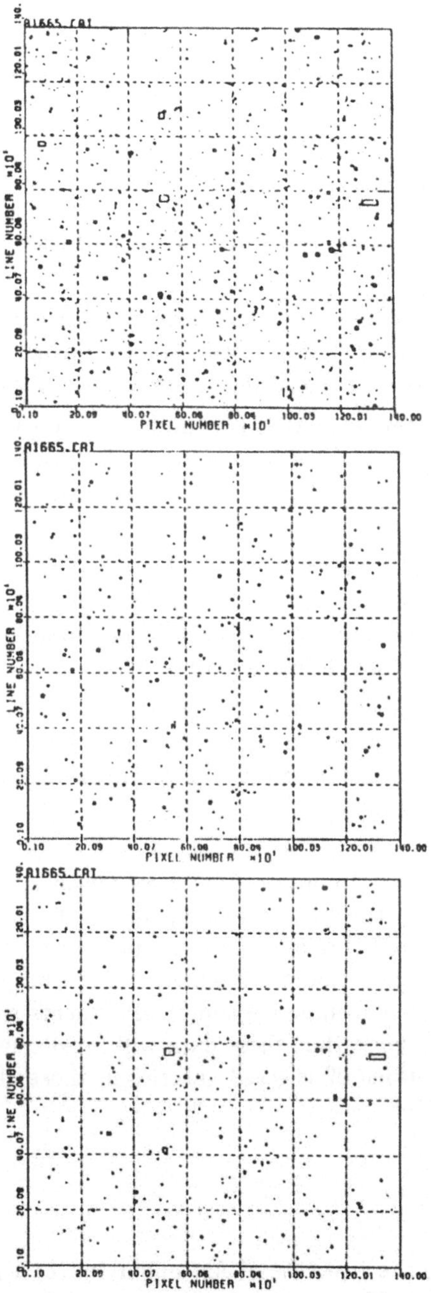

Fig. 3. Maps of the investigated clusters of galaxies. (a) all detected objects, (b) stars, (c) galaxies.

4 Present status

The project began in 1984, shortly after the PDS 1010 G microdensitometer in Monte Porzio started operation. After inspection of the plates and selection of the clusters to be studied, initial trials were made to determine the optimal values of spot size and step size for scanning. Spot sizes of 20 μm and 15 μm were chosen for the 098-04 and 127-02 emulsions, respectively. The scan step size was chosen to be equal to the spot size. The Rome Observatory software was updated and adapted to this program, then all plates were scanned.

The accuracy of the object detection procedure and the star-galaxy separation was checked by comparison between the object list and visual inspection of both the original plates and digitized images displayed on a video terminal.

So far, catalogues have been generated for 16 clusters. Each catalogue contains the following data for each galaxy: Coordinates, total magnitude, intensity profile, number of pixels in the object, sizes of major and minor axis, position angle and object classification (star or galaxy).

For illustration, we present here data on two clusters, A 1661 and A 1665. These are both distant clusters (distance class 6) of richness class 2 and 3, respectively. Abell (1980) gave diameters of 12' and 13', respectively, for these clusters. Both appear on one 127-02 plate (PS 9985) of 120 min exposure.

In *Fig. 2* we plot Δm *vs.* magnitude for all objects in the field of A 1661. As the figure shows, there is a clear separation between stars and galaxies over an interval of two magnitudes, even for such a distant cluster. *Fig. 3* presents maps of detected objects in the field of the cluster A 1665.

The magnitudes appearing in the figures, as well as those in the final catalogues, are the machine magnitudes. To obtain proper magnitudes, we need zero points for calibration. In order to do this, at least one galaxy with known magnitude in the F-band (or r-band) measured through a known aperture is required. The literature search resulted in some data, usually photometry of the brightest cluster members. For some other bright cluster members, good measurements are available in different photometric systems. We hope that these data will be also useful for our purposes. We plan to make observations of selected galaxies in those clusters, where no data are available so far.

Acknowledgments

PF is greatly indebted to the Institute of Astronomy of the University of Rome "La Sapienza" and the Vatican Observatory for their hospitality and support, enabling him to carry out the present project. PH is supported by the Natural Sciences and Engineering Research Council of Canada. GP thanks the Jagiellonian University Observatory for its hospitality during his stay there. We are grateful to the Mt. Wilson and Palomar Observatories for providing observing time and the loan of plate material.

References

Abell, G.O., 1980. Preprint.
Butcher, H., Oemler, A., 1985. *Astrophys. J. Suppl.*, **57**, 665.
Di Chio, P., Nanni, D., Pittella, G., Trevese, D., Vignato, A., 1983. In *Clustering in the Universe*, ed. Gerbal, D., Mazure, A., Éditions Frontières, Gif-sur-Yvette, p. 145.
Dressler, A., 1984. *Ann. Rev. Astr. Astrophys.*, **22**, 185.
Hickson, P., 1977a. *Astrophys. J.*, **217**, 16.
Hickson, P., 1977b. *Astrophys. J.*, **217**, 964.
Nanni, D., Pittella, G., Trevese, D., Vignato, A., 1980. *SPIE*, **264**, 48.
Oegerle, W.R., Hoessel, J.C., Ernst, R.M., 1986. *Astr. J.*, **91**, 697.
Oemler, A., 1974. *Astrophys. J.*, **194**, 1.
Pittella, G., 1987. *Acta Cosmologica*, **15**, 119.
Pittella, G., Vignato, A., 1979. *Mem. Soc. astr. Ital.*, **50**, 537.
Stobie, R.S., 1980. *SPIE*, **264**, 208.
Thuan, T.X., Gunn, J.E., 1976. *Publ. astr. Soc. Pacific*, **88**, 543.

An Application of Projected Distance Cross-Correlation for Abell Clusters

J. Hollósi
Space Telescope Science Institute
Baltimore, MD, USA, and
Roland Eötvös University
Budapest, Hungary

G. Efstathiou
Institute of Astronomy
Cambridge, England, UK

Abstract

A new method is applied for the estimation of the spatial two-point correlation function of Abell clusters. This method is based on the projected distance cross-correlation function w_{SA} between the Struble and Rood catalogue and the Abell catalogue. The σ-correlation function is determined, the best fit is $w_{SA}(\sigma) = \sigma_{SA}/\sigma$, where $\sigma_{SA} = 2.3\,\text{Mpc}$. The spatial cross-correlation function can be derived from $w_{SA}(\sigma)$ by an integral transformation. In our case, we obtain $\xi_{SA}(r) = (r/r_{SA})^{-\gamma}$, where $\gamma = 1.88$ and $r_{SA} = 40\,\text{Mpc}$. A further normalization, based on a comparison of angular correlations, is needed to obtain the autocorrelation function since the Struble and Rood catalogue is not a fair subsample of the Abell catalogue. Thus we get $\xi_{AA}(r) \approx (r/r_{AA})^{-\gamma}$, where $\gamma = 1.88$ and $r_{AA} = 33\,\text{Mpc}$. We note that the estimation is very sensitive to the estimated distance limits of the Abell catalogue. This result agrees well with $\xi_{AA}(r) = (r/30\,\text{Mpc})^{-2}$ obtained from the Limber equation.

1 Motivation

The spatial correlation of Abell clusters is an important cosmological problem because the Abell catalogue (1958) is the best defined catalogue of clusters. The problem is that we do not know the spatial positions of these clusters, we have only angular coordinates for most of them. The Struble and Rood catalogue which contains Abell clusters with measured redshifts was published in 1987. This gave a chance to check the previous estimations of the spatial correlation function.

2 Data

The two catalogues were used for our calculations:

- the Abell catalogue (1958) including 2712 clusters
- the Struble and Rood catalogue (1987) including 588 clusters *with measured redshifts.*

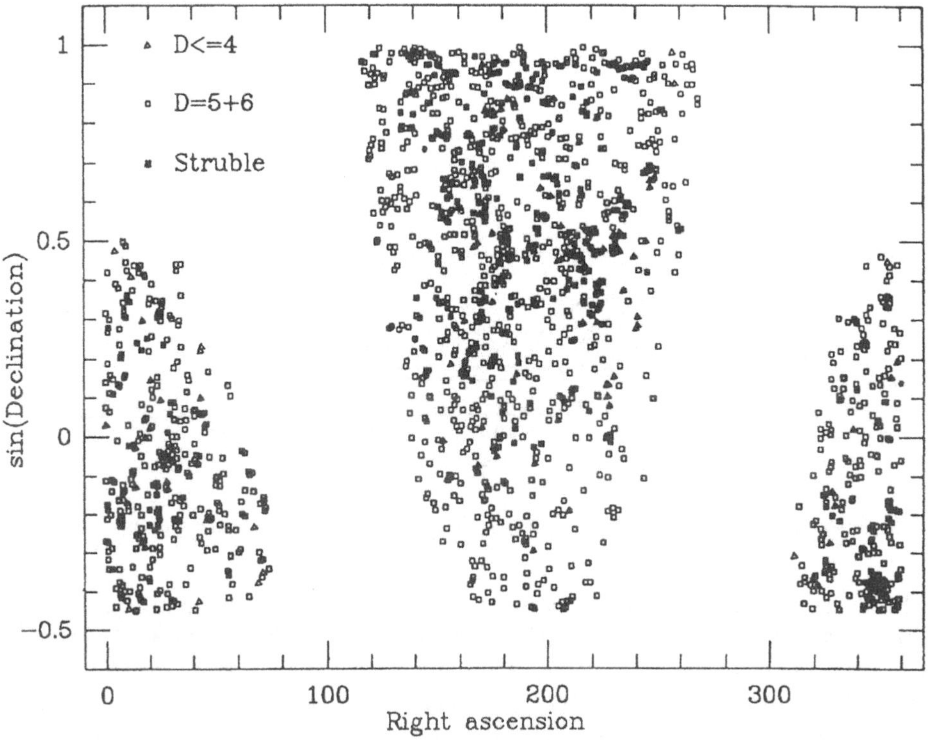

Fig. 1. The Abell catalogue and its subset, the Struble and Rood catalogue with measured redshifts

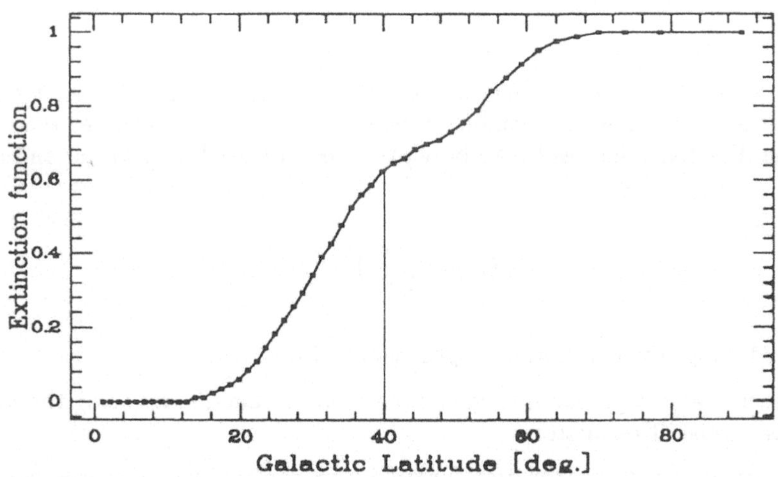

Fig. 2. Extinction function of the Abell catalogue.

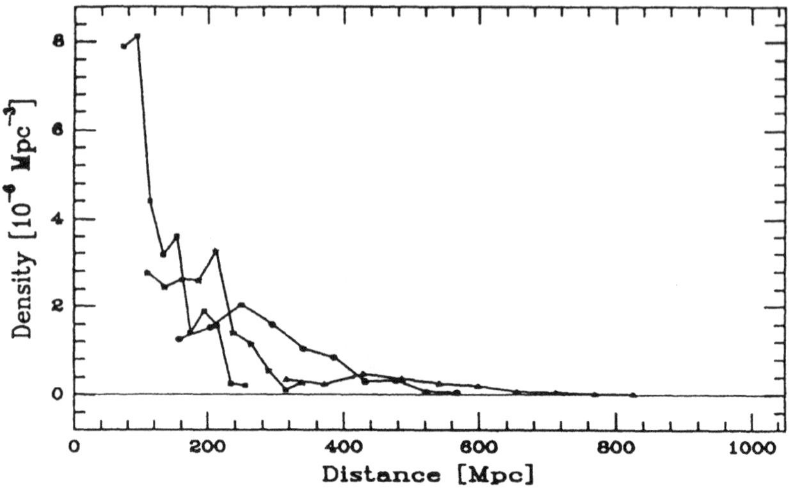

Fig. 3. Densities of the Struble and Rood clusters in various distance classes.

The cluster distributions in projected spherical coordinates are shown in *Fig. 1*. The smoothed extinction function of the catalogues, which is the effect of galactic obscuration, is shown in *Fig. 2*. (For the original extinction curves, see *Fig. 3* in Tóth *et al.*, these proceedings, p. 201). Only the high latitude samples ($|b^{II}| \geq 40°$) of the catalogues with 1418 and 310 rich clusters, respectively, were used in order to avoid regions of poor statistics. The extinction curve was also applied to the random catalogues.

3 Methods to estimate the spatial correlation

Let us summarize the previous estimation methods:

We know a nearly fair subsample of the Abell catalogue with measured redshifts. It contains all clusters in the distance classes $D \leq 4$, thus we can get their spatial correlation directly. The result can be written as a power law (Bahcall and Soneira 1983):

$$\xi(r) \approx \left(\frac{r}{r_0}\right)^{-\gamma} , \tag{1}$$

where $\gamma \approx 1.8$ and the correlation length $r_0 \approx 24$ Mpc.

We can calculate an '*approximate*' spatial correlation function by using the $m_{10} - Z$ relation as a redshift estimator.

We can derive the spatial correlation function from the angular correlation which can be estimated easily. The Limber equation, which is the connection between them, can be inverted at small angular separation by assuming a smooth selection function.

Table 1. Estimation of distance limits by Peebles and from the Struble and Rood catalogue.

	D≤ 3	$D = 4$	$D = 5$	$D = 6$
Seldner and Peebles [Mpc]	0 – 194	194 – 292	292 – 440	440 – 664
Struble and Rood [Mpc]	0 – 240	100 – 320	150 – 500	300 – 720
Number of clusters	38	57	565	758
Density $[10^{-6}\,\mathrm{Mpc}^{-3}]$	4.5	2.8	8.1	3.1

3.1 Distance limits of Abell's distance groups

Figures 3 and *4* show the spatial density of clusters for each Abell distance class in the Struble and Rood catalogue and the normalized densities, respectively, to check the given limits of the distance classes. It shows clearly that Abell's categorization, which was based on the $m_{10} - Z$ relation, has large errors and leads to strong overlaps.

3.2 Estimation of distance limits

In addition to the original distance limits of Abell we know an estimation by Peebles (1977), given in Table 1. It can be compared with *Fig. 4* where the measured redshifts verify them more or less. The densities of the distance classes can be computed considering these distance limits. We find that the densities are far from constant but this scatter can be explained by the uncertainities of the distance limits and the overlaps. We assume that the selection function is equal to 1 up to 600 Mpc and 0 above. From the present data we cannot derive more.

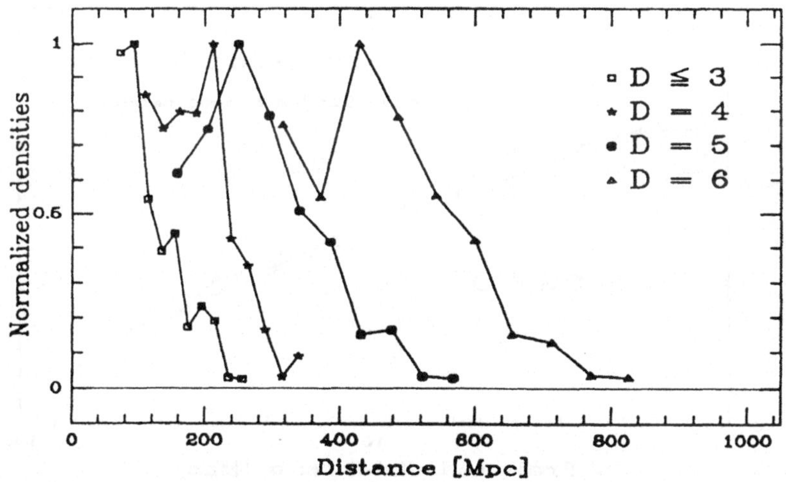

Fig. 4. The same as before, but with normalized curves, The figure shows well the distance limits of the Abell distance classes.

3.3 The projected distance cross-correlation function

Let σ denote the projected distance of two clusters. This means, that if one of them has a measured redshift and its distance is y, and the angular separation between them is θ, then $\sigma = y \cdot \theta$.

Now we define $w(\sigma)$, the projected distance cross-correlation function as follows: dn is the expected number of A-clusters in the solid angle element $d\Omega$ at separation σ from an S-cluster. The letters A and S refer to the Abell catalogue and the Struble and Rood catalogue, respectively.

$$w_{SA} = \overline{n} \cdot (1 + w_{SA}(\sigma)) \cdot d\Omega \tag{2}$$

The next formula gives the best method to estimate the σ-correlation function avoiding edge effects (the $\langle\ \rangle_\sigma$ symbol denotes the number of pairs whose projected distances are σ):

$$w_{SA} \approx \frac{\langle D_S D_A \rangle_\sigma}{\langle D_S R_A \rangle_\sigma} - 1\,. \tag{3}$$

Our best power law fit,

$$w_{SA}(\sigma) = \frac{\sigma_{SA}}{\sigma}\,, \tag{4}$$

where $\sigma_{SA} = 2.3\,\mathrm{Mpc}$, is shown in *Fig. 5*.

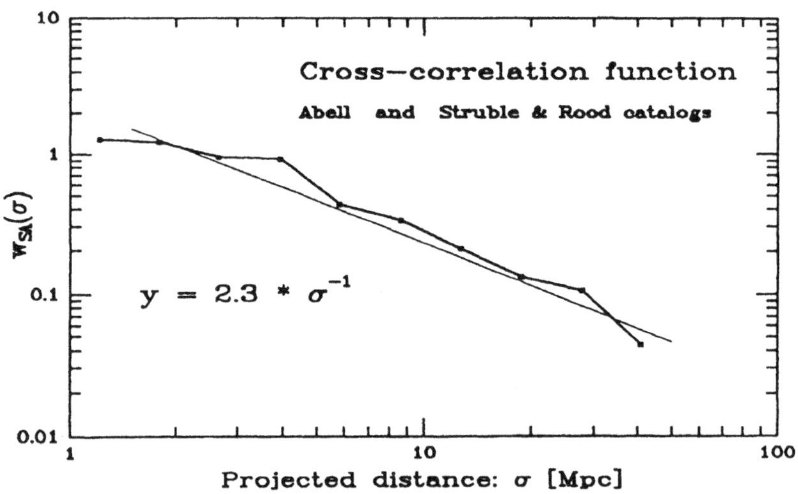

Fig. 5. Projected distance cross-correlation function between the Abell and Struble and Rood catalogues.

3.4 Relation between $w_{SA}(\sigma)$ and $\xi_{SA}(r)$

The σ-correlation function can be expressed in terms of the spatial cross-correlation function, similar to the Limber equation

$$w_{SA}(\sigma) = \frac{\int_0^\infty \int_0^\infty \rho_S\, p_S(y)\, \rho_A\, p_A(x)\, x^2\, \xi_{SA}(r)\, dx\, dy}{\int_0^\infty \rho_S\, p_S(y)\, dy \cdot \int_0^\infty \rho_A\, p_A(x)\, x^2\, dx}, \tag{5}$$

where $r^2 = x^2 + y^2 - 2xy \cdot \cos(\sigma/y)$ (*Fig. 6*), ρ_S and ρ_A are the mean densities of clusters, p_S and p_A are the selection functions. The formula is too complicated for inversion, so we use the following approximation:

$$r^2 \approx (y-x)^2 + \frac{x}{y} \cdot \sigma^2 . \tag{6}$$

3.5 The inverse equation

The inverse equation was derived by Lilje and Efstathiou (1988). At small separation,

$$\xi_{SA}(r) = -\frac{1}{\pi B r} \cdot \frac{d}{dr} \int_r^\infty \frac{\sigma\, w_{SA}(\sigma)}{(\sigma^2 - r^2)^{\frac{1}{2}}}\, d\sigma . \tag{7}$$

The constant B is determined by the redshift distribution of the Struble and Rood catalogue and the selection function of the complete Abell catalogue:

$$B = \frac{\int_0^\infty p_A(x)\, p_S(x)\, x^2\, dx}{\int_0^\infty p_S(y)\, dy \cdot \int_0^\infty p_A(x)\, x^2\, dx} = 7.4 \cdot 10^{-4} . \tag{8}$$

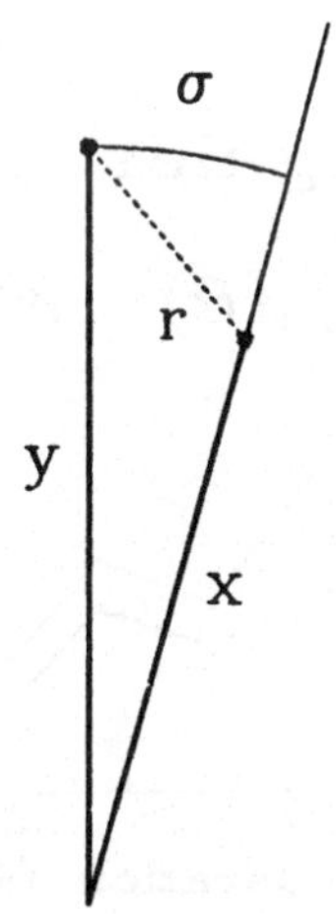

Fig. 6. Relation between the spatial and projected distance cross-correlations.

The result of the integral-transformation is

$$\xi_{SA}(r) \approx \left(\frac{r}{r_{SA}}\right)^{-\gamma} , \tag{9}$$

where $\gamma = 1.88$ and $r_{SA} = 40$ Mpc (*Fig. 7*).

Note that this is the spatial cross-correlation function and not the auto-correlation.

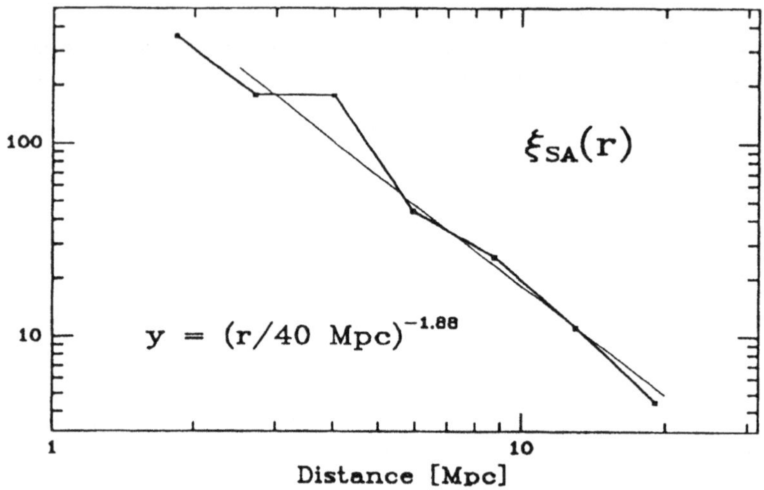

Fig. 7. Spatial cross-correlation function between the Abell and Struble and Rood catalogues.

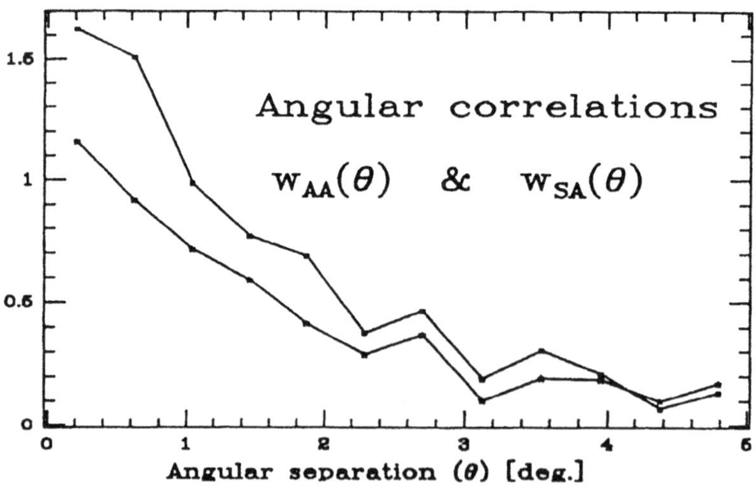

Fig. 8. Comparison of the angular correlations of the Abell and Struble and Rood catalogues.

3.6 The spatial cross-correlation function

A special normalization is needed for the spatial cross-correlation function because the Struble and Rood catalogue is not a fair subsample of the whole Abell catalogue. This means that the picking method of Struble and Rood clusters may have some virtual correlation effects which we should avoid. That effect can be seen in *Fig. 8* which shows the angular auto- and cross-correlation functions.

We found that their ratio is nearly constant, ≈ 1.45 up to 4 degrees (*Fig. 9*). Accepting the simple assumption that the ratios of the angular and spatial correlation

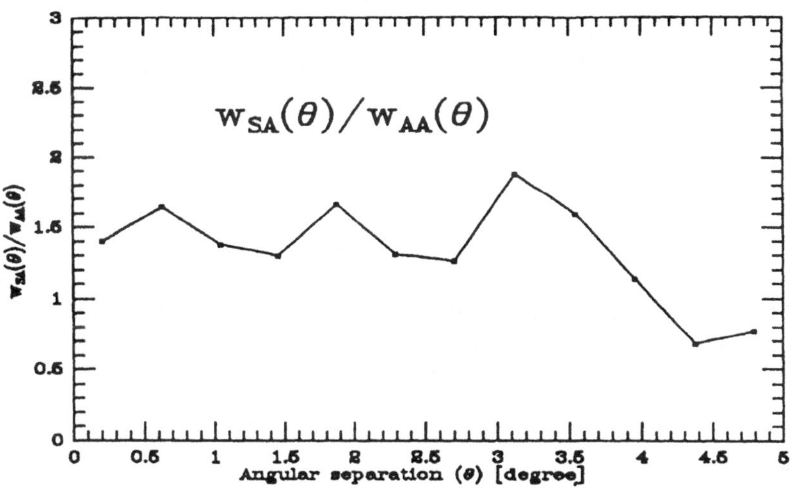

Fig. 9. Ratio of the angular correlation functions of the Abell and Struble and Rood catalogues.

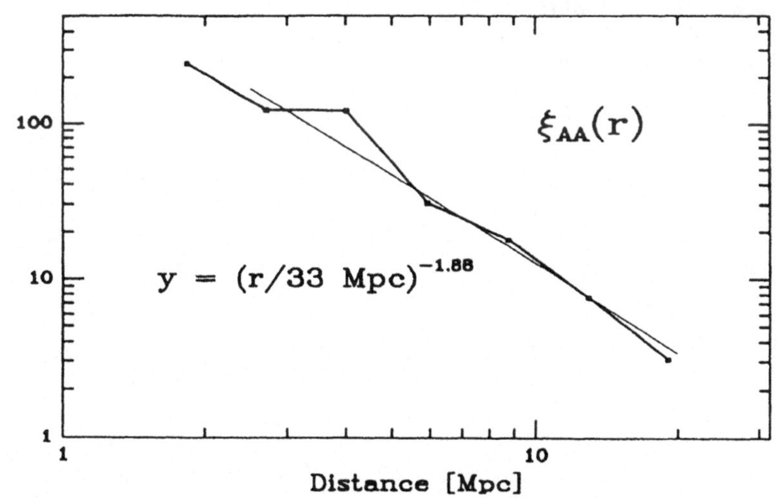

Fig. 10. Spatial correlation function of the Abell clusters.

functions are about the same, we can correct the spatial cross-correlation function to the autocorrelation function of the Abell clusters. This assumption can be derived from the fact that connections of correlation functions are linear.

4 Conclusion

Our best power law fit to the spatial correlation function of the whole Abell catalogue is the following (*Fig. 10*):

$$\xi_{AA}(r) \approx \left(\frac{r}{r_{AA}}\right)^{-\gamma}, \tag{10}$$

where $\gamma = 1.88$ and $r_{AA} = 33$ Mpc. We note that the amplitude depends on the distance limit of the Abell catalogue. The dependence has the following form:

$$r_{AA} \propto D^{\frac{3}{\gamma}}, \tag{11}$$

thus from the $\approx 10\,\%$ uncertainty of D:

$$r_{AA} = 33 \pm 5\,\text{Mpc}\,. \tag{12}$$

Our result is in a good agreement with the inversion of the Limber equation. Although this estimation gives a somewhat stronger correlation than the previous estimation of Bahcall and Soneira based on the nearest 104 clusters, the exponent is nearly the same.

References

Abell, G.O., 1958. *Astrophys. J. Suppl.*, **3**, 211.
Bahcall, N.A., Soneira, R.M., 1983. *Astrophys. J.*, **270**, 20.
Lilje, P.B., Efstathiou, G., 1987. *Mon. Not. R. astr. Soc.*, **231**, 635.
Peebles, P.J.E., 1977. *Astrophys. J.*, **215**, 703.
Seldner, M., Peebles, P.J.E., 1977. *Astrophys. J.*, **215**, 703.
Struble, M.F., Rood, H.J., 1987. *Astrophys. J. Suppl.*, **63**, 543.

Three-Point Correlations of Abell Clusters

G. Tóth and J. Hollósi
Space Telescope Science Institute
Baltimore, MD, USA, and
Roland Eötvös University
Budapest, Hungary

A.S. Szalay
Roland Eötvös University
Budapest, Hungary, and
Johns Hopkins University
Baltimore, MD, USA

Abstract

We estimate the irreducible three-point correlations of Abell clusters using all distance class 5+6 clusters with latitude $|b^{II}| \geq 40°$. We find that these clusters satisfy a relation between the two- and three-point correlation functions:

$$\zeta(r,s,u) \approx Q\big(\xi(r)\xi(s) + \xi(s)\xi(u) + \xi(u)\xi(r)\big)$$

similar to that for galaxies. The value of Q has large uncertainties:

$$Q = 0.9 \pm 0.5 \, ,$$

with a strong discrepancy between the northern and southern hemispheres, $Q_{north} \approx 0.3$, while $Q_{south} \approx 1.7$ [1]. Higher order terms seem to be absent in ζ. Several error estimation methods are applied.

1 Definition of the three-point correlation function

Spatial:

Let V_1, V_2 and V_3 be three volume elements in space. Let us denote their separations by r_{12}, r_{23} and r_{31} (*Fig. 1*a). With the mean density of clusters being ρ the expected number of triplets in the volume elements is $n_{exp} = \rho^3 V_1 V_2 V_3$ for a uniform distribution. In case of a correlated distribution there is an excess of triplets, thus the number of them will be:

$$n = \rho^3 V_1 V_2 V_3 \big(1 + \xi(r_{12}) + \xi(r_{23}) + \xi(r_{31}) + \zeta(r_{12}, r_{23}, r_{31})\big) \, , \qquad (1)$$

where ξ and ζ are the spatial two- and three-point correlation functions, respectively.

[1] Work in progress indicates that this discrepancy is less significant, when one takes into account that the distribution of triplets is non-Poissonian; strongly correlated.

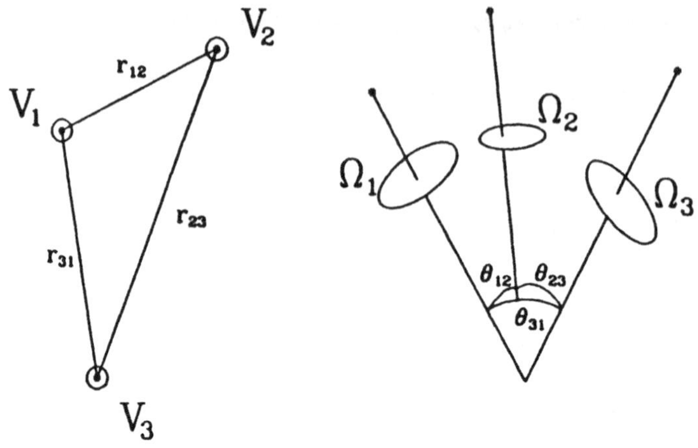

Fig. 1a, b. Definition of the three-point correlation functions.

Angular:

Let Ω_1, Ω_2 and Ω_3 be three solid angle elements on the surface of the unit sphere with separations θ_{12}, θ_{23} and θ_{31} (*Fig. 1b*). With the mean surface density of clusters being η, the expected number of triplets in the angle elements is $n_{exp} = \eta^3 \Omega_1 \Omega_2 \Omega_3$ for a uniform distribution. In case of a correlated distribution there is an excess of triplets, thus the number of them will be

$$n = \eta^3 \Omega_1 \Omega_2 \Omega_3 \left(1 + w(\theta_{12}) + w(\theta_{23}) + w(\theta_{31}) + z(\theta_{12}, \theta_{23}, \theta_{31})\right) , \quad (2)$$

where w and z are the angular two- and three-point correlation functions, respectively.

2 Motivation

Comparison with galaxies:

The distribution of galaxies were studied in detail by Groth and Peebles (1977). They determined the three- and even the four-point correlations so as to get quantitative results on the distribution of galaxies. An interesting relation was found between the spatial two- and three-point correlation functions:

$$\zeta_{gal}(r, s, u) \approx Q_{gal} \left(\xi_r \xi_s + \xi_s \xi_u + \xi_u \xi_r\right) , \quad (3)$$

where $Q_{gal} = 0.8 \ldots 1.3$ for various catalogues.

Comparison with models:

There are several models (both numerical and analytical) of clustering and they can be tested by comparing their correlation functions with the correlation function of an observed catalogue. A well known analytical model (Kaiser 1984, Bardeen *et al.* 1986, Politzer and Wise 1984) is the biasing of density fluctuations, where we assume that there were primordial Gaussian fluctuations of the mass density, and visible objects were formed only at places where the density reached a certain value ('biasing'). In a more general model (Szalay 1988) we have an arbitrary non-linear relation between luminosity and mass density. The three-point correlation function can be expanded in terms of the two-point correlation function, and the leading terms are:

$$\zeta(r,s,u) = Q\left(\xi_r\xi_s + \xi_s\xi_u + \xi_u\xi_r\right) + Q^3\xi_r\xi_s\xi_u + Q'\left(\xi_r^2\xi_s + \xi_r^2\xi_u + \ldots\right) . \quad (4)$$

Clusters are more likely to have preserved their initial distribution below the correlation length than galaxies, so the relation between the 2nd and 3rd coefficients can be checked for this.

3 Data

A magnetic tape of the Abell catalogue prepared by the Bulgarian Academy of Sciences is used. Our copy was obtained from UC Berkeley. We have discovered that somewhere in the copying processes an error occured: every 57th cluster is missing from the catalogue. Subsequently we found that this error is by no means specific to our tape. Several major institutes also had the faulty catalogue on their computer, thus we warn everybody to check his catalogue for the error. Abell's (1958) original paper was used to complete the data.

The catalogue contains 2712 clusters altogether (see *Fig. 1* in Hollósi and Efstathiou, *these proceedings*), the $D = 5+6$ (distance groups 5 and 6 and richness class $R \geq 1$) sample is used with declinations greater than $-27°$ and galactic latitudes $|b^{II}| \geq 40°$. Two subsamples are chosen for error estimation :

(HL)	High Latitude with $\lvert b^{II}\rvert \geq 40°$	(1323 clusters)
(NC)	North Cap with $b^{II} \geq +40°$	(844 clusters)
(SC)	South Cap with $b^{II} \leq -40°$	(479 clusters)

Within these geometrical boundaries only clusters of the statistical sample are present (the only exception is A915), thus no further geometrical constraints are necessary to restrict ourselves to the statistical sample. The areas of HL, NC and SC are 3.46, 2.24 and 1.22 sterad, respectively.

The extinction function is determined from the distribution of surface density $\eta(b^{II})$ from the data at different galactic latitudes. The clusters are counted in 50 equal area bins in the north galactic cap. The result is smoothed by a simple averaging process, and saved in a file to be used for generating random catalogues (*Fig. 2*). The same function is used for the south galactic cap.

4 How to Estimate the three–point correlation function

Triplet counts in space:

The usual way of calculating the three-point correlation function $\zeta(r,s,u)$, *e.g.* for galaxies, is to determine the number of triplets with given separations both in the data and random catalogues: $N_D(r,s,u)$ and $N_R(r,s,u)$. Assuming that we have already determined the two-point correlation function ξ, we can estimate ζ:

$$\zeta(r,s,u) = \frac{N_D(r,s,u)}{N_R(r,s,u)} - \xi_r - \xi_s - \xi_u - 1\,. \tag{5}$$

Unfortunately only ≈ 300 Abell clusters have measured redshifts, thus we cannot apply this method to this subsample, because $N_D(r,s,u)$ would be too small.

From the angular three-point correlation function:

For two-point correlations the angular function $w(\theta)$ can be expressed in terms of the spatial function $\xi(r)$ and vice versa, so we can estimate $\xi(r)$ from $w(\theta)$ even if we do not know the redshifts.

For three-point correlations we can express the angular function z in terms of the spatial function ζ, but the inversion is unknown.

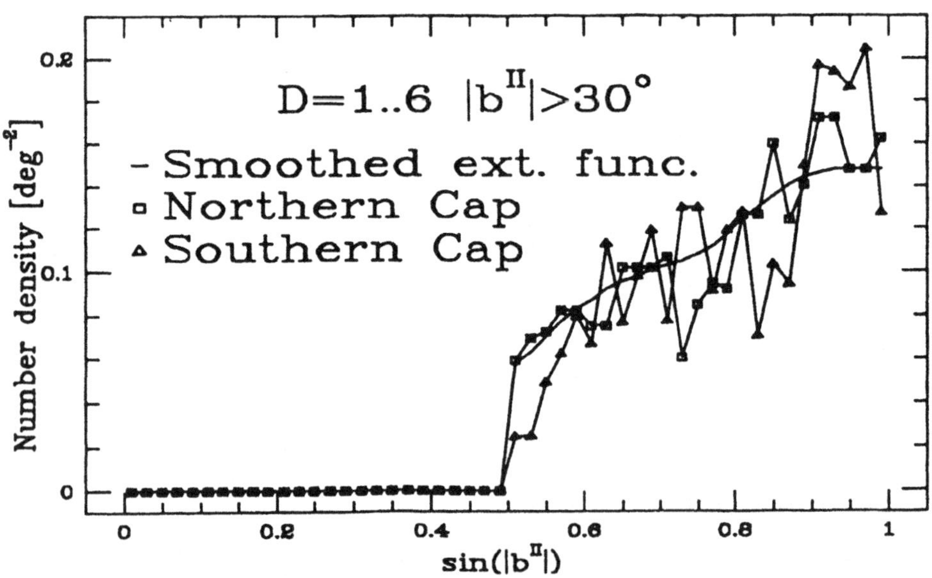

Fig. 2. Extinction function of the Abell clusters.

Expansion of the three-point correlation function:

Let us expand ζ into the symmetric terms of ξ up to the third order:

$$\begin{aligned}
\zeta(r,s,u) = \; & Q_0 \\
& +Q_1 \cdot (\xi_r + \xi_s + \xi_u) \\
& +Q_{11} \cdot (\xi_r\xi_s + \xi_s\xi_u + \xi_u\xi_r) \\
& +Q_2 \cdot (\xi_r^2 + \xi_s^2 + \xi_u^2) \\
& +Q_{111} \cdot (\xi_r\xi_s\xi_u) \\
& +Q_{12} \cdot (\xi_r^2\xi_s + \xi_r^2\xi_u\xi_s^2\xi_u + \xi_s^2\xi_r\xi_u^2\xi_r + \xi_u^2\xi_s) \\
& +Q_3 \cdot (\xi_r^3 + \xi_s^3 + \xi_u^3)
\end{aligned} \tag{6}$$

Now the coefficients $Q_{\underline{k}}$ are to be determined instead of ζ, and this is possible, because the terms can be projected separately into angular terms, as we will see below. On the other hand we can easily answer the questions asked in Sect. 2 from the coefficients.

5 Projection of the expansion

Using the notation $\underline{k} \in \{0, 1, 11, 2, 111, 12, 3\}$ the expansion above can be written in a shorter form:

$$\zeta(r,s,u) = \sum_{\underline{k}} Q_{\underline{k}}\, \xi_{\underline{k}}(r,s,u)\,, \tag{7}$$

where $\xi_{\underline{k}}(r,s,u) = \xi_r^{k_1}\xi_s^{k_2}\xi_u^{k_3} + \text{symm.}$
Assuming that the selection function $P(r)$ is known, the expansion can be projected onto an equation for angular correlation functions:

$$z(a,b,c) = \sum_{\underline{k}} A_{\underline{k}} w_{\underline{k}}(a,b,c)\,. \tag{8}$$

The connection between $Q_{\underline{k}}\xi_{\underline{k}}$ and $A_{\underline{k}}w_{\underline{k}}$ is:

$$A_{\underline{k}} w_{\underline{k}}(a,b,c) = \frac{\rho^3}{\eta^3}\int_0^\infty dr_1\, r_1^2 P(r_1)\int_0^\infty dr_2\, r_2^2 P(r_2)\int_0^\infty dr_3\, r_3^2 P(r_3)\, Q_{\underline{k}}\xi_{\underline{k}}(r,s,u)\,, \tag{9}$$

where

$$\begin{aligned}
r^2 &= r_1^2 + r_2^2 - 2r_1r_2\cos a \\
s^2 &= r_2^2 + r_3^2 - 2r_2r_3\cos b \\
u^2 &= r_3^2 + r_1^2 - 2r_3r_1\cos c
\end{aligned} \tag{10}$$

Adopting $\xi(r) = (r/r_0)^{-\gamma}$ and $P(r) = 1$ for $d \le r \le D$ we can integrate analytically. The calculation is similar to the projection of $\xi(r)$ onto $w(\theta)$, but it is more complicated. The results are:

$$\begin{aligned}
w_0 \quad &(a,b,c) = 1 \\
w_1 \quad &(a,b,c) = w_a + w_b + w_c \\
w_{11} \quad &(a,b,c) = w_a w_b + w_b w_c + w_c w_a \\
w_2 \quad &(a,b,c) = \tfrac{w_a^2}{a} + \tfrac{w_b^2}{b} + \tfrac{w_c^2}{c} \\
w_{111} \quad &(a,b,c) = \tfrac{w_a w_b w_c}{a+b+c} \qquad \text{(for } \gamma = 2 \text{ only)} \\
w_{12} \quad &(a,b,c) = \tfrac{w_a^2 w_b + w_a^2 w_c}{a} + \tfrac{w_b^2 w_c + w_b^2 w_a}{b} + \tfrac{w_c^2 w_a + w_c^2 w_b}{c} \\
w_3 \quad &(a,b,c) = \tfrac{w_a^3}{a^2} + \tfrac{w_b^3}{b^2} + \tfrac{w_c^3}{c^2}
\end{aligned} \tag{11}$$

The angular coefficients are proportional to the spatial coefficients, but the ratios are different for the different terms due to the various moments of $P(r)$ involved. Let us denote the ratios by $R_{\underline{k}} = A_{\underline{k}}/Q_{\underline{k}}$. The numerical results for $\gamma = 2$ and for the distance limits $d = 250$ Mpc and $D = 600$ Mpc are:

$$\begin{aligned}
R_0 &= 1 \\
R_1 &= 1 \\
R_{11} &= 1.036 \\
R_2 &= 5.9 \\
R_{111} &= 11.3 \\
R_{12} &= 5.65 \\
R_3 &= 63.4
\end{aligned}$$

a, b and c are measured in degrees. Note that the spatial terms and the corresponding angular terms are similar, but all higher order terms are strongly amplified in the angular coefficients.

If we know *the angular two- and three-point correlation functions* we can determine the coefficients $A_{\underline{k}}$ from the angular expansion (Eqn. 7), and then the spatial coefficients $Q_{\underline{k}} = A_{\underline{k}}/R_{\underline{k}}$.

6 Estimation of angular correlations

Angular two-point correlation:

20 random catalogues were genarated with the proper areas and extinction functions for the HL, NC and SC samples. The number of pairs in 12 linear bins is counted up to 5° separation for the data samples and the random catalogues as well. The estimator of the angular correlation for each bin is

$$w(\theta) = \frac{N_D(\theta)}{N_R(\theta)} - 1 , \tag{12}$$

where $N_R(\theta)$ and $N_D(\theta)$ are the number of pairs in a bin of $[\theta - \frac{\Delta\theta}{2}, \theta + \frac{\Delta\theta}{2}]$ for the random and data catalogues, respectively. N_R can be estimated with an analytic formula, because for such small separations the non-linear effects of the boundary and the density gradient are negligible. Thus we can calculate the expected value of pair frequency analytically instead of generating several hundreds of random catalogues to eliminate the statistical scattering:

$$N_R(\theta) = \frac{1}{2} A \cdot \overline{\eta^2} \cdot 2\pi\theta\Delta\theta \ , \tag{13}$$

where A is the area of the sample, $\overline{\eta^2}$ is the mean squared surface density, $2\pi\theta\Delta\theta$ is the area where the second point can be if the first one is fixed. The formula is divided by 2, because otherwise all pairs would be counted twice. It is still necessary to determine the accurate coefficient of θ from fitting to the pair-frequency function of the random catalogues, because the presence of boundaries reduces the number of pairs.

The best power law fit for the HL sample is

$$w(\theta) = \frac{1}{\theta} \ . \tag{14}$$

The estimated angular correlation function is determined up to 30° and compared with the projected spatial correlation function and its analytic approximation (*Fig. 3*). The angular correlation functions of the different samples agree fairly well (*Fig. 4*).

Angular three-point correlation:

All triplets with separations up to 5° are counted and put into bins according to the length of the sides. Let $n_D(a,b,c)$ and $n_R(a,b,c)$ be the number of triplets in

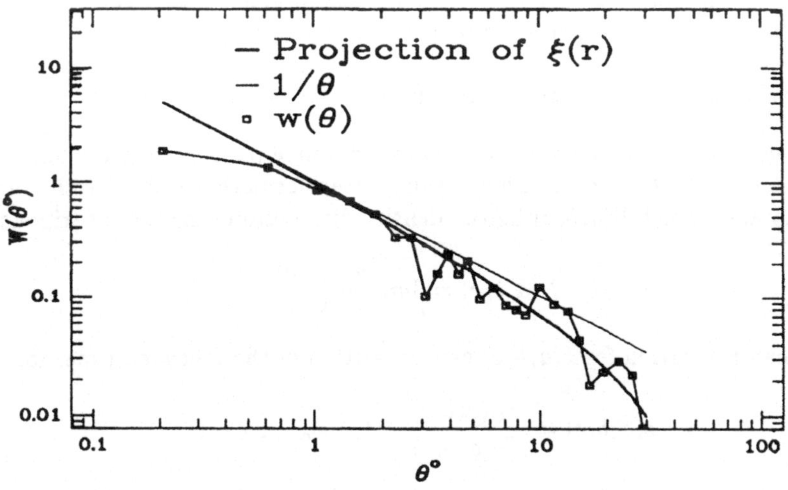

Fig. 3. Smooth curve: projection of $\xi(r) = (r/30\,\mathrm{Mpc})^{-2}$ analytically.
Straight line: best power law fit: $1/\theta$.
Points: angular two-point correlation function of the HL sample.

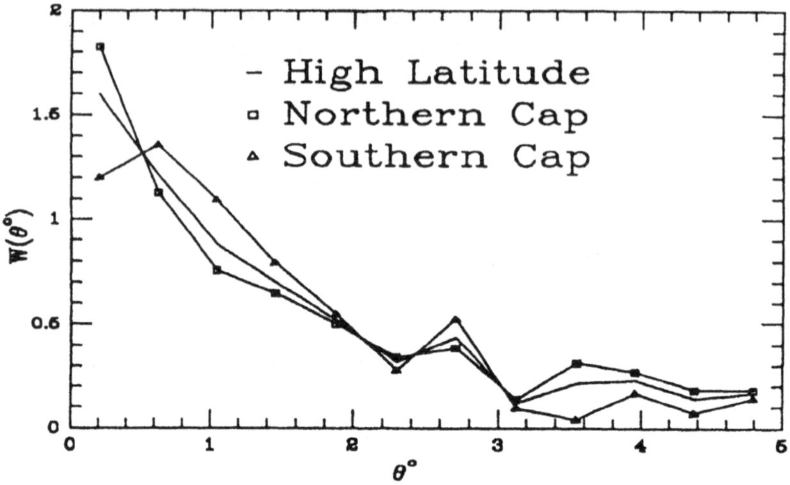

Fig. 4. Comparison of the angular two-point correlation functions of the various samples.

a bin of $V(a,b,c) = [a - \frac{\Delta}{2}, a + \frac{\Delta}{2}] \times [b - \frac{\Delta}{2}, b + \frac{\Delta}{2}] \times [c - \frac{\Delta}{2}, c + \frac{\Delta}{2}]$ for the data and random catalogues, respectively. Δ is the size of the bin. The estimator for the angular three-point correlation function is

$$z(a,b,c) \approx \frac{n_D(a,b,c)}{n_R(a,b,c)} - 1 - w_a - w_b - w_c \, . \tag{15}$$

This estimation, however, has some disadvantages. The equation is accurate for infinitesimal Δ only, but reducing Δ is limited because of the relatively small number of triplets in the data (≈ 30000). There are also problems with generating enough random catalogues to get a good estimation for $n_R(a,b,c)$.

7 Transformation into triplet counts

Our purpose is to determine the expansion of the three-point correlation function and we do not need $z(a,b,c)$ itself. Thus a simple transformation will solve the problems mentioned above. Let $P(a,b,c)$ be the density distribution function of triplets:

$$P(a,b,c) = \lim_{\Delta \to 0} \frac{n(a,b,c)}{\Delta^3} \, . \tag{16}$$

The estimator equation for $z(a,b,c)$ can be written in the following exact form:

$$z(a,b,c) = \frac{P_D(a,b,c)}{P_R(a,b,c)} - 1 - w_a - w_b - w_c \, . \tag{17}$$

Now we will rearrange this equation and integrate over the volume of a bin; thus an exact equation can be obtained for the data triplet counts independently of the size of the bins:

$$P_D(a,b,c) \quad = \quad P_R(a,b,c)\big(1 + w_a + w_b + w_c + z(a,b,c)\big)$$

$$\begin{aligned} n_D(a,b,c) &= \iiint\limits_{V(a,b,c)} da\,db\,dc P_R(a,b,c)\Big(w_0 + w_1 + \sum_{\underline{k}} A_{\underline{k}} w_{\underline{k}}(a,b,c)\Big) \\ n_D(a,b,c) &= \sum_{\underline{k}} B_{\underline{k}} n_{\underline{k}}(a,b,c)\,, \end{aligned} \tag{18}$$

where

$$B_0 = A_0 + 1\,, \quad B_1 = A_1 + 1\,, \quad B_{\underline{k}} = A_{\underline{k}} \quad (\underline{k} \neq 0,1)\,,$$

$$n_{\underline{k}} = \iiint\limits_{V(a,b,c)} da\,db\,dc P_R(a,b,c) w_{\underline{k}}(a,b,c)\,. \tag{19}$$

Having determined the $n_{\underline{k}}(a,b,c)$ functions from Eqn. 19 we can fit them to $n_D(a,b,c)$ with the $B_{\underline{k}}$ parameters in Eqn. 2. From $B_{\underline{k}}$ we can calculate $A_{\underline{k}}$ and $Q_{\underline{k}}$ easily. For $n_{\underline{k}}$ we still need $P_R(a,b,c)$, *the probability density of random triplets*, and we have to *integrate* in Eqn. 19.

8 Distribution of random triplets

Since we use small separations, the non-linear effects of the boundary and the density gradient are negligible, so uniform distribution is a good approximation for the random catalogues. For a uniform distribution of points the distribution of triplets can be calculated analytically in a similar way to that which we used for random pairs. The result is

$$P_R(a,b,c) = A\overline{\eta^3}\frac{8\pi abc}{\sqrt{2\left(a^2b^2 + b^2c^2 + c^2a^2\right) - a^4 - b^4 - c^4}}\,, \tag{20}$$

where η is the surface density and A is the area of the sample.

The accurate coefficient in this equation for a finite area might be slightly smaller due to the presence of the boundaries. The coefficient is corrected after the integration of Eqn. 19.

Since $P_R(a,b,c)$ is singular at $a + b = c$, the terms in Eqn. 19 must be integrated carefully. We used a special Monte Carlo integration.

9 Fit to the triplet data

We determine $n_D(a,b,c)$ in 239 bins with size $\Delta = 5°/12$, the sides of the triplets are $a \le b \le c \le 5°$. A weighted least-squares method is used for fitting in Eqn. 18:

$$\chi^2 = \frac{1}{239}\sum_{a,b,c}\frac{1}{\sigma_D^2(a,b,c)}\Big(n_D(a,b,c) - \sum_{\underline{k}} B_{\underline{k}} n_{\underline{k}}(a,b,c)\Big)^2\,, \tag{21}$$

where $\sigma_D(a,b,c)$ is the scatter of the triplet count in the $V(a,b,c)$ bin. Poisson errors $\sigma_D^2(a,b,c) = n_D(a,b,c)$ were assumed.

10 The results

When we fit all terms, the third order terms have very small coefficients. Neglecting them and fitting only up to second order we get:

$$Q_0 = -0.048 \quad Q_1 = 0.045 \quad \mathbf{Q_{11} = 0.94} \quad Q_2 = -0.002 \qquad (\chi^2 = 1.52)$$

Only the third term seems to be significant, so let us fit this one only:

$$\mathbf{Q_{11} = 0.93 \qquad (\chi^2 = 1.53)}$$

Note that reducing the number of fitted terms did not increase χ^2 too much. If we had only Poisson errors, χ^2 should be equal to 1.

11 Error estimate

Fitting on subsamples:

Q_{11} (hereafter Q) is determined for the NC and SC subsamples as well:

$$Q_{NC} = 0.31 \qquad\qquad Q_{SC} = 1.74$$

The general behaviour of the coefficients for the NC and SC sample is similar to the coefficients for the HL sample (the third order terms are less significant than the lower order terms, and the third coefficient is outstanding), but the results are different.

Perturbation of the triplet distribution:

We test whether the above mentioned discrepancy is due to the Poisson errors or not. The triplet counts are perturbed with 1σ noise and Q is calculated for every perturbed data set:

$$Q_{HL} = 0.91 \ldots 0.94 \quad Q_{NC} = 0.31 \ldots 0.4 \quad Q_{SC} = 1.7 \ldots 1.75$$

The lack of overlapping shows clearly that the discrepancy cannot be explained by Poisson errors.

The relation χ^2 vs. Q:

Figure 5 shows well that $Q \approx 1$ gives the best compromise with $\chi^2 \approx 3$ for the NC and SC samples.

Examination of the deviation from the fitted function:

We tested whether the rest (after subtracting the fitted function) has a Gaussian distribution. The rest is calculated for each bin first, then divided by $1\sigma = \sqrt{n_D(a,b,c)}$. The statistics of these numbers is shown in *Fig. 6*. Although the curve is flatter than the Gaussian belonging to $\chi^2 = 1$ (smooth curve), it is not too far from a Gaussian.

11 Conclusions

Our main result is that there seems to be a connection between the spatial three-point and two-point correlation functions for clusters similar to that for galaxies:

$$\zeta(r,s,u) \approx Q\big(\xi(r)\,\xi(s) + \xi(s)\,\xi(u) + \xi(u)\,\xi(r)\big) \ , \tag{22}$$

where $Q = 0.9 \pm 0.5$. There is a strong discrepancy between the northern and southern galactic caps. It might be due to the non-Poisson distribution of triplets, or due to the differences between the hemispheres in the completeness of the catalogue. If we accept that $Q = 1$ and if there is no cubic term in ζ, we have to conclude that

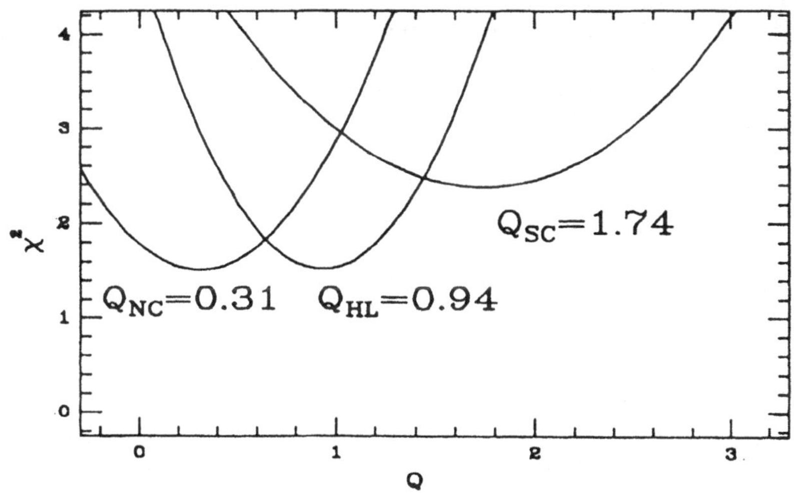

Fig. 5. Relation between χ^2 and Q for the three samples.

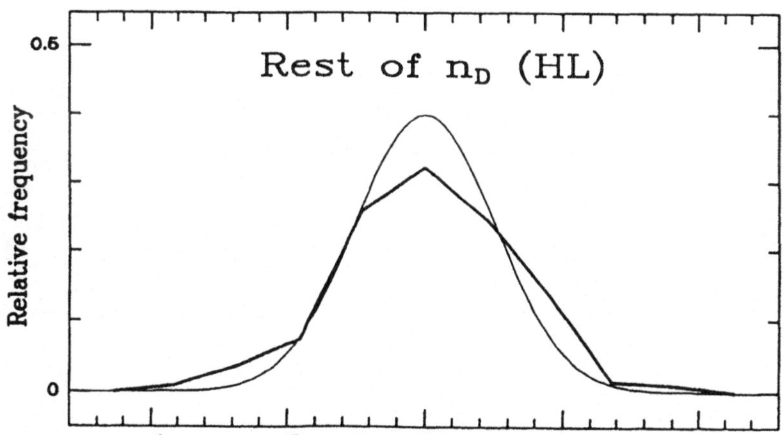

Fig. 6. Deviation of the triplet counts from the fit.

Gaussian biasing cannot be responsible for a major amplification of the correlations. These problems are under investigation (Hollósi *et al.* 1988).

Finally, we mention that we are planning to carry through the whole process for other cluster and galaxy catalogues, too.

References

Abell, G.O., 1958. *Astrophys. J. Suppl.*, **3**, 211.
Bardeen, J.R., Bond, J.R., Kaiser, N., Szalay, A.S., 1986. *Astrophys. J.*, **304**, 15.
Groth, E.J., Peebles, P.J.E., 1977. *Astrophys. J.*, **217**, 385.
Hollósi, J., Szalay, A.S., Tóth, G., 1988. In preparation.
Kaiser, N., 1984. *Astrophys. J.*, **284**, L9.
Peebles, P.J.E., 1980. *The Large Scale Structure of the Universe*, Princeton University Press, Princeton.
Politzer, D., Wise, M.B., 1984. *Astrophys. J.*, **285**, L1.
Struble, M.F., Rood, H.J., 1987. *Astrophys. J. Suppl.*, **63**, 543.
Szalay, A.S., 1988. *Astrophys. J.*, in press.

Part IV

Superclusters and Intergalactic Dust

The Hydra-Centaurus Supercluster

Guido Chincarini
Università degli Studi di Milano, and
Osservatorio Astronomico di Brera
Milano, Italy

Paolo Vettolani
Istituto di Radioastronomia CNR,
Bologna, Italy

Abstract

We discuss the status of our work in the region $10^h < \alpha < 14^h$ and $-50° < \delta < -20°$. An estimated overdensity of about 2 for the visible mass could, coupled to the streaming motion detected by Lynden-Bell *et al.*, give $\Omega_0 \simeq 0.07$.

1 Introduction

The theoretical investigations by Fall (1975) and Peebles (1980 and references therein), among others, showed that an overdensity $\Delta\rho/\rho$ perturbs the Hubble flow by an amount $\Delta v/v$ given by the equation

$$\frac{\Delta v}{Hr} = \frac{1}{3}\frac{\Delta\rho}{\rho}\Omega_0^{0.6} . \tag{1}$$

There are at least two consequences:

i: On local scales, we expect a component of shear motion, if the distribution of the density perturbation is not isotropic and thus defines a preferential direction in space;
and

ii: we must define as the fundamental cell of the Universe, *i.e.* a region of space with the mean characteristics of the cosmos and therefore undistinguishable from any other region of equal volume, one which has no motion with respect to the microwave background and an internal energy density equal to the mean energy density of the Universe. Is the size of such a cell about 1/10 of the horizon or as large as the horizon itself? Furthermore, since a concept of inertial frame of reference in the above is included, shall we relax the above concept of cell and investigate the concept of random scale motion among otherwise identical parts of space?

From the observational point of view it is clear, indeed, that the strong local asymmetry in the distribution of mass due to the Virgo cluster slows down locally the general expansion. On the other hand, measurements of the microwave background show a dipole component in the direction $l = 269°$, $b = 28°$. Since the Virgo cluster is in the direction $l = 284°$, $b = 75°$, other density irregularities must perturb the Hubble flow in such a way that by combining the various motions on larger and larger scales we end up with a vector pointing in a direction which is in agreement with the observed dipole anisotropy of the microwave background.

The region of Hydra-Centaurus is becoming one of the most interesting targets of observational cosmology. Indeed, there is a long standing evidence that somewhere in this direction there is a large concentration of mass which perturbs the Hubble flow on a large scale.

Following the work by Rubin *et al.* (1976), it became clear that, unless we dismiss their results as due to some observational bias, we have to take into account the possibility of a perturbation in the general direction of Hydra-Centaurus. This point was stressed in a preliminary form by Chincarini (1982) who also noticed that the components of the motions were rather close to the supergalactic plane. The obvious reasoning was that the motion of the local group with respect to the microwave background had to be explained as the result of the various motion components generated by known local perturbations. Following a lecture series held in Rio de Janeiro, it was decided to begin a redshift survey in this region of the sky using the new facilities set up at the National Observatory of Brazil. More recently, observations were obtained at ESO/La Silla by Vettolani *et al.* and in South Africa by Fairall.

A detailed study on the search for motion with respect to the microwave background was published by Tammann and Sandage (1984, and it was clearly shown that all the data available indicated a density perturbation in the direction of Hydra-Centaurus. The most interesting results on this matter, however, came from the work of Burstein *et al.* (1986 and subsequent papers) who showed that there is a clear indication of motion of a large region of local space in the direction $l = 307°$, $b = 9°$. This region is located somewhat south of the Hydra-Centaurus supercluster toward Pavo-Indus.

In this paper we describe some of the results obtained so far in the region we are surveying.

2 Redshift surveys in the Hydra-Centaurus region

In the Hydra-Centaurus supercluster (*Fig. 1*), clusters of various richness are present; these are: the Antila cluster, the Centaurus cluster, the Hydra cluster and Klemola 27. A few areas have been surveyed for redshifts by Hopp and Materne (1985), Da Costa *et al.* (1986, 1987), Fairall *et al.* (1988). Clusters of galaxies in this region have been studied also by Melnick and Moles (1987) and by Lucey *et al.* (1986). As discussed below, important surveys in a region south of the one presented in *Fig. 1*, in Pavo-Indus, have been carried out by Fairall (1987) and more recently by Dressler (1988).

The analysis of the ESO-Uppsala catalogue (Lauberts 1982) using the percolation algorithm shows that between the Hydra and the Centaurus region the galaxian density is very low.

The redshift surveys demonstrate that the Hydra supercluster is connected to the

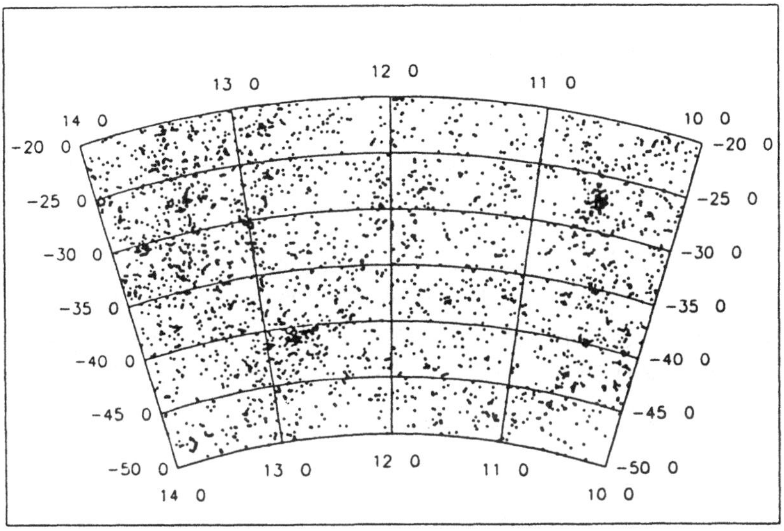

Fig. 1. Equal-area projection of the 3018 galaxies of the ESO-Uppsala catalogue in the survey region.

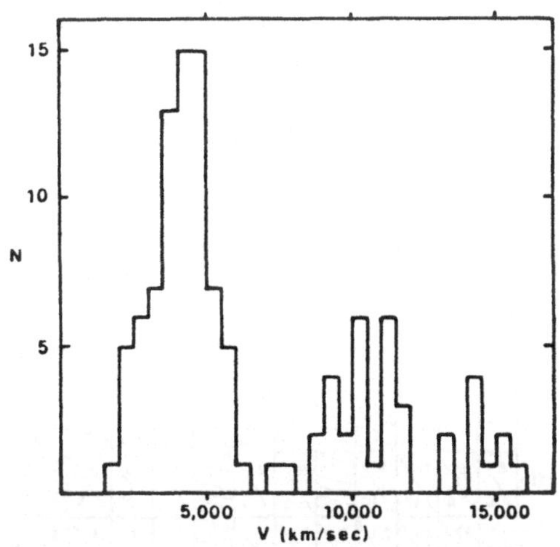

Fig. 2. The distribution in redshift in the region of Hydra-Centaurus observed by Da Costa *et al.* (1986).

Centaurus supercluster only through a low density filamentary [1] structure at about 3 000 km s^{-1}. The Hydra-Centaurus supercluster redshift distribution shows a peak at about 4 500 km s^{-1}. Other overdensities are detected at about 10 000 km s^{-1} and 15 000 km s^{-1} (*Fig. 2*) (Da Costa *et al.* 1987; Fairall *et al.* 1988). The redshift distri-

[1]Here and in some previous work we use the word "filamentary" in a rather broad sense and not in the strict meaning of the word given in the Webster dictionary. In a similar way, we discuss the Hydra-Centaurus supercluster as the complex with redshifts in the range 2 500 – 5 000 km s^{-1}, leaving to more detailed discussions the subgrouping in redshift space.

bution continues with similar characteristics in the Pavo-Indus region, again showing a peak at about 4 600 km s^{-1} (Fairall 1987) (*Fig. 3*).

Fairall *et al.* (1988) compiled a redshift catalogue of 484 objects with magnitudes brighter than $B_T = 14\overset{\text{m}}{.}5$. At the distance of the Hydra-Centaurus supercluster we are, therefore, sampling the luminosity function at a somewhat fainter magnitude than the break. In sample A, in which 251 galaxies have magnitudes measured on the ESO plates by Lauberts, redshifts are known for 354 galaxies and the survey is 73 % complete. In sample B the redshift is now known for 596 galaxies so that the survey is 50 % complete. The redshift completeness map is given in *Fig. 4*.

The distribution of galaxies with redshift 2 500 km s^{-1} $< v <$ 4 500 km s^{-1} (*Fig. 5*) shows a low density connection between the regions of Hydra and Centaurus. Most

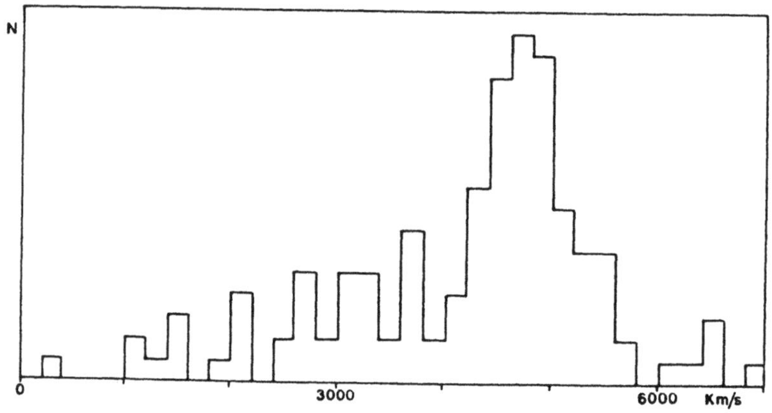

Fig. 3. The distribution in redshift in the region of Pavo-Indus observed by Fairall (1987).

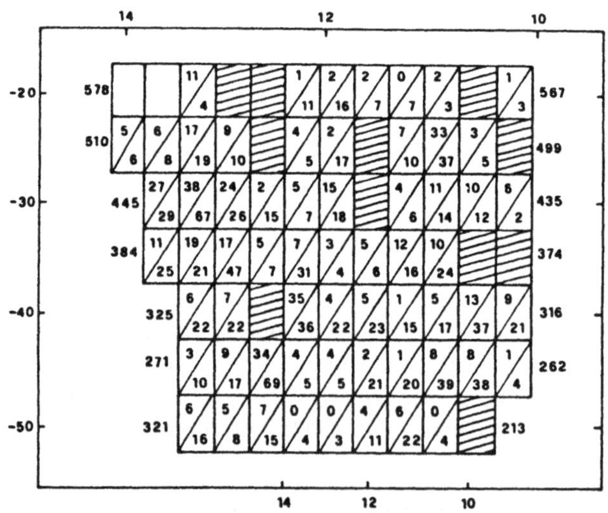

Fig. 4. Redshift completeness map for sample A (see text).

of the galaxies linking the two regions are at redshifts of about 3 000 km s^{-1} and at a declination of about $-37°$. On the other hand, in the redshift range 4 500 km s^{-1} $< v <$ 6 500 km s^{-1} (*Fig. 6*), we do not detect galaxies in the region $11^h < \alpha < 12^h30^m$ and $-40° < \delta < -20°$. Inspection of the redshift completeness map (*Fig. 4*) shows that the northern part of the void is poorly sampled and we cannot exclude the existence of a northern boundary at about $\delta = -20°$.

The agglomerate of galaxies in Centaurus is formed by various subgroupings. Two concentrations of galaxies are centered at about 3 050 km s^{-1} and 4 500 km s^{-1} (see also Lucey *et al.* 1986) and the concentration at 3 050 km s^{-1} seems to be formed by two subgroups separated by about 7°.

A small and rather well bounded void is visible in the cone diagram of *Fig. 7*. The cone diagram, α *vs.* redshift, covers the declination range $-50° < \delta < -20°$ so that the projected void is about 30° long in declination and about 6°.1 in right ascension at a redshift of about 4 200 km s^{-1}. We will call this a pipe-shaped void.

The survey by Dressler (1988), bounded by the galactic coordinates $-35° < b < +45°$ and $290° < l < 350°$, shows a similar distribution in redshift space (his *Fig. 2*) with the difference that the peaks at 10 000 km s^{-1} and 15 000 km s^{-1} are less prominent when compared to the 4 500 km s^{-1} peak.

It is quite evident from the above redshift surveys that a density enhancement over a very large region is present in the redshift range 3 000 – 5 000 km s^{-1} with a peak at about 4 500 km s^{-1}. Substructure is indicated.

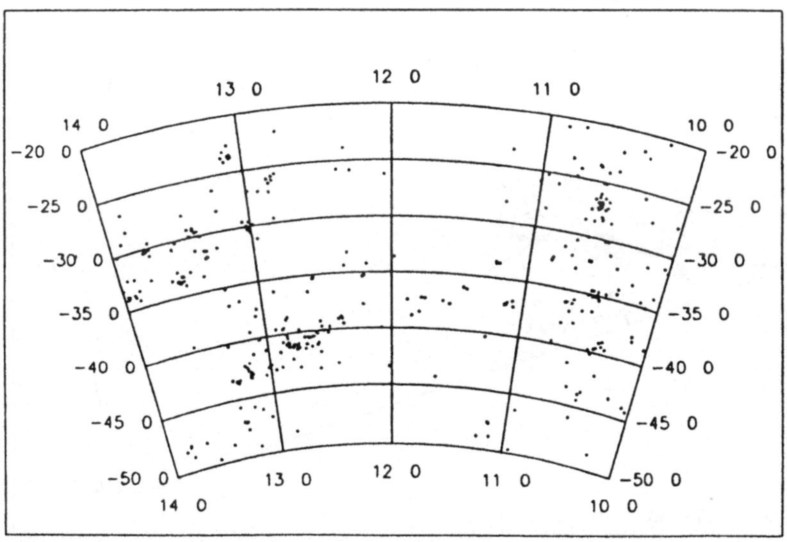

Fig. 5. Distribution of galaxies from sample B with radial velocities between 2 500 km s^{-1} and 4 500 km s^{-1} in equal-area projection.

3 A possible value for Ω_0?

Lynden-Bell *et al.* (1987), using a sample of 385 elliptical galaxies, show that the motions are best described by a flow toward a mass concentration centered on $l = 307°$, $b = 9°$ at a redshift of about $(4\,350 \pm 350)\ \mathrm{km\,s^{-1}}$. The streaming motion at

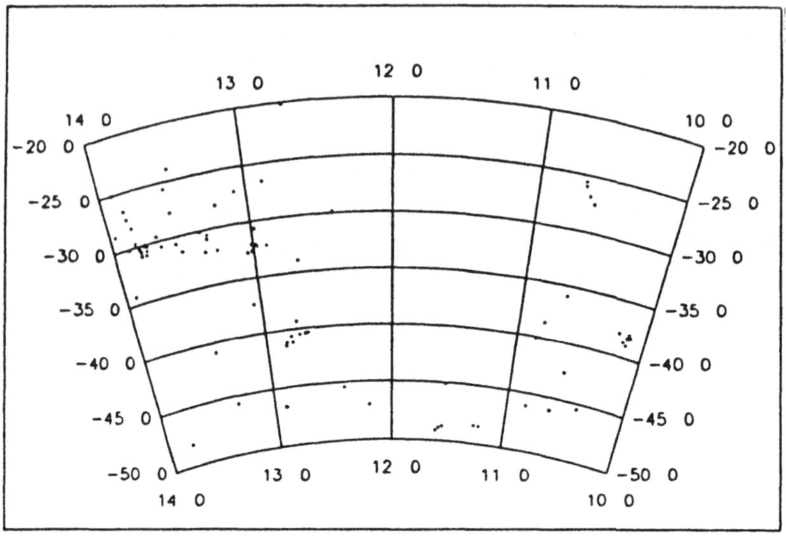

Fig. 6. Distribution of galaxies from sample B with radial velocities between $4\,500\ \mathrm{km\,s^{-1}}$ and $6\,500\ \mathrm{km\,s^{-1}}$ in equal-area projection.

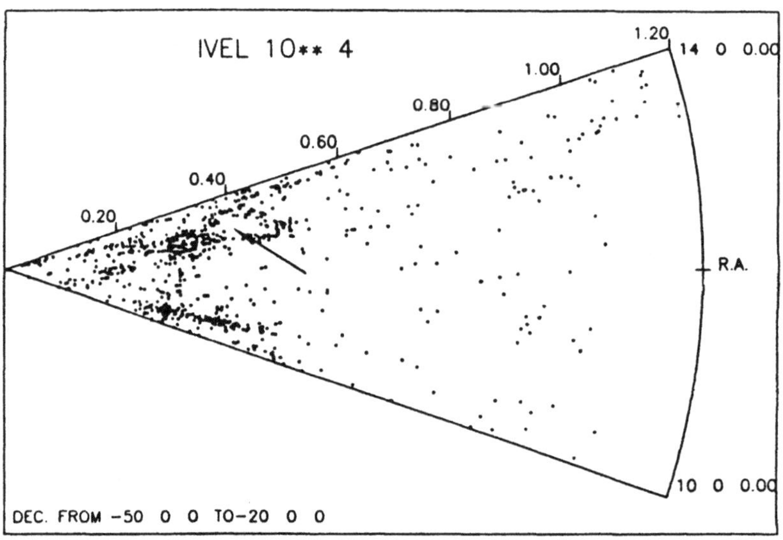

Fig. 7. Wedge diagram, compressed in declination, for the whole area. The arrow points to a pipe-shaped void.

the Sun is $(570 \pm 60)\ \mathrm{km\,s^{-1}}$.

The redshift histograms can also be used as a measure of overdensity with respect to the mean density of the Universe. Indeed, a measure of the density perturbation is given by the relation:

$$\frac{\Delta\rho}{\rho} = \frac{\int_{z_1}^{z_2} \mathrm{N_{obs}}(z)\,dz}{\int_{z_1}^{z_2} \langle \mathrm{N} \rangle_{\mathrm{expected}}(z)\,dz} - 1\,, \tag{2}$$

where $\langle \mathrm{N} \rangle_{\mathrm{expected}}$ is the number of objects we would observe in the solid angle defined by the sample, assuming uniform distribution of galaxies and counts limited at the same limiting magnitude as the sample and smoothed all over the sky. These counts define the mean density of the Universe and are known (they must be corrected for the Virgo cluster, see also Olowin *et al.* 1988).

Our sample (Fairall *et al.* 1988) covers a solid angle of 0.43 sterad and gives an excess density

$$\frac{\Delta\rho}{\rho} = 2.37\,. \tag{3}$$

The sample by Dressler (1988) extends over an area of 0.85 sterad (since the redshifts have not been published an estimate has been made using his *Fig. 2*) and gives

$$\frac{\Delta\rho}{\rho} = 1.2\,. \tag{4}$$

Assuming a shear motion model (Eqn. 1) and a mean overdensity of 1.8, we obtain

$$\Omega_0 = 0.07\,. \tag{5}$$

Are we really overestimating the density fluctuations by about a factor 10 using visible matter?

On the other hand, the effect of the Perseus-Pisces supercluster (see Da Costa *et al.* 1986) and/or the presence of a weaker perturbation (located further away) could favor a larger value of Ω_0.

References

Burstein, D., Davies, R.L., Dressler, A., Faber, S.M., Lynden-Bell, D., Terlevich, R.J., Wegner, G.A., 1986. In *Galaxy Distances and Deviations from the Hubble expansion*, eds. Madore, B. and Tully, R.B., Reidel, Dordrecht.

Chincarini, G., 1982. In *The Large Scale Structure of the Universe*, Rio de Janeiro Summer School, University of Oklahoma, preprint.

Da Costa, L.N., Nunes, M.A., Pellegrini, P.S., Willmer, C., Chincarini, G., Cowan, J.J., 1986. *Astr. J.*, **91**, 6.

Da Costa, L.N., Willmer, C., Pellegrini, P.S., Chincarini, G., 1987. *Astr. J.*, submitted.

Dressler, A., 1988. Preprint.

Fairall, A.P., Vettolani, G., Chincarini, G., 1988. *Astr. Astrophys.*, submitted.

Fall, S.M., 1975. *Mon. Not. R. astr. Soc.*, **172**, 23p.

Hopp, U., Materne, J., 1985. *Astr. Astrophys. Suppl.*, **61**, 93.

Lucey, J.R., Currie, M.J., Dickens, R.J., 1986. *Mon. Not. R. astr. Soc.*, **222**, 417.

Lynden-Bell, D., Faber, S.M., Burstein, D., Davies, R.L., Dressler, A., Terlevich, R.J., Wegner, G., 1987. *Astrophys. J.*, submitted.
Melnick, J., Moles, M., 1987. *Rev. Mex. Astr. Astrophys.*, **14**, 72.
Olowin, R., de Souza, R.E., Chincarini, G., 1988. *Astr. Astrophys.*, in press.
Peebles, P.J.E., 1980. *The Large Scale Structure of the Universe.* Princeton University Press, Princeton.
Rubin, V.C., Ford, W.K., Thonnard, N., Roberts, M.S., Graham, J.A., 1976. *Astr. J.*, **81**, 687.

On the Orientation of Double Galaxies

Piotr Flin
Jagiellonian University Observatory
Kraków, Poland

Abstract

The double galaxies in the Local, the Perseus, and the Coma/A1367 superclusters were studied. It is shown that these objects reveal the same tendency of alignment as the other galaxies belonging to the supercluster. The galaxy planes tend to be perpendicular to the plane of the parent supercluster and perpendicular to the radius-vectors from the centre of the supercluster. The angle between the rotation axes of galaxies in pairs, denoted as β, was also determined. The ambiguity connected with the unknown sense of galaxy tilt and spin in the method presented was not removed, the range of the β-angle is $0° - 180°$. The distribution of the β-angle is highly non-random. There is a statistically significant excess of small absolute values of the β-angle and a deficit near perpendicular configurations.

1 Introduction

The search for alignments of galaxy rotation axes for the members of three superclusters with known spatial geometry has been performed (Flin and Godlowski 1986, Flin 1987) with positive results. The rotation axes of galaxies tend to align with the plane of the parent supercluster and the galaxy planes tend to be perpendicular to their radius vectors. It is interesting to check, whether similar alignments can be observed if we restrict ourselves to double galaxies. In the case of double galaxies it is worthwhile to consider the β-angle, determined by the difference between the rotation axes of the galaxies constituting the pair. Previous work on the orientation of galaxies in pairs was discussed in detail by Noerdlinger (1979) and Helou (1984). The issue was also taken up by Karachentsev (1981). The classic approach consists in studying the distribution of differences between the position angles of component galaxies, sometimes supplemented by additional information, *e.g.* the winding of spiral arms. This method was applied by most of the previous authors; the resulting acute angle gives usually an isotropic distribution. Helou (1984) had at his disposal data containing true galaxy spins, with the differences between spins ranging from 0° to 180° and he found anisotropy. Preferred is the anti-parallel alignment of spins.

Naturally, such spins are the most valuable ones, however, the number of galaxies with determined true spins is rather limited. Therefore, in the present paper a different approach to the question of relative orientation of paired galaxies is proposed. The investigated angle is that between the rotation axes of a pair, without solving the ambiguity connected with the sense of rotation and the sign of the galaxy tilt. This gives the investigated range of β-angle $0^\circ - 180^\circ$ and permits to construct a large sample, which is important for any statistical investigation.

The paper is laid out in the following manner: observational data are presented in Sect. 2, the description of the method applied is given in Sect. 3, Sect. 4 is devoted to the alignment of galaxies in pairs in the three investigated superclusters, while Sect. 5 presents the result of the investigation of the β-angle. The conclusions constitute Sect. 6.

2 Observational data

In order to be included into the present study two galaxies had to be situated in the northern hemisphere and listed as a pair in Arigo *et al.*'s (1978) sample of double galaxies or in the notes to the UGC, as well as in the main part of the UGC (Nilson 1973). Moreover, the radial velocities, the major and minor axes, and the position angles of both components must be known. The galaxy coordinates came from the UGC. The radial velocities of galaxies were taken from the literature. The values of the major and minor axes and position angles were taken from the UGC or from Arigo *et al.* Sometimes they were determined on the Palomar Sky Survey prints by the present author who also calculated the position angles of each component from the relative values presented by Arigo *et al.* For performing the transformation, the position angle of the line connecting the centres of both components was taken from the UGC notes or from Tifft (1980) or determined by the author.

The Helou (1984) sample of 30 pairs (one non-UGC pair is excluded from the analysis) with differences between the radial velocities of the components $\Delta V \leq 250$ km s^{-1} served to check the correctness of the applied method.

The Perseus sample contained 56 pairs situated within the supercluster, *i.e.* $22^h \leq \alpha \leq 4^h$, $21\overset{\circ}{.}5 \leq \delta \leq 45^\circ$, $4000 \leq V_r \leq 8500$km s^{-1} and with $\Delta V \leq 250$km s^{-1}. The 33 pairs attributed to the Coma/A1367 supercluster are located within $11^h \leq \alpha \leq 13.6^h$, $16^\circ \leq \delta \leq 37^\circ$ and $4000 \leq V_r \leq 9000$ km s^{-1}, with $\Delta V \leq 250$ km s^{-1}. The fact that the region considered is more extended than the supercluster itself is due to the scarcity of data, but should not affect further considerations too much. From the Arigo *et al.* sample, 54 pairs with $V_r \leq 2600$ km s^{-1} and $\Delta V \leq 250$ km s^{-1} were extracted and considered as belonging to the Local Supercluster.

The samples of pairs are not complete from the statistical point of view, however, as the Helou (1984) sample, they are homogeneous with respect to the parameters considered.

The axial ratios q_o, as given in the UGC, were corrected to standard photometric diameters using the prescription of Fouqué and Paturel (1985), and they served for calculating the galaxy inclination:

$$i = \cos^{-1}\left(\frac{q_o^2 - q_t^2}{1 - q_t^2}\right)^{\frac{1}{2}} . \qquad (1)$$

It was assumed that q_t changes with morphological types from 0.25 to 0.15.

The question arises, how reliable the data given in the UGC are. There are a number of papers which deal with this problem. The Arigo *et al.* measurements are totally independent from the UGC. *Fig. 1* and *2* represent the comparison of position angles and axial ratio measurements in these two sources. From the inspection of the figures it follows that the position angles deviate slightly more for $b/a > 0.6$. Note that the Arigo *et al.* data were obtained using the value of the pair position angle, which certainly increases the errors. The angular deviation for the overall sample is 13°.4. The differences of axial ratios, as given in *Fig. 2*, are obtained from data which were both reduced to standard photometric diameters; the "standard", however, is different for each sample. The standard deviation is about 0.1. The mean value of the b/a

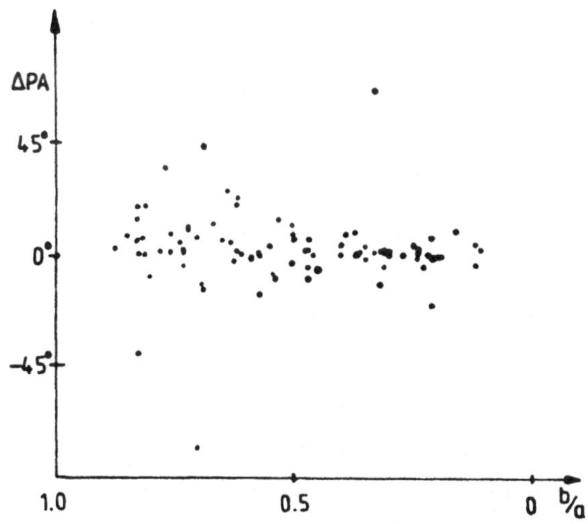

Fig. 1. The differences between position angles in Arigo *et al.* and the UGC relative to the axial ratios.

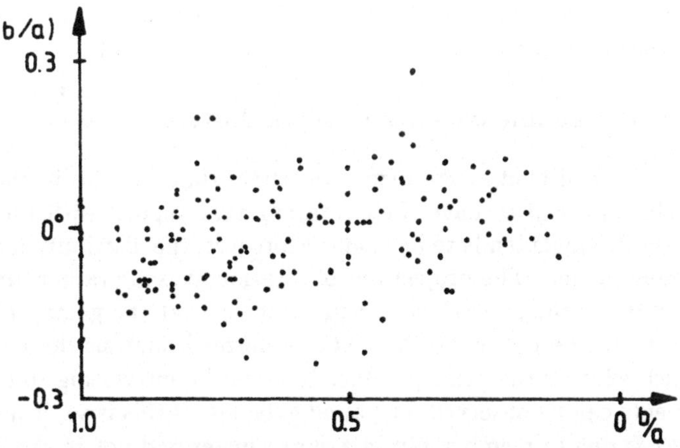

Fig. 2. The differences of the axial ratios in Arigo *et al.* and the UGC relative to the axial ratios.

differences for $b/a < 0.4$ is above the zero line, while for $b/a > 0.6$ it is below the zero line, which probably means that in at least one set of data the axial ratios were overcorrected.

3 Method of analysis

The analysis is performed exactly in the manner presented in detail in the previous paper (Flin and Godlowski 1986). It is assumed that the galaxy rotation axis is perpendicular to the galaxy plane, which due to the ambiguity in the galaxy tilt gives two possible solutions for each galaxy. Both solutions are taken into account in the calculations. The coordinates of each galaxy are transformed from the equatorial coordinate system into a coordinate system connected with each parent supercluster. The value of the galaxy position angle is also expressed in the new coordinate system.

Two angular distributions are analysed: the polar angle δ_D between the rotation axis of the galaxy and the plane of the supercluster and the azimuthal angle η between the projection of the rotation axis on the supercluster plane and the x-axis, pointing towards supergalactic $l = 0, b = 0$. This analysis permits to study the alignment of galaxies with the plane of the parent supercluster.

The β-angle between the rotation axes of the components was calculated using the well-known formula of spherical trigonometry:

$$\cos\beta = \sin b_1 \sin b_2 + \cos b_1 \cos b_2 \cos(l_1 - l_2) , \qquad (2)$$

where b and l are the coordinates of the galaxy rotation axes. For each pair eight possible solutions were obtained and included, as equally probable, into further calculations. In comparison to previous work, where only the analysis of position angles was performed, the allowed range of the β-angle is twice as large, but there is a symmetry of the angles β and $180° - \beta$. Galaxies seen "face-on" are, however, included into the analysis. The present analysis permits only to test the isotropy of the distribution, and in the case of anisotropy to check whether the rotation axes of paired galaxies are more nearly parallel or perpendicular. It does not allow to separate parallelism and anti-parallelism, as it was done in the work of Helou (1984).

4 Orientation of double galaxies in superclusters

It was shown (Flin and Godlowski 1986, Flin 1987) that galaxies in the three investigated superclusters tend to have their rotation axes aligned with the supercluster plane, that is the δ_D-angle tends to be small. Moreover, the distribution of the η-angle is also a non-random one. The projections of rotation axes are rather directed toward the main structure of the supercluster, which means that the galaxy planes tend to be perpendicular to the radius vector. The performed analysis does not permit to determine which effect is the principal one. It would be interesting to check whether similar alignments can be observed for paired galaxies. If this is so, it means that the dominant effect is due to membership in a supercluster and not to duplicity.

In order to test the distribution of rotation axes in pairs with respect to the main plane of the parent supercluster, the two angles, δ_D and η, were determined. The

Table 1. The result of the χ^2-test

Supercluster		no. of solutions	δ_D-angle degrees of freedom	δ_D-angle χ^2-value	η-angle degrees of freedom	η-angle χ^2-value
LSC	all double	2550 108	13	13.1	15	17.5
Coma	all double	140 132	10	12.7	13	30.4
Per	all double	642 224	13	7.8	16	28.0

distributions of both angles were compared with the distributions of the respective angles for the general sample of galaxies belonging to each supercluster. The statistical hypothesis was that both samples do not differ significantly from each other, *i.e.*, they come from the same population. The check was performed using the χ^2-test for angular data (Batschelet 1981). The contingency tables are given in Table 1. From Table 1 it follows that the distribution of the δ_D-angle is always the same for double galaxies as for the general sample. In the case of the η-angle, the similarities between the distributions, though not as large as for the δ_D-angle, are statistically significant

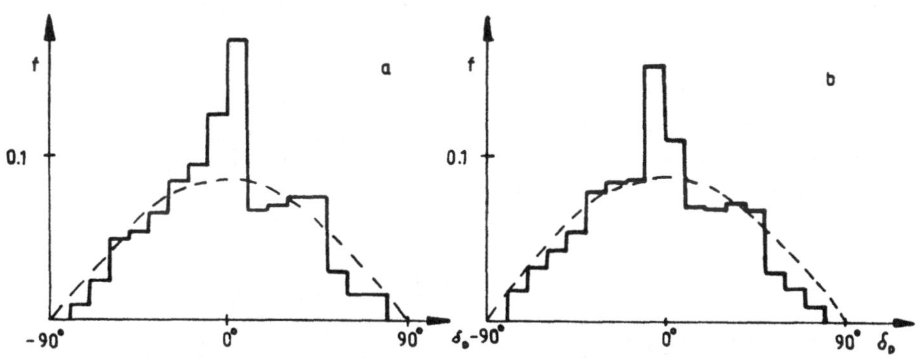

Fig. 3. The distribution f of the δ_D-angle for (a) double galaxies and (b) all bright ($m \leq 14.5$) galaxies in the Perseus supercluster (broken lines denote isotropic distributions).

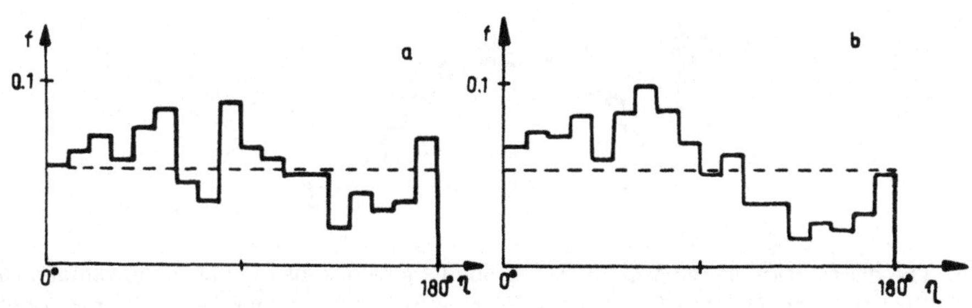

Fig. 4. The same as *Fig. 3*, but for the η-angle.

(at the significance level $\alpha = 0.01$). The distributions of the δ_D- and η-angles for single bright galaxies and for double galaxies in *Figs. 3* and *4* show that the two populations show noticeable similarities.

The performed analysis permits to state that the rotation axes of galaxies in pairs tend to be aligned with the plane of the parent supercluster, and that the projections of the rotation axes onto the supercluster plane tend to point towards the superclusters.

5 Analysis of the β-angle

The resulting distributions are shown in *Fig. 5*, separately for each investigated sample, and for the total sample of pairs belonging to superclusters. The isotropy of the distributions was checked by application of the Rayleigh test for angular data (Batschelet 1981); for each sample the probability of isotropy P was less than 0.001. Moreover, the number of solutions falling into bins with 15° width was compared with the number predicted from the random distribution, and the comparison is shown in Table 2. An excess of solutions is observed in the bins $0° - 15°$ and $165° - 180°$, while there is a manifest lack of solutions in the bins $75° - 105°$. In the sample containing the pairs in all superclusters the differences are about 2σ. It should be stressed that the observed anisotropy is not due to the ambiguity in the determination of the β-angles. Such an ambiguity could only blur the existing anisotropy, due to the symmetrisation of the β- and $(180° - \beta)$-angles, but it is not able to produce it.

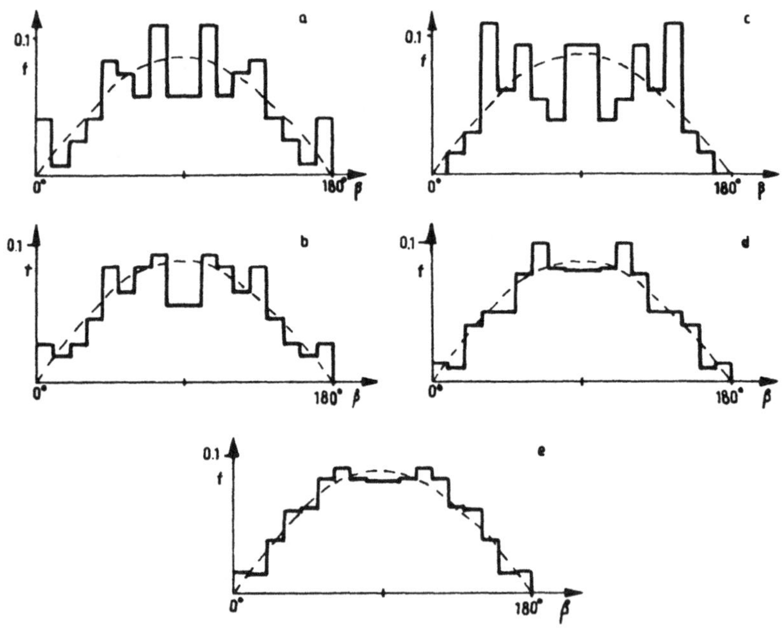

Fig. 5. The distribution of the β-angle in the double galaxies for (a) the Helou sample, (b) the LSC, (c) the Coma/A1367 SC, (d) the Perseus SC, and (e) composite plot for double galaxies in superclusters (the broken lines denote isotropic distributions).

Table 2. Check of isotropy of the β-angle

β-angle range	Helou obs	Helou exp	LSC obs	LSC exp	Coma obs	Coma exp	Per obs	Per exp	pairs in SC obs	pairs in SC exp
$0° - 15°$ and $165° - 180°$	24	8	16	7	4	4	16	15	36	27 ± 5.1
$15° - 30°$ and $150° - 165°$	12	24	16	22	8	13	40	45	64	79 ± 8.9
$30° - 45°$ and $135° - 150°$	52	38	36	34	44	20	68	71	148	126 ± 11.2
$45° - 60°$ and $120° - 135°$	44	50	48	45	24	26	90	93	162	164 ± 12.8
$60° - 75°$ and $105° - 120°$	48	58	60	52	20	31	128	108	208	191 ± 13.8
$75° - 90°$ and $90° - 105°$	60	62	40	56	28	33	106	116	174	205 ± 14.3

6 Conclusions

It is shown that the applied method is a useful tool for studying the mutual orientation of rotation axes of paired galaxies. It has the advantage that it includes objects for which the true spins are not known and thus permits to study large samples, so important in statistics. Moreover, in comparison to the position angle analysis it allows to include galaxies seen "face-on". Two galaxies forming a pair when seen "face-on" or nearly "face-on" have rotation axes close to each other, *i.e.* the β-angle is small. To drop these galaxies off the analysis is to introduce a systematic shift in the observed distribution. This is probably the reason that random distribution was so frequently obtained. The effect for the Helou sample is the same as in the present study, which validates the applied method; due to the limitation of the method, however, parallel and anti-parallel alignments cannot be separated.

The main results of the present study are:

i. the rotation axes of galaxies forming pairs tend to be aligned with the plane of the parent supercluster,

ii. the projections of rotation axes of paired galaxies onto the supercluster plane tend to point toward the centre of the structure, *i.e.* the galaxy planes tend to be perpendicular to the radius vector from the centre of the structure,

iii. the distribution of the β-angle is a non-random one, there is a strong tendency for rotation axes of paired galaxies to form a small $|\beta|$-angle and to avoid angles $90° \pm 15°$.

The observational results should be compared with predictions resulting from theoretical considerations of galaxy origin. From (i) and (ii) it follows that the alignments of paired galaxies and of single ones within superclusters are similar. This supports the hypothesis of a common origin of galaxies belonging to superclusters. The existence of an anisotropy and the direction of alignment (parallelism of the rotation axes with the supercluster main plane) are in agreement with a dissipative scenario of galaxy origin.

The lack of isotropy in the distribution of the β-angle is an argument against both the capture mechanism and the tidal torque hypothesis as explanations of the origin of double galaxies. The anti-parallelism of spins detected by Helou (1984) is against the turbulence scenario, exactly as points (i) and (ii). Thus, all observational evidence strongly favours the non-turbulence "top-down" scenario of galaxy origin.

References

Arigo, R., Czuhai, K., Hibbard, E., Noerdlinger, P., Wisner, K., 1978. *Astrophys. J.*, **223**, 410.

Batschelet, E., 1981. *Circular Statistics in Biology*, Academic Press, London and New York.

Flin, P., 1987. In *IAU Symp. No. 130, The Structure of the Universe*, eds. Audouze, J., Szalay, A. Kluwer, Dordrecht, in press.

Flin, P., Godlowski, W., 1986. *Mon. Not. R. astr. Soc.*, **222**, 525.

Fouqué, P., Paturel, G., 1985. *Astr. Astrophys.*, **150**, 192.

Helou, G., 1984. *Astrophys. J.*, **284**, 471.

Karachentsev, I.D., 1981. *Astrofizika*, **17**, 692.

Nilson, P., 1973. *Nova Acta Regiae Societatis Scientiarum Upsaliensis ser. V.A.* Vol. 1.

Noerdlinger, P.D., 1979. *Astrophys. J.*, **229**, 877.

Tifft, W.G., 1980. *Astrophys. J.*, **239**, 445.

Visual Light and Infrared Observations as Complementary Sources of Data on Intergalactic Dust

Bogdan Wszołek and **Konrad Rudnicki**
Jagiellonian University Observatory
Kraków, Poland

Paolo de Bernardis and **Silvia Masi**
Istituto di Fisica dell' Università
Roma, Italy

Abstract

The significance of investigations of intergalactic dust in absorption, as well as of complementary observations in emission, is discussed. Infrared observations, particularly interesting for the Okroy Cloud, are cited. A program for further investigations is suggested.

1 General remarks

As is well known, intergalactic dust (IGD) is difficult to study observationally. If we leave out purely theoretical publications as well as all observational contributions which proved to contain substantial observational errors or misinterpretations of the results, and sometimes even reflect systematic side-effects instead of physical reality, then there remain about 40 papers giving observational evidence of the existence of IGD (Rudnicki 1986) — within 70 years of research, initiated by the "prehistorical" paper of Lundmark and Lindblad (1917). Those 40 papers deal mostly with the search for intergalactic continuous extinction. The effects found are very small indeed. An advantage of optical investigations of absorption attributed to IGD is that there are methods which can distinguish between interstellar (within our own Galaxy) and intergalactic extinction. This applies to Hoffmeister's (1962) method of searching for the effect, and also Kwast's (1974) Θ-method, particularly in its modified form (Zabierowski 1985), which is convenient for numerical use.

On the other hand, if IGD absorbs light, it must also emit radiation. A major difficulty is the absence of methods in the infrared domain which distinguish between IGD and radiation originating within our Galaxy. The unknown temperature of IGD clouds makes the problem even more complicated. As of today, no methods like those of Hoffmeister or Kwast have been conceived for infrared observations.

2 Preliminary results

The search for IGD, based on its emission effects, is thus rather a matter of the future, when relevant methods will have been developed. Therefore, for the time being, we have attempted to use the infrared IRAS data just as additional evidence for the existence of IGD clouds, well established otherwise. Since IGD radiation is obviously cooler than 300 K while at least some parts of it should have a temperature higher than 3 K, we assumed, to start with, a value of 30 K and looked for corresponding emission in the IRAS 100 μm survey. So far we have found a distinct maximum in infrared emission in the Okroy Cloud (Murawski 1983), centered exactly on the cloud centre as established in the original report of Okroy (1965). This maximum can

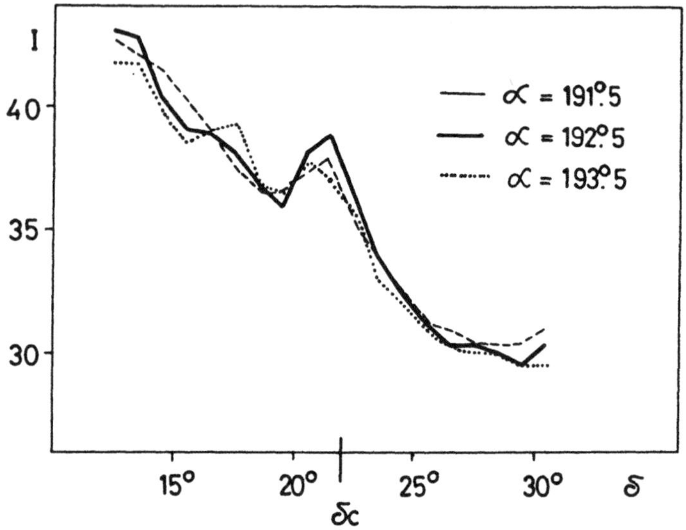

Fig. 1. Profiles of 100 μm radiation (IRAS data) in the Okroy Cloud at fixed right ascensions. δ_c denotes the cloud centre as given in Okroy's original paper. The units here as well as in the next figures are IRAS units: 7.5×10^{-15}W cm^{-2}sr$^{-1}\mu$m^{-1}.

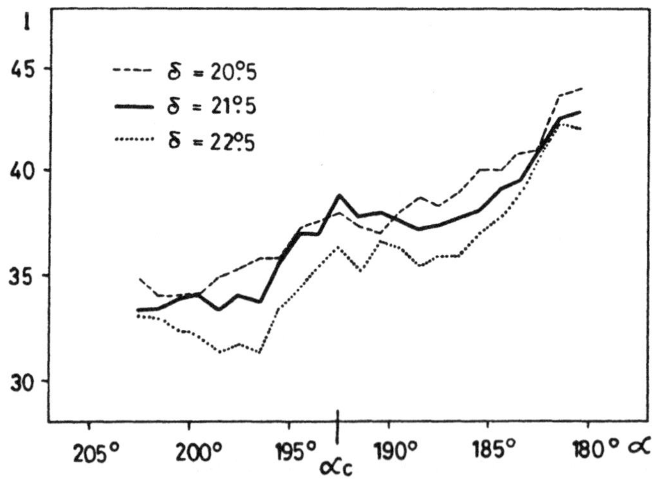

Fig. 2. As in Fig. 1, but for profiles at fixed declinations. α_c denotes the cloud centre.

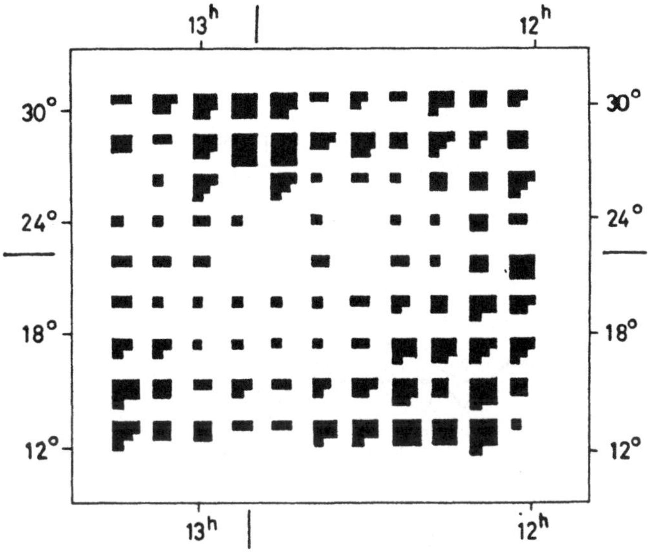

Fig. 3. Galaxy density map with proportional blackening (including all galaxies from the Zwicky Catalogue), a fragment of Okroy's original map. The centre of the cloud is indicated by lines on the margins.

be seen in the declination profiles (right ascension fixed) at three neighbouring right ascensions (*Fig. 1*); but only in one right ascension profile (declination fixed) (*Fig. 2*). This leads to the conclusion that the cloud is elongated in right ascension, which is in agreement with Okroy's original drawing (*Fig. 3*). Of interest may also be a sudden emission drop from one side of the Coma cluster, accompanied by a similarly sharp increase of galaxy density towards the centre of the Coma Cluster. The conclusion, that some outer parts of the Coma Cluster might extend further into the direction of lower declination but is concealed from sight by the overlapping cloud, is a tempting one.

A similar check for the Rudnicki-Baranowska Cloud (Kwast 1974) has not yielded any such distinct effects (*Figs. 4* and *5*).

A noticeable infrared radiation maximum has been found in the neighbourhood of the Okroy Cloud (*Figs. 6* and *7*). Whether this can be attributed to an IGD cloud, or perhaps linked to the peripheral parts of the Okroy Cloud itself, is a matter for further investigations.

3 A tentative program for further investigations

From the results obtained so far one can take the following hints for further investigations:

(1) to plot all regions of the sky with a deficit of galaxies if it is manifest in all magnitudes;

(2) to investigate those regions using Kwast's Θ-method and look for Hoffmeister's effects to distinguish the actual IGD clouds from clouds within our Galaxy as well as from voids between clusters;

(3) to look for infrared and radio emission in all regions where the existence of distinct IGD clouds has been confirmed by (2);

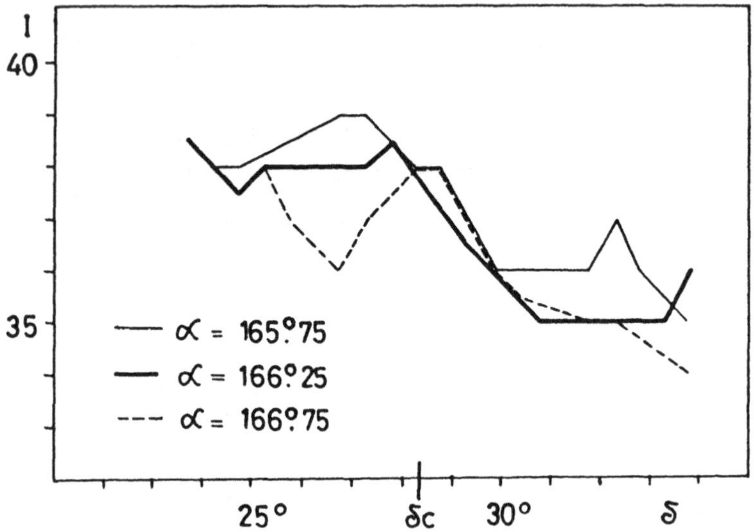

Fig. 4. Profiles of 100 μm radiation at fixed right ascensions, as in Fig. 1, for the Rudnicki-Baranowska Cloud. In the central section, a part of the profile is omitted due to the absence of data. The centre of the cloud as seen in absorption is denoted with δ_c.

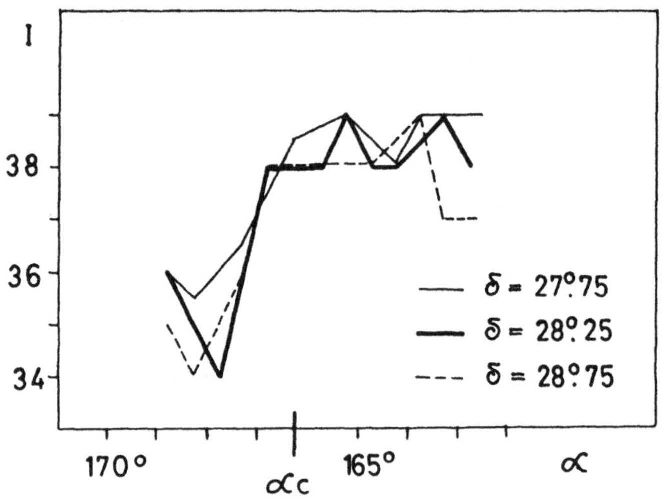

Fig. 5. As in Fig. 4, but for profiles at fixed declinations. The centre of the Rudnicki-Baranowska Cloud is denoted with α_c.

(4) to determine the infrared and radio emission spectra for all these clouds (they are probably different for different clouds);

(5) to determine the physical and chemical properties of IGD, and their possible relation to the uniformly distributed dust and gas in intergalactic space;

(6) to conceive methods of discovering IGD clouds directly by spectral analysis of their radiation;

(7) to determine the contribution of IGD radiation to the cosmic background radiation, and herewith to falsify or verify the hypothesis of Rana (1979, 1980a, 1980b) that the 3 K relic radiation is just of IGD origin.

The programme we outlined here, even the points (1) to (4) only, is very laborious indeed due to the extensive observational evidence needed, as well as to the broad range

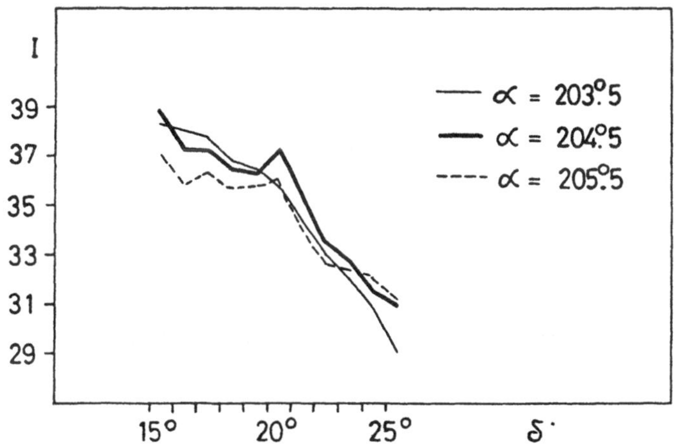

Fig. 6. A maximum of 100 μm radiation (IRAS data) near the Okroy Cloud; profiles at fixed right ascensions.

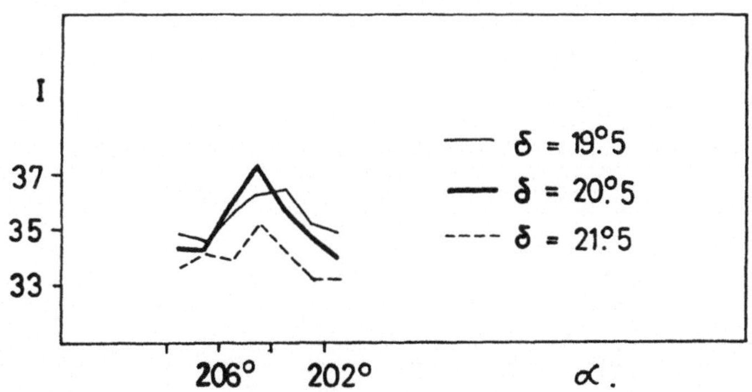

Fig. 7. As in Fig. 6, but for profiles at fixed declinations.

of methods to be applied. It can be successfully realized only through collaboration of a number of astronomical institutions.

References

Hoffmeister, C., 1962. *Z. Astrophys.*, **55**, 46.
Kwast, T., 1974. *Acta Cosmologica*, **2**, 65.
Lundmark, K., Lindblad, B., 1917. *Astr. Nachr.*, **205**, 161.
Murawski, W., 1983. *Acta Cosmologica*, **12**, 7.
Okroy, R., 1965. *Astr. Tsirk. No. 320*, 4.
Rana, N.C., 1979. *Astrophys. Space Sci.*, **66**, 173.
Rana, N.C., 1980a. *Astrophys. Space Sci.*, **67**, 201.
Rana, N.C., 1980b. *Astrophys. Space Sci.*, **71**, 123.
Rudnicki, K., 1986. *Gamow Cosmology, Proc. Int. School Phys. "Enrico Fermi", Course 86, eds. Melchiorri, F., Ruffini, R.*, North-Holland, Elsevier, Amsterdam, p. 480.
Zabierowski, M., 1983. *Astrophys. Space Sci.*, **117**, 179.

Part V

Quasars

Dieter Engels

Volker Gericke

Christian de Vegt

Klaus Beuermann

Quasar Search on Objective Prism Plates

H.J. Hagen, D. Engels,
D. Groote, D. Reimers
Hamburger Sternwarte
Hamburg-Bergedorf, F. R. Germany

A machine-based search for quasar candidates down to $B \sim 18^{\mathrm{m}}.5$ on objective prism plates is conducted at Hamburg. Our aim is to provide several hundred bright QSOs for the purpose to make follow-on studies on the physics of QSOs and their environment. We are especially interested in peculiar objects like BALs, QSO pairs, and gravitational lenses (Engels *et al.* 1988).

The survey plates are taken with the Schmidt telescope on Calar Alto equipped with a $1^\circ.7$ objective prism. The dispersion of the spectra on the plates is $\sim 140\,\mathrm{nm\,mm^{-1}}$ at Hγ. The Kodak IIIa-J emulsion used limits our range to $z \sim 3.3$.

The plates are scanned in Hamburg with a PDS 1010 G microdensitometer controlled by a PDP 11/24 computer. For data storage a 170 MByte disc drive and a 1600 bpi magnetic tape drive are currently available. Machine driver software and all data processing programme have been written in Hamburg and run under RSX 11M plus on the PDP 11/24 (Hagen 1987).

Complete plate scans are made perpendicular to the direction of dispersion in a low-resolution mode. The background is removed automatically. The current scan time is about 16 hours per plate. The scan results in 30 000–50 000 low-resolution spectra (*Fig. 1a*).

Candidates are selected on the basis of the presence of emission lines or a blue continuum. These candidates are scanned in a high-resolution mode and the resulting spectra (*Fig. 1b*) are evaluated by eye. For each field two plates are used to discriminate against plate faults.

First QSO confirmations of the candidates were made through slit spectroscopy with the 3.5 m telescope on Calar Alto/Spain (*Fig. 1d*). We find typically 25–30 quasars per field (25 deg^2) with a limiting magnitude $\sim 18^{\mathrm{m}}.5$.

References

Engels, D., Groote, D., Hagen, H.J., Reimers, D., 1988. In *Proceedings of the workshop on Optical Surveys for Quasars*, ed. Osmer, P., *Publ. astr. Soc. Pacific*, in press.
Hagen, H.J., 1987. *PhD thesis*, University of Hamburg.

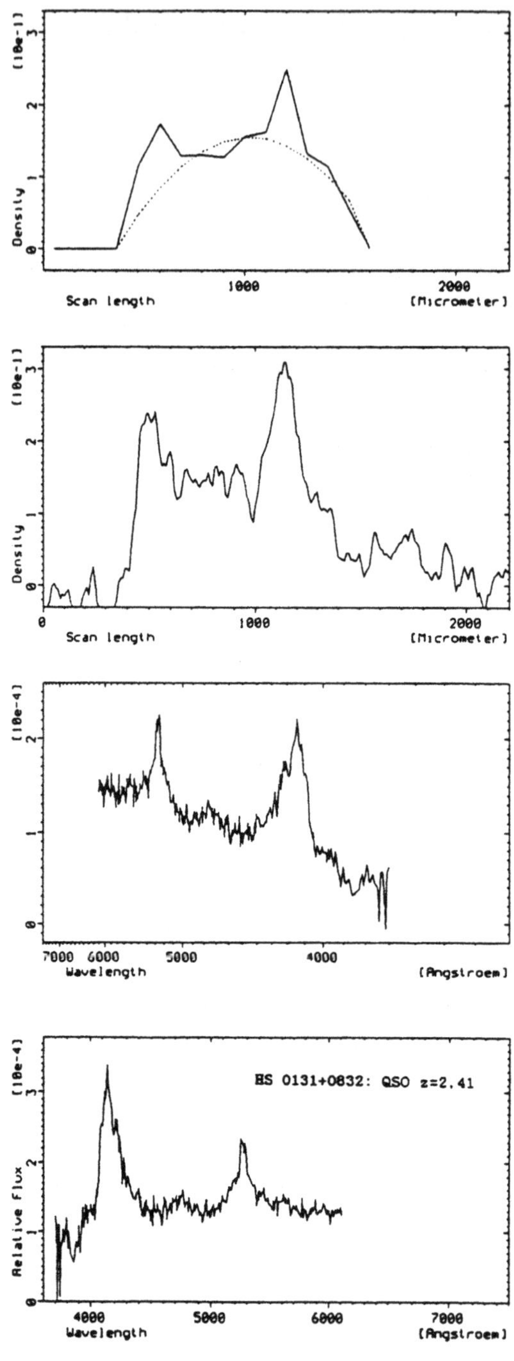

Fig. 1. a) Low-resolution spectrum of a QSO candidate, b) high-resolution spectrum, c) slit-spectrum convoluted with the prism dispersion and d) slit spectrum of the confirmed QSO

A Search for Homogeneous Samples of Quasars

Volker Gericke
Astronomisches Institut
Westfälische Wilhelms-Universität
Münster, F.R. Germany

Abstract

An automated search for quasars, based on MRSP software, is applied to low-dispersion objective prism plates. Two criteria are used. The first criterion is the UV excess of low and medium redshift quasars. The second criterion is the presence of emission lines. 25 000 non-overlapping spectra in the ESO/SRC field No. 411 were analyzed; 600 show UV excess, 2 500 spectra have apparent emission lines. Both criteria taken separately lead to a large number of "false" objects. In the combined feature space known quasars occupy a characteristic region. Among the objects from this region in the limited magnitude range $18^{m} - 19^{m}$ 600 quasar candidates were found. Redshifts were determined from the low-dispersion objective prism spectra for 255 of these objects. The spatial distribution of 158 Ly α quasars is discussed.

1 Introduction

When normal galaxies are used to study large scale structures in the universe, only data up to redshifts $z = 1$ are available so far. With quasars it is possible to investigate large scale distributions of luminous matter at distances up to $z = 4$ and beyond.

To date, two quasar catalogues exist; both list the majority of the approximately 3 500 known quasars distributed over the whole sphere (Hewitt and Burbidge 1987, Véron-Cetty and Véron 1987). The catalogues contain very inhomogeneous data, because they include the results of different authors using different techniques to find and to identify quasars. Some spectra are obtained through slit spectroscopy, others from objective prism plates, leading to different degrees of reliability of the redshifts listed. In some cases the origin of the quoted redshifts remains unclear.

Several *automated* quasar surveys, intended to increase the number of known objects, are in progress. The first automated procedures were employed by Clowes *et al.* (1984) and Hewett *et al.* (1985). More recent work based on very low dispersion objective prism spectra is reported by Beuermann and Clowes (1988). A survey using somewhat higher dispersion spectra was started in Hamburg and is presented by Hagen *et al.* (1988). Automated procedures applied to low-dispersion objective prism spectra in Cambridge have led to remarkable results (Foltz *et al.* 1987). The present contribution concerns the automated search for quasars as part of the MRSP.

The ESO/SRC field No. 411 was investigated, with MRSP software applied to a film copy of the direct atlas J-plate and to the film copy of a low dispersion objective prism plate ($240\,\mathrm{nm\,mm^{-1}}$) taken in this region. The preprocessing of direct plates and corresponding objective prism plates is described by Horstmann (1988) and Schuecker (1988). The criteria used for the automated quasar search are discussed in Sects. 2 to 4.

2 The UV-excess criterion

For the UV-excess excess criterion of low and medium redshift quasars, the center of intensity X_C of each spectrum is calculated:

$$X_C = \frac{\sum_{i=1}^{n} I_i X_i}{\sum_{i=1}^{n} I_i} . \tag{1}$$

The summations are taken over all pixels n of each spectrum. I_i is the intensity at pixel i; x_i is its position.

X_C values and magnitudes determine the feature space shown in *Fig. 1*. Small values of X_C correspond to red objects, high values to blue ones. Faint objects are located in the lower part of the diagram. Most objects lie on a 'main sequence'. Quasars which are found in the abovementioned catalogues are marked by squares.

Many of the quasars are found outside the 'main sequence', but the QSOs extend into the region of (normal) stars. This is due to the fact that the excess criterion

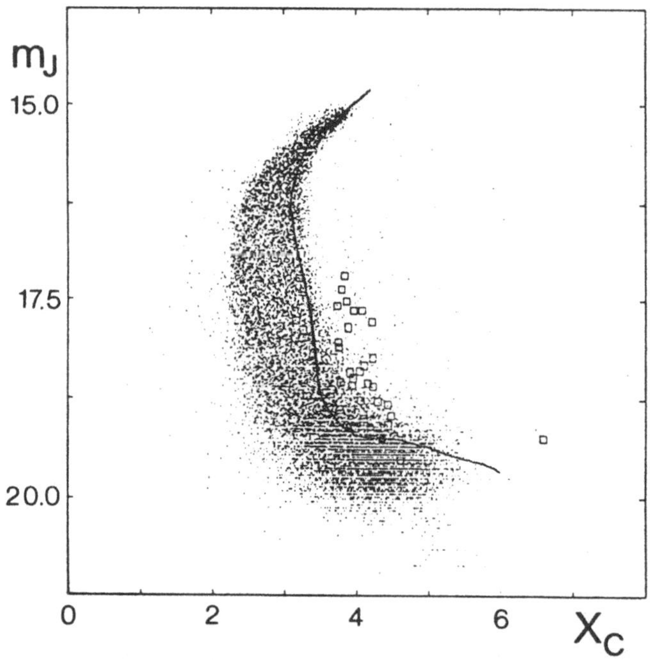

Fig. 1. PDS m_J magnitudes *vs.* center of intensity X_C. Large values of X_C correspond to blue objects. Quasars listed in the catalogues of Hewitt and Burbidge (1987) and Véron-Cetty and Véron (1987) are shown as squares.

is continuous, so that blue stars and relatively red quasars overlap. In low-density spectra the mixing results from blurring due to noise. A small percentage of the effect may also be due to possible misidentifications in the catalogues, with stars in the range G – M listed as quasars.

A random sample was taken from the region on the blue side of the linear discriminant, indicated in the diagram. The discriminant was determined by setting knots interactively and fitting a cubic spline function. The sample includes few quasars. From a total of 596 blue starlike objects only 82 are high probability quasar candidates as indicated by their clearly non-stellar continua and emission lines.

No quasars were found on the red side of the feature space.

3 The emission line criterion

To locate possible emission lines, the spectra between 350 nm and 520 nm are transformed to a linear wavelength scale. The fact that through this process the noise becomes nonrandom (wavelength dependent) is tolerated, though it restricts the recognition of emission lines to *strong* lines. The spectral intensities are corrected for atmospheric and instrumental extinction and for the sensitivity of the emulsion using the curves from Clowes *et al.* (1980). By these procedures the widths of the lines are directly comparable.

For each spectrum the positions $P_i^{\max}$ and the intensities $I(P_i^{\max})$ of all maxima are calculated, *i.e.* in the beginning of the calculations each peak is considered a possible emission line. The maxima are found from second derivatives calculated using Lagrangian polynomials. Best results are obtained for polynomials with 5 knots separated by $H = 6.8$ nm. The second derivative of the central knot is calculated by

$$I_3'' = \frac{1}{24H^2}\left(-2I_1 + 32I_2 - 60I_3 + 32I_4 - 2I_5\right) . \tag{2}$$

I_k is the intensity of the kth knot. The constants are weighting factors, determined from the theory of Lagrangian polynomials. Calculating the second derivative acts as a filter of width H, because the value for the central knot is determined by the values of the side knots. The values of the second derivatives at the edges of the spectra are calculated with similar formulae, but lower accuracy. In this way the second derivative is obtained for each pixel.

During the next step the height and the width of each emission feature is determined. To calculate the height the continuum must be known. The points of inflection P_i^1 and P_i^2 below each maximum are determined; I_i^m, the mean value of the intensities at the two points, is used to determine the continuum I_i^C in the vicinity of each emission feature:

$$I_i^C = I_i^m - (I(P_i^{\max}) - I_i^m) . \tag{3}$$

The relative height I_i^r of an emission feature is

$$I_i^r = \frac{I(P_i^{\max}) - I_i^C}{I_i^C} . \tag{4}$$

The distance W_i between P_i^1 and P_i^2 is a measure of width:

$$W_i = |P_i^1 - P_i^2| \,. \tag{5}$$

In the corresponding feature space (*Fig. 2*), the intensity of the highest emission peak I_M of each spectrum is plotted *vs.* the width W_M of the same emission feature. Known quasars from the catalogues are again marked by squares. Peaks occur at multiples of the filter width H. The first peak at $W_M = 6.8\,\mathrm{nm}$ contains all emission-like structures with W_M of the order of H. This peak is largely due to noise, though it includes also the narrow line quasars. The second peak at $W_M = 13.6\,\mathrm{nm}$ includes all emission features with W_M of the order of $2H$. They are often due to real emissions, continuum features and emission lines. A third, very weak peak can be seen around $W_M = 20.4\,nm$. These features are caused by changes in the continuum.

The existence of more than one peak shows that the height of a spectral feature alone is not a sufficient characteristic for an emission line.

A sample taken from the region around $W_M = 13.6\,\mathrm{mn}$ which is marked by the rectangle, leads to about 2 500 objects with apparent emission lines. Only a few of them are quasars. Objects with strong absorption breaks are found here because they feign emission lines. The most prominent absorption objects in this sample are G- and K-type stars, with a strong continuum hump between the G-band and the Ca II-break. Very red stars with narrow red contina simulate single-emission line objects.

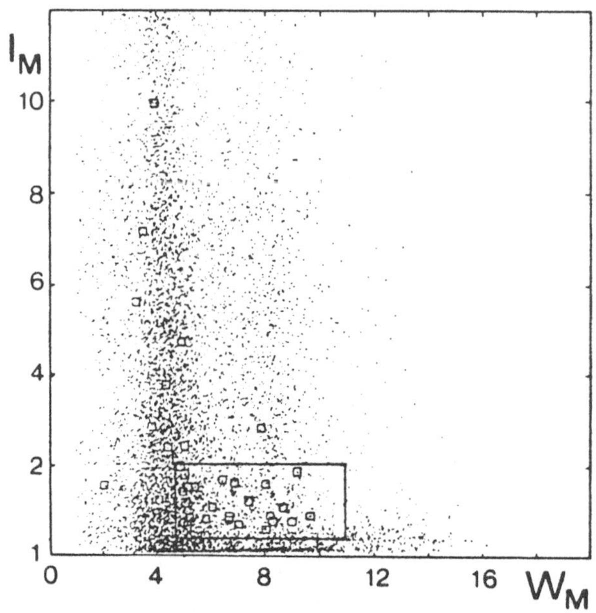

Fig. 2. Height I_M of the maximum emission feature *vs.* its width W_M. Quasars listed in the catalogues are shown as squares.

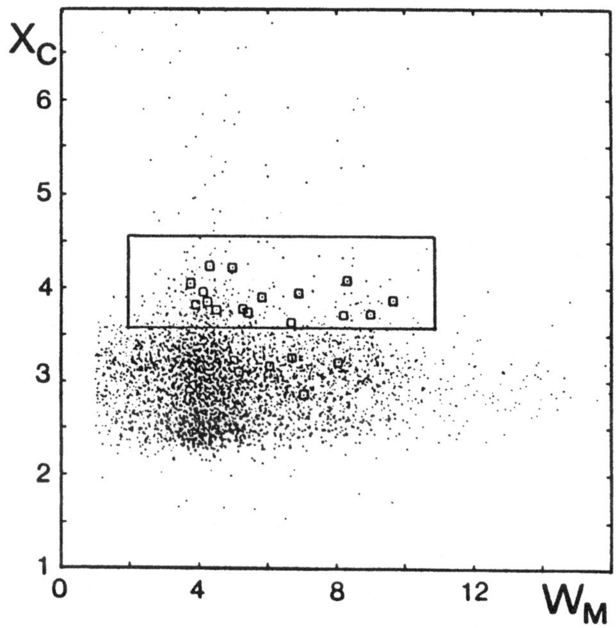

Fig. 3. Center of intensity X_C *vs.* width W_M of the highest peak in the magnitude range $18^m - 19^m$. Quasars listed in the catalogues are shown as squares.

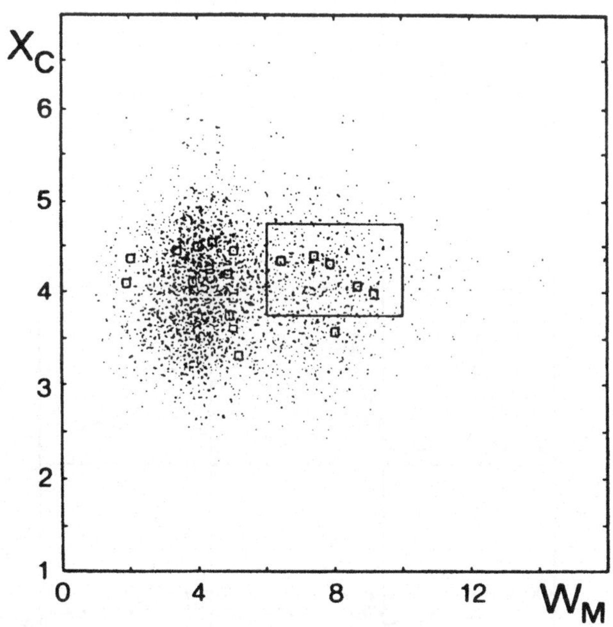

Fig. 4. Center of intensity X_C *vs.* width W_M of the highest peak in the magnitude range $19^m - 20^m$. Quasars listed in the catalogues are shown as squares.

4 The combined feature space

The UV-excess criterion and the emission line criterion taken separately do not suffice to unambiguously identify quasars. A more efficient way to locate quasars is the combination of two or more criteria. The two criteria discussed above can be combined in a new feature space with center of intensity X_C *vs.* width W_M of the highest emission feature. In *Fig. 3* the combined feature space includes all objects in the magnitude range $18^m - 19^m$. The separation between quasars and other starlike objects is clearly more apparent than in the cases where only one criterion was used. A sample of 600 objects is taken from the region extending above $X_C = 392\,\mathrm{nm}$ (marked by a rectangle). Among these are only 5 % which can be *excluded* as quasars from a visual inspection of their objective prism spectra.

Figure 4 shows the combined feature space in the magnitude range between $19^m - 20^m$. The separation between quasars and other starlike objects is not as good as in the case of *Fig. 3* because of the higher noise in the spectra. About 600 objects are selected from the region around $W_M = 13.6\,\mathrm{nm}$, marked by the rectangle. Only a few of the quasar candidates can be excluded to be quasars. But for faint objects it is much more difficult to confirm objects as quasars, because of the weak continua and the numerous spurious emission features.

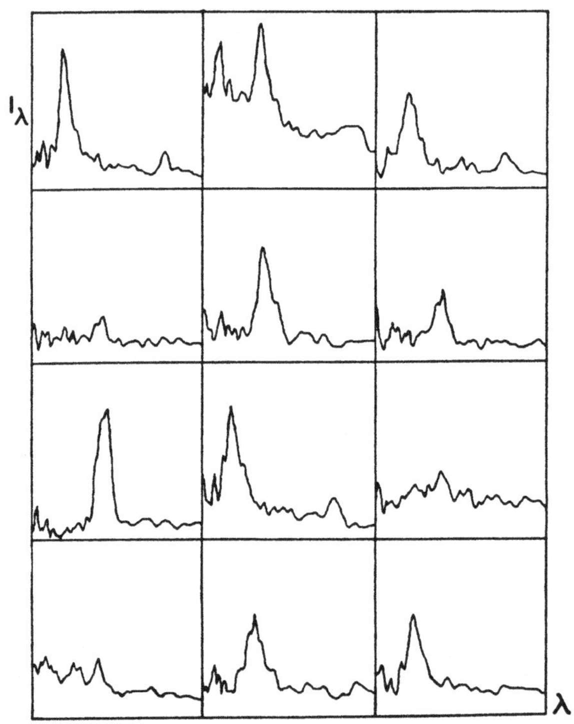

Fig. 5. Objective prism spectra of Ly α quasars found in this survey, intensity-corrected and on a linear wavelength scale.

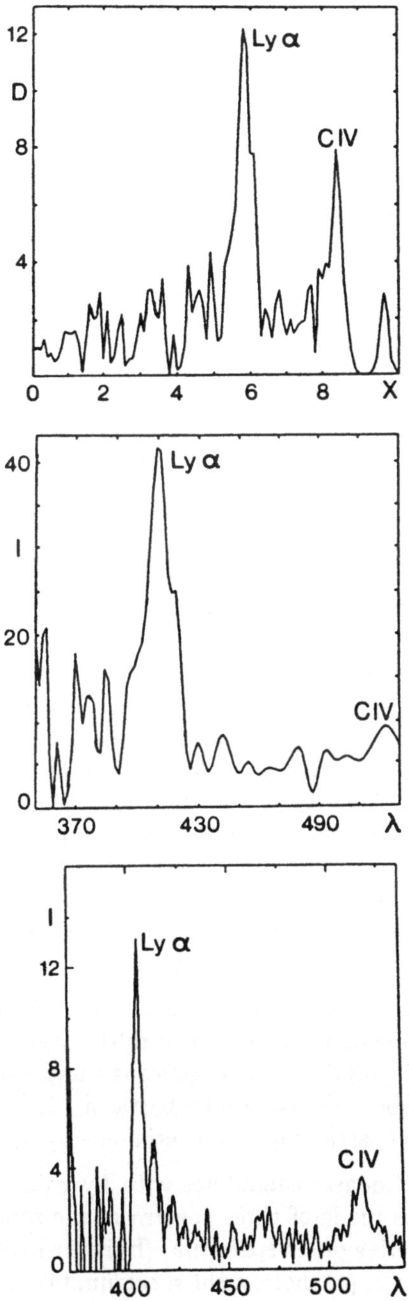

Fig. 6. Spectra of the quasar 0054-29, from top to bottom:
a) density spectrum (objective prism plate)
b) intensity corrected spectrum on a linear wavelength scale (objective prism plate)
c) slit spectrum (EFOSC, ESO 3.6 m-telescope).

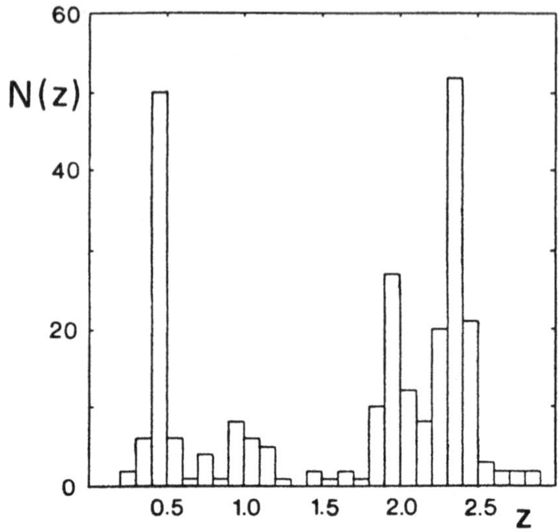

Fig. 7. Histogram of z-values of 255 quasars found in this survey.

5 A census of quasars in field No. 411.

For the quasar candidates in the magnitude range $18^m - 19^m$ redshifts were obtained interactively from the objective prism spectra. The identification of Ly α emission is quite reliable, especially, because in most cases other lines such as N V, Si IV or C IV are also present. The identification of the Mg II emission is more difficult, because no other prominent emission lines appear at the same redshift. When only one strong emission feature was present, the underlying continuum was used as additional criterion; when the strong line appeared on a very blue continuum (attributable to the Lyman continuum), the feature was identified as Ly α, when the continuum appeared flat, the emission-line was assigned to Mg II. Structural criteria, high and narrow for Ly α, low and more diffuse for Mg II, were also used.

255 redshifts were obtained with an average of 2.0 lines per spectrum. Among the objects are 158 with Ly α emission. In these spectra the average is 2.4 lines. These data constitute a well-defined, physically homogeneous sample of quasars in the redshift range $1.9 < z < 2.9$ and absolute magnitudes between -25^m and -27^m (K-corrections are negligible in this range, assuming exponential energy distribution in the spectra).

Figure 5 shows spectra of quasar candidates with Ly α lines found in this survey. In *Fig. 6* a-c a comparison is made of three spectra of the same quasar. *(a)* shows the uncorrected density objective prism spectrum, *(b)* is the intensity corrected spectrum on a linear wavelength scale, *(c)* shows a slit spectrum taken with EFOSC at the ESO 3.6 m-telescope.

6 Quasar statistics

The z-distribution of all quasars is presented in *Fig. 7.* Most quasars are found at z values near 0.4, where the Mg II emission line appears in the central part of the

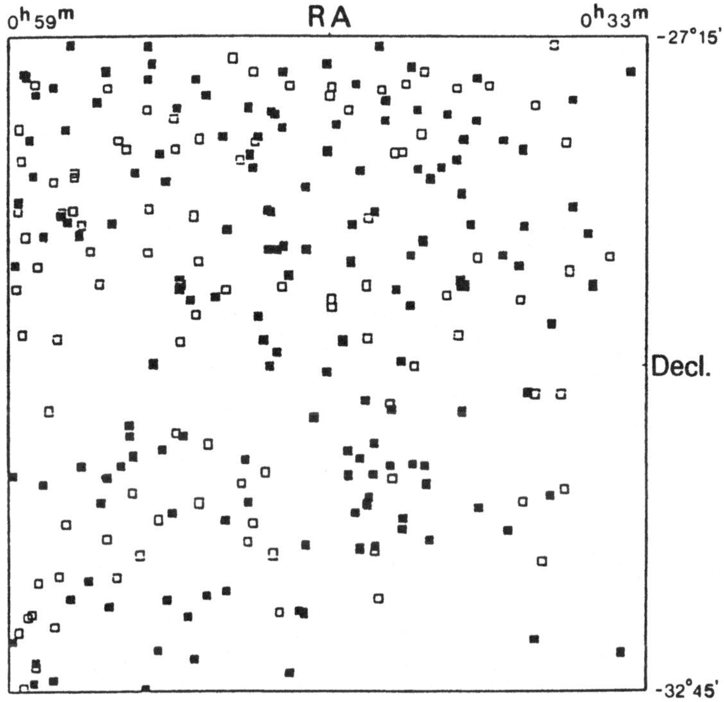

Fig. 8. Distribution of the 255 quasars in ESO/SRC field No. 411. Quasars with Ly α lines are shown as filled squares, all other quasars as open squares.

spectral window, and around $z = 2.4$ with Ly α. For other redshifts, only weak emission lines can be expected, so that it is difficult to find quasars at these values. The peak around $z = 0.4$ may be too high, because Mg II quasars are more uncertain. Because the rest wavelength of Mg II is much larger than the rest wavelength of Ly α, the range of observable z-values for Mg II quasars is much smaller. This explains the narrowness of the peak at small redshifts in the z-histogram as compared to peak at large z-values.

In *Fig. 8* the two-dimensial distribution of all quasars with measured redshifts in field No. 411 is shown. There seems to be an indication of clustering for the Ly α quasars, *e.g.* a concentration in the south-west quadrant.

z-histograms of the Ly α quasars only, arranged in $55' \times 55'$ cells, show an indication of clustering in the same region and at a depth of $z = 2.4$ (*Fig. 9*). Other, minor concentrations are suggested. The relevant positions in the histograms are shaded. The extent of the major "cluster" is of the order $100\,h^{-1}$ Mpc.

The two-dimensional distribution of the quasars was also studied using the angular correlation function $w(\theta)$ given by Hewett (1982). *Fig. 10* shows $w(\theta)$ for the Ly α quasars. The resolution used is 6'. There may be clustering at angles smaller than $1\overset{\circ}{.}3$ and on a characteristic scale of $100\,h^{-1}$ Mpc. Because of the small number of objects large noise is expected. In order to test whether this could lead to spurious effects,

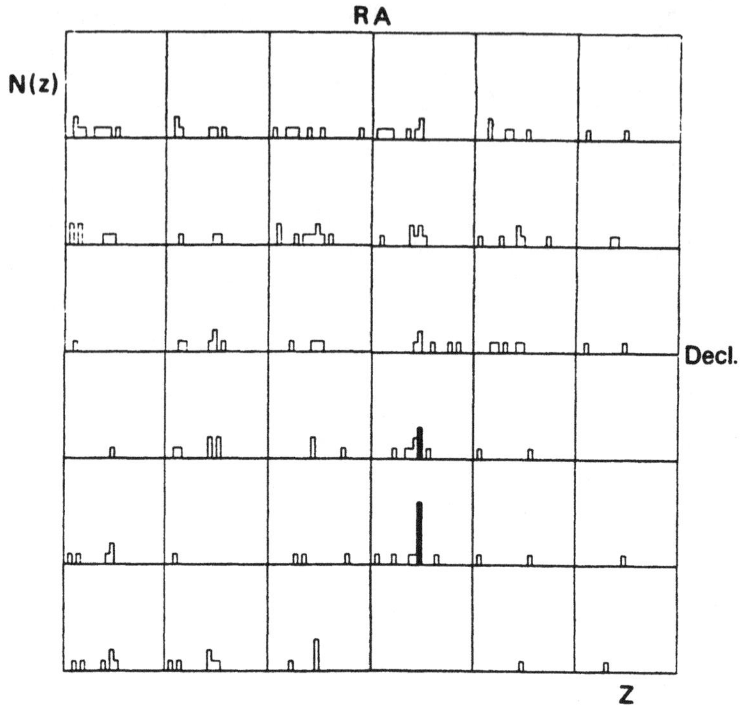

Fig. 9. Histograms of z-values of Ly α quasars in field No. 411, segmented into 36 sections. Ranges: $0 \leq N \leq 10$, $1.8 \leq z \leq 3$, bin size $\Delta z = 0.05$.

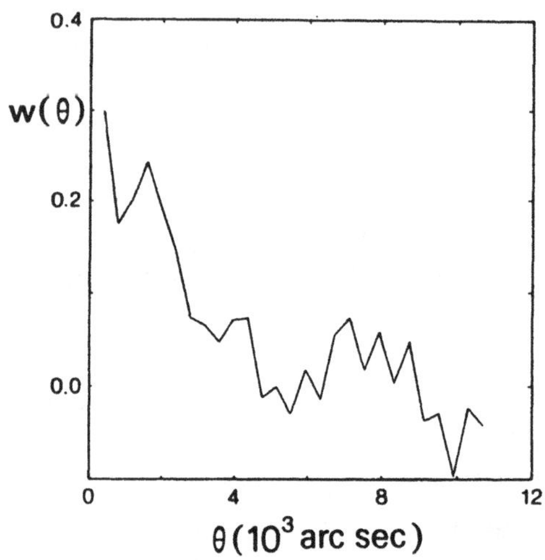

Fig. 10. Two-point correlation function for Ly α quasars.

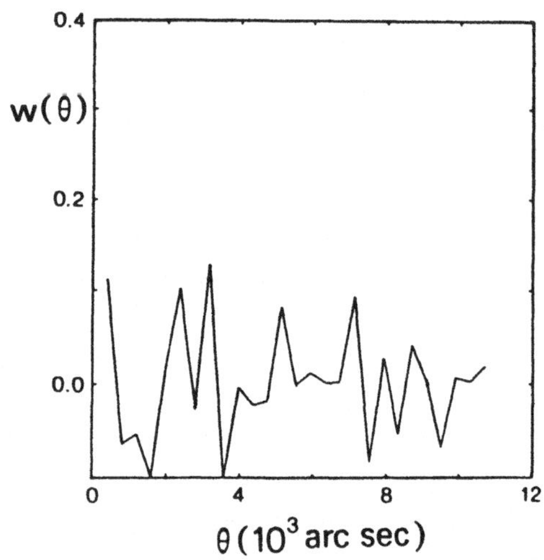

Fig. 11. Two-point correlation function for a random distribution.

the correlation function for random distribution, using the same average number of objects per cell, was determined. The result is shown in *Fig. 11*. Mg II quasars show marginal clustering. This may be attributed to the small number of objects (70), the larger number of misidentifications and/or the absence of clustering. The number of quasars which are neither Ly α nor Mg II quasars is too small for a test.

Data of three more fields have recently become available from the MRSP. With the expected number of about 600 Ly α quasars, we will be able to derive clustering properties from a more reliable homogeneous sample of quasars.

References

Beuermann, K., Clowes, R.G., 1988. *These proceedings*, p. 246.
Clowes, R.G., Smith, M.G., Wallace, P.T., Cannon, R.D., Savage, A., Boksenberg, A., 1980. *Mon. Not. R. astr. Soc.*, **193**, 415.
Clowes, R.G., Cooke, J.A., Beard, S.M., 1984. *Mon. Not. R. astr. Soc.*, **207**, 99.
Foltz, C.B., Chaffee, F.H., Hewett, P., MacAlpine, G.M., Turnshek, D.A., Weymann, R.J., Anderson, S.F., 1987. *Astr. J.*, **94**, 1423.
Hagen, H.E., Engels, D., Groote, D., Reimers, D., 1988. *These proceedings*, p. 233.
Hewett, P.C., 1982. *Mon. Not. R. astr. Soc.*, **201**, 867.
Hewett, P.C., Irwin, M.J., Bunclark, P., Bridgeland, M.T., Kibblewhite, E.J., He, X.T., Smith, M.G., 1985. *Mon. Not. R. astr. Soc.*, **213**, 971.
Hewitt, A., Burbidge, G., 1987. *Astrophys. J. Suppl.*, **63**, 1.
Horstmann, H., 1988, *These proceedings*, p. 111.
Schuecker, P., 1988. *These proceedings*, p. 142.
Véron-Cetty, M.-P., Véron, P., 1987. *A Catalogue of Quasars and Active Nuclei*, 3rd edition, ESO scientific report No. 5.

Slit Spectroscopy of Candidates from Automated Quasar Detection on UKSTU Prism Plates[1]

K. Beuermann
Institut für Astronomie und Astrophysik
Technische Universität
Berlin, F.R. Germany, and
Max-Planck-Institut für Extraterrestrische Physik
Garching bei München, F.R. Germany

R.G. Clowes
Royal Observatory, Blackford Hill
Edinburgh, Scotland, UK

Abstract

A search for quasars is being performed in ESO/SERC fields 86, 119, and 120 at and near the South Ecliptic Pole (just outside the LMC). This area will be extensively covered by the ROSAT all-sky soft X-ray survey. The present search aims at providing a data base of optically selected extragalactic objects which may later be compared with the sample of X-ray selected objects. The quasar candidates were identified on objective prism plates from the UK Schmidt telescope (UKSTU) by visual means in field 86 and by employing Automated Quasar Detection (AQD) in fields 119 and 120. Slit spectra were subsequently obtained for 39 candidates in field 119 and 3 candidates in field 86, using the ESO/MPI 2.2 m telescope at La Silla, Chile. The results demonstrate that an automated quasar search can be successfully be performed in these comparatively crowded star fields.

1 Introduction

Visual and automated searches of objective prism plates represent a well-established technique to identify a large number of quasar candidates (*e.g.* Clowes and Savage 1983, Barbieri and Cristiani 1986). Visual searches are useful in crowded fields but suffer from subjective selection criteria. Automated Quasar Detection (AQD) is the generic name for a system of software and procedures developed at the Royal Observatory in Edinburgh (ROE) that allows quasar candidates to be automatically discovered from measurements of objective prism plates (Clowes *et al.* 1984, Clowes 1986). It is based on the COSMOS fast plate-measuring machine at ROE (MacGillivray and Stobie 1984) and typically uses plates from the UK Schmidt telescope (UKSTU).

The present program was initiated in view of the ROSAT all-sky soft X-ray survey (Trümper 1984). This survey will cover the areas near the North and South Ecliptical

[1]Based on observations collected at the European Southern Observatory, La Silla, Chile, with the 2.2 m telescope of the Max-Planck Society

Poles (NEP and SEP) with the highest exposure integral. Here, the survey will be deeper than on the average over the sky. Furthermore, X-ray sources near the poles will be in view longer than near the equator with a maximum of 180 days at the poles. For the brighter X-ray sources near the poles, X-ray variability studies may, therefore, be performed in the course of the ROSAT survey.

We have initiated an optical search for quasars and AGNs at and near the South ecliptic pole. The SEP ($\alpha, \delta = 6^{\mathrm{h}}, -66\overset{\circ}{.}5$) is located about 3° from 30 Doradus at the fringes of the hydrogen envelope of the LMC. The star density in this region (ESO/SERC field 86) is still comparatively high and crowding was considered too severe for AQD to be employed. Instead, a visual search for emission-line AGNs and quasars was performed in two sections of field 86, using the TV blink comparator at ROE.

Because of the high star density in field 86, we decided to employ AQD in two fields near the SEP, ESO/SERC fields 119 and 120. The fields were selected for three reasons: (1) The star density is sufficiently low to permit the use of AQD. (2) The fields receive a reasonably high exposure during the ROSAT survey since the plate centers are separated from the SEP by only about 6° and 8°, respectively. (3) The fields are located sufficiently far outside the hydrogen envelope of the LMC to permit soft X-ray studies of extragalactic background objects. The H I column densities in fields 119 and 120 range from $1.7 \cdot 10^{20}$ to $6.0 \cdot 10^{20}$ H atoms cm^{-2} which corresponds to optical depths $\tau = 1$ against photoabsorption in cold matter for soft X-rays of 0.22 and 0.36 keV, respectively.

The aim of the present initial study is (1) to test the quasar selection criteria, (2) to study the contamination of the quasar sample by blue stars of LMC membership, and (3) to study the completeness of quasar detection as a function of star density. To this end, slit spectoscopy of so far 42 candidates was performed and first results of this program are presently reported.

2 Automated Quasar Detection (AQD)

AQD was developed at ROE to remove from an inherently powerful technique the drawbacks of visual scanning which include unknown and time-varying selection criteria, wastage of information, and tedium. A long period of software development was necessary, but the final result realizes the full potential and allows large numbers of quasars to be discovered in large areas of the sky with pre-defined and constant criteria, recording of all important data, and relatively little tedium. The initial version of AQD was described by Clowes *et al.* (1984) and the current version by Clowes (1986).

The following options are available for finding quasars, or indeed, for finding other classes of objects with appropriately distinctive spectra:

- emission lines
- absorption lines
- continuum discontinuities
- ultraviolet excess

- red excess

Of course, emission lines and ultraviolet excess are the most productive options for quasars, and the others are intended for the rarer types such as BAL and high redshift quasars. The options of emission lines and ultraviolet excess only were used for field 119. Furthermore, galaxy-like images were downgraded, thereby effectively eliminating nearby AGNs.

The limiting values of the spectral selection criteria were deliberately set quite low. Consequently many candidates were selected, most of which will not be quasars. The purpose here is to make all potential candidates available so that the grading of candidates and re-selections can then be performed on this much amller database according to the needs of particular applications. The standard grading process has been used for field 119. In this, star-like images, ultraviolet excess, and emission lines all make positive contributions to the grade. Features that atre apparently stellar (usually for F stars) make negative contributions. The procedure presently used discriminated against galaxy-like images. Our sample will, therefore, not contain Seyfert galaxies which are potential powerful X-ray sources. In the future, because the aim is to discover AGNs and not only quasars, it will be necessary to devise a slightly different scheme.

In the following, we list some important features of AQD. A more thorough discussion may be found in Clowes (1986):

- The usual magnitude range is $B \sim 17.0 - 20.5$.
- AQD excels at detecting quasars in the range $z \sim 1.8 - 3.0$, but can, in fact, detect quasars at all $z < 3.0$.
- Many quasar candidates are so obvious that while spectroscopy is important for establishing the identifications of lines and for accurate redshifts, it is not strictly necessary for their confirmation.
- AQD is well suited for projects that require large numbers of quasars and/or coverage of large areas of sky.
- The photographic requirements are easily satisfied. A minimum of one objective-prism plate and a sky-survey direct plate are required.
- All objects have celestial coordinates accurate to ~ 1 arcsec.
- The CPU requirements are comparatively small on a small VAX (usually ~ 40 hours on a VAX 11/780 for one UK Schmidt field).
- Losses from overlapping spectra may be made negligible by also processing a second prism plate with the dispersion direction rotated by 90° relative to the first.
- The maximum area which can be measured by COSMOS in a single measurement is $286.7 \times 286.7\,\mathrm{mm}^2$, which for a UKSTU plate is ~ 28.6 square degrees.

3 Selection of quasar candidates in fields 86, 119, and 120

Table 1 lists the prism plates presently available. In field 86, a visual search in two sections, covering a total of 5 square degrees, yielded a total of 74 quasar candidates.

Table 1. UKSTU objective prism plates of ESO/SERC fields 86, 119, and 120. All plates were taken on IIIa-J emulsion without filter.

ESO/SERC Field	Plate Number	Status
86	UJ9076P	visual search
119	UJ10739P	COSMOS, AQD
119	UJ11514P	COSMOS
120	UJ10725P	
120	UJ10743P	
120	UJ11484P	COSMOS

The emission-line criterion only was used. *Fig. 1a* shows an example of the direct plate and the prism plate as photographed from the TV screen. The figure illustrates the problems encountered by crowding and by overlap of the prism spectra near the SGP. *Fig. 1b* displays the corresponding slit spectrum.

In field 119, quasar candidates were identified by AQD using a combination of emission lines and ultraviolet excess as selection criteria. As a consequence, the sample may contain more blue stars than a comparable sample based on the presence of strong emission lines only. So far, only one plate was processed completely (UJ 10739P) and no effort was made to resolve overlaps which affect about 40 % of all spectra in field 119. This will be possible in the future by using plate UJ 11514P taken with the prism rotated by 90°. In the present form, the quasar search is therefore definitively incomplete. Furthermore, as noted above, it cannot presently qualify as an AGN search because we discriminated against galaxy-like images.

In this first step of the program, a total of 64 new quasar candidates with blue magnitudes between 16.4 and 19.9 were identified in field 119. This field contains two previously known quasars, PKS 0506-61 and PKS 0522-611 with $V = 16.9$ and $V = 18.1$, respectively, which were both rediscovered in the survey.

4 Slit spectroscopy

Slit spectroscopy of a number of the selected candidates from field 119 was performed on 20–26 October 1987 using the ESO/MPI 2.2 m telescope at La Silla, Chile, equipped with the B&C spectrograph and a thinned, backside illuminated CCD chip (RCA 501 EX) with 30 μm pixel size as detector. All spectra were taken at low spectral resolution with ESO grating number 13 (450 Å/mm), covering the wavelength range from 3380 Å to 10220 Å (FWHM resolution 1.5 pixels or 20 Å). The slit width was 2 arcsec and the seeing on 20–24 October was better than 1.5 arcsec, in part better than 1 arcsec. Flux calibration was achieved by observing standard stars three times a night (Wolf 1346, BD 17°4708, 40 Eri B).

Of the 64 quasar candidates in field 119 slit spectra were taken of 39. In field 86, lower priority and bad weather left us with only 3 good spectra. Slit spectra of the candidates in field 119 were taken starting with the brighter candidates. Of the total of 42 objects observed spectroscopically, 20 tuned out to be quasars with redshifts between $z = 0.38$ and 2.70, 2 were extragalactic H II regions with $z = 0.050$ and 0.065,

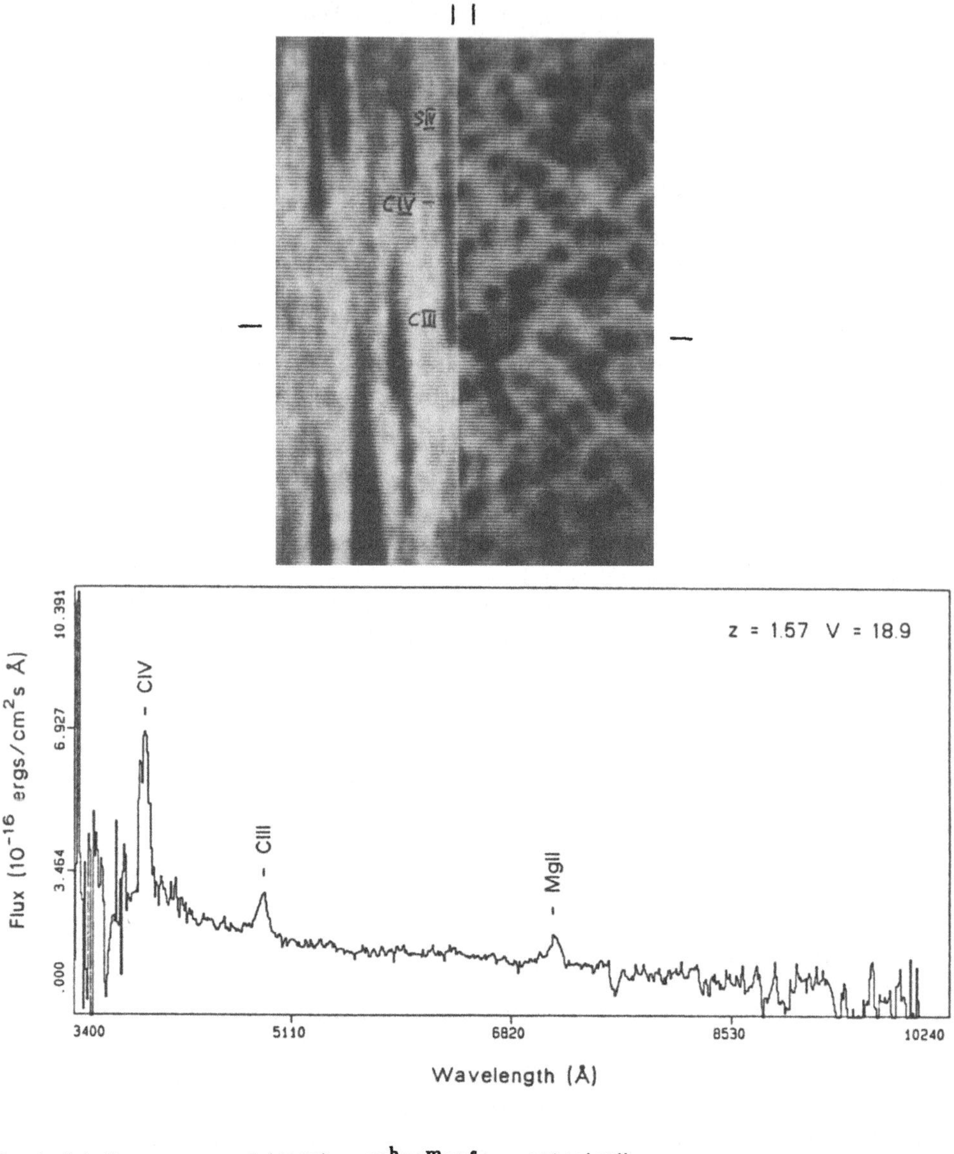

Fig. 1. (a) Quasar at α, δ (1950) = $05^h53^m57^s.9$, $-66°21'48''$ at a separation of 37 arcmin from the South Ecliptical Pole. The direct image (ESO/SERC J-plate) and the prism plate are shown as photographed from the screen of the TV comparator at ROE. The object is identified by tick marks, west is on top, north to the left. The picture illustrates the problems encountered by crowding and overlap of prism spectra near the SEP. (b) slit spectrum of the object yielding $z = 1.57$, V = 18.9.

18 were stars, 2 of uncertain nature. There is no definite brightness limit for these objects but typically they are brighter than $V = 19$.

Fig. 2a shows the distribution of the objects identified in field 119 as a function of apparent visual magnitude, separately for stars, quasars, and extragalactic H II regions. *Fig. 2b* shows the z-distribution of all non-stellar objects. The maximum near $z \simeq 2$ for prism-selected quasars (Hewitt and Burbidge 1987) is also borne out in this study. *Fig. 2c* shows the distribution in absolute V–magnitude. Following Véron-Cetty and Véron (1987), M_v was calculated using $H_0 = 50\,\mathrm{km\,s^{-1}\,Mpc^{-1}}$ and a continuum optical spectral index of F_ν equal to 0.7. *Fig. 2c* demonstrates that all objects identified as extragalactic except for the two nuclear H II regions qualify as quasars, having $M_v < -24.0$. With respect to the ROSAT X-ray survey it is noteworthy that the X-ray emitting AGNs cluster at lower z-values and that spectroscopically a large fraction of them qualify as Seyfert 1 nuclei. As noted in Sect. 2, our present procedure discriminates against these objects. We plan, however, to extend our search to low-z AGNs by using different options in selecting the candidates.

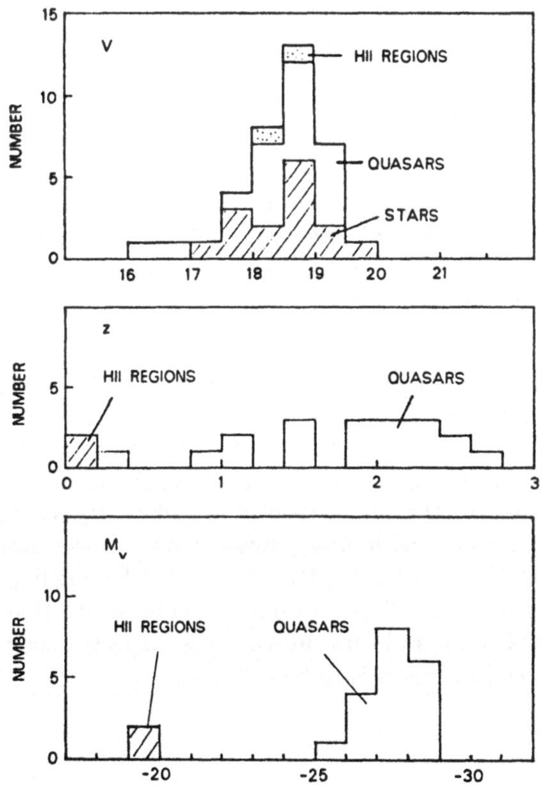

Fig. 2. Brightness- and z-distributions of objects identified by slit spectroscopy in field 119. (a) Distribution in apparent V magnitude, (b) z-distribution of all extragalactic objects, (c) distribution in absolute visual magnitude for all extragalactic objects. $H_0 = 50\,\mathrm{km\,s^{-1}\,Mpc^{-1}}$, $q_0 = 0$, and a nonthermal spectral index of 0.7 were assumed.

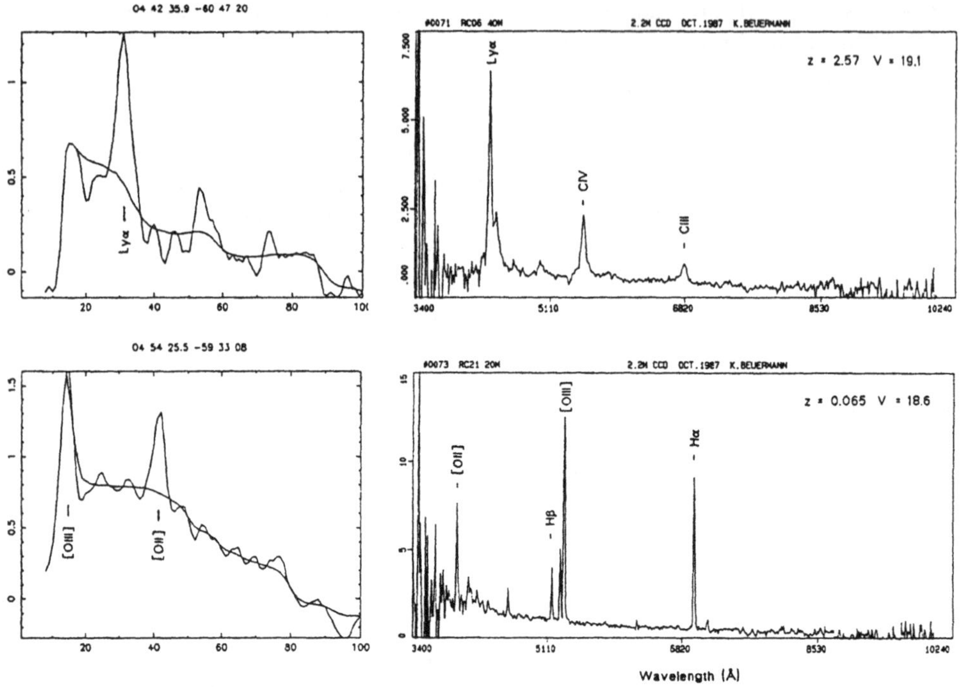

Fig. 3. Comparison of prism and slit spectra for selected objects. In the prism spectra, pixel 10 corresponds to the cutoff of the IIIa-J emulsion at 5380 Å, pixel 39 is at 4009 Å, and pixel 66 at 3397 Å. Ordinates are in relative units for the prism spectra and in 10^{-16} ergs cm^{-2} s^{-1} $Å^{-1}$ for the slit spectra. (a) strong-line quasar with $z = 2.57$, (b) extragalactic H II–region with $z = 0.065$

In *Fig. 3.*, we compare selected prism and slit spectra of objects identified in field 119. Panel (a) shows a strong-line quasar with $z = 2.57$, V = 19.1 , panel (b) an extragalactic nuclear H II region. Such objects are easily identified in prism spectra and do not necessarily require slit spectroscopy for confirmation. Quasars of somewhat lower redshift are easily recognized as being extragalactic because Ly α, Si IV, and C IV may be identified in the prism spectra if sufficiently strong. Problems arise for objects of lower redshift with weak emission lines if the dominant line is near 4000 Å($z \simeq 1.5$ for C IV, $z \simeq 1.0$ for C III, and $z \simeq 0.4$ for Mg II), because these may be mixed up with blue stars. These examples demonstrate that liberal use of the selection criteria is called for and that subsequent slit spectroscopy is necessary in the maximum degree of completeness is to be achieved.

5 Discussion

The present study demonstrates that Automated Quasar Detection (AQD) may successfully be employed in the comparatively crowded star fields near to but outside the LMC. Our results for field 119 are preliminary because plate UJ 10739P was not optimal and the plate taken with the prism rotated by 90° (UJ 11514P) has not yet been fully processed. Contamination by blue stars was not found to be excessive

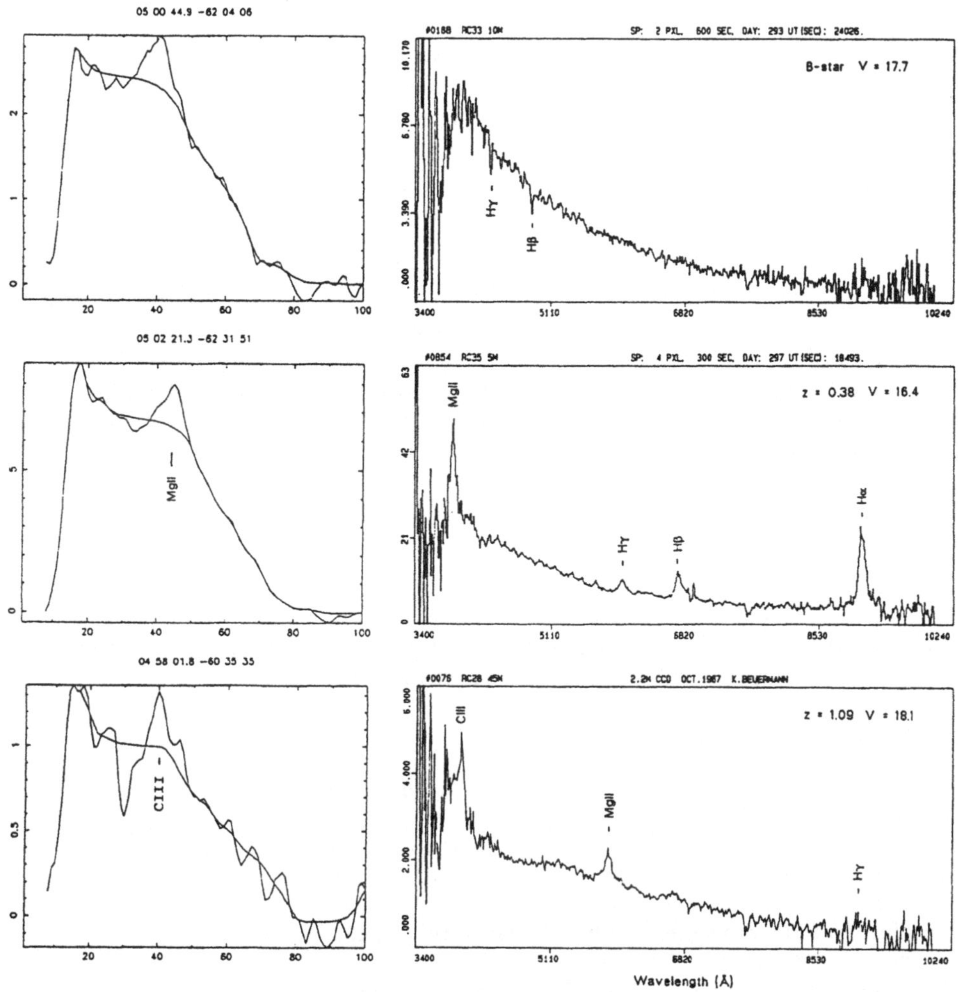

Fig. 3. cont.
(c) B star, (d) quasar with Mg II line near 3900 Å, and (e) quasar with C III line near 4000 Å.

considering the proximity of the LMC and the liberal selection criteria.

The number of identified quasars above an effective brightness limit of about $V = 19$ is 0.7 per degree^2. Considering the fraction of candidates for which slit spectra are not yet available and the loss due to non-recovered overlaps, the corrected rate is approximately 1.5 quasars per degree^2 brighter that $V = 19$. This number is comparable to that found in independent surveys (Clowes and Savage 1983, Barbieri and Cristiani 1986) and further demonstrates that AQD operates successfully as the star densities encountered in field 119.

While appropriate selection criteria can almost entirely discriminate against blue stars, application of such criteria will cause a loss of completeness among weak-line quasars at certain redshifts below 1.8. For 7 out of 20 identified quasars, some other line than Ly α was the dominant line within the sensitive wavelength range of the J–emulsion. We have attempted to keep the success rate at $z < 1.8$ reasonably high by not systematically discriminating against blue objects with a possible line in the prism spectrum near 4000 Å. Comparison of the prism and slit spectra shows that our criteria are reasonable because on the one hand about 1/3 of all identified quasars have $z < 1.8$ while on the other hand the load on observing time by including blue stars is still tolerable (40 % of all spectra taken).

Continuation of the present program will provide a sample of quasars and AGNs near the SEP which can be further studied prior to the launch of ROSAT and provide an important data base for comparison with the sample of X-ray emitting extragalactic objects. The brighter optical quasars may also provide information on the interstellar medium in the fringes of the LMC. The survey plates will ultimately also allow stellar X-ray emitting objects in the outer parts of the LMC to be identified and their optical properties to be studied.

Acknowledgments

We thank the staffs of the UK Schmidt telescope and the COSMOS group for their support. KB thanks the staff at ROE for help in using the TV comparator and all staff at ESO, La Silla, for their support at the ESO/MPI 2.2 m telescope during MPI time. Data processing used, in part, the UK starlink facilities.

References

Barbieri, C., Cristiani, S., 1986. *Astr. Astrophys. Suppl.*, **63**, 1.
Clowes, R.G., 1986. *Mitt. Astr. Ges.*, **67**, 174.
Clowes, R.G., Savage, A., 1983. *Mon. Not. R. astr. Soc.*, **204**, 365.
Clowes, R.G., Cooke, J.A., Beard, S.M., 1984. *Mon. Not. R. astr. Soc.*, **207**, 99.
Hewitt, A., Burbidge, G., 1987. *Astrophys. J. Suppl.*, **63**, 1.
MacGillivray, H.T., Stobie, R.S., 1984. *Vistas Astr.*, **27**, 433.
Trümper, J., 1984. *Physica Scripta*, T7, 209.
Véron-Cetty, M.-P., Véron, P., 1987. *ESO Scientific Report No. 5*.

The Use of Quasars for the Construction of a Future Extragalactic Inertial Reference Frame

C. de Vegt
Hamburger Sternwarte
Hamburg-Bergedorf, F.R.Germany

Abstract

Quasars will provide optimal candidates for the construction of a future inertial extragalactic reference frame in astronomy. The selection of suitable candidate objects depends on their individual optical and/or radio properties. Optical and Radio quasar survey projects will provide a sound basis to achieve this goal.

Most aspects of this astrometric key-project have been discussed extensively in the recent literature. Instead of duplicating this effect here, a small selection of references is given which covers the main topics of this exciting field of research.

1 General discussion of reference frames

Hughes, J. (ed.), 1986. Reference Frames. In: *Highlights of Astronomy*, **7**, ed. Swings, J.P., p. 51.

de Vegt, C., 1986. The extragalactic optical/radio reference frame. A progress report, IAU Comm. 24 Working Group. In *IAU Symp. 109, Astrometric Techniques*, eds. Eichhorn, H.K., Leacock, R.J., p. 173.

de Vegt, C., 1984. Interrelation of present optical and radio reference frames. In *IAU Symp. 110, VLBI and Compact Radio Sources*, eds. Fanti, R., Kellermann, K., Setti, G., p. 357.

de Vegt, C., Johnston, K.J., 1988. Reference frames in astrometry. In *IAU Symp. 128, The Earth's Rotation and Reference Frames for Geodesy and Geodynamics*, eds. Babcock, A.K., Wilkins, G.A., p. 25.

2 Catalogs of candidate sources and general observation programs

Argue, A.N. *et al.*, 1984. A catalogue of selected compact radio sources for the construction of an extragalactic radio/optical reference frame. *Astr. Astrophys.*, **130**, 191.

de Vegt, C., Johnston, K.J., Russell, J., Hughes, J.A., Jauncey, D., Nicholson, G., 1988. Toward the establishment of a radiooptical reference frame. In *Proc. HIPPARCOS INCA Conference, Sitges, Jan. 1988*, in press.

Johnston, K.J., Florkowski, D., de Vegt, C., 1988. Precise VLA astrometry of HIPPARCOS radio link stars. In *Proc. HIPPARCOS INCA Conference, Sitges, Jan. 1988*, in press.

de Vegt, C., Hughes, J.A., Jauncey, D., Johnston, K.J., Nicholson, G., 1988. An observational program for a unified optical radio reference frame. In *IAU Coll. 100, Fundamentals of Astrometry, Cel. Mechanics*, eds. Eichhorn, H.K., Pakvor, I., in press.

3 Results of radio and optical observation programs

de Vegt, C., Gehlich, U.K., 1982. Precise optical positions of radio sources in the FK4-system

II. Results from 28 sources on the northern hemisphere and a preliminary comparison of the optical-radio reference frame. *Astr. Astrophys.*, **113**, 213.
de Vegt, C., 1982. Radio sources for precise astrometry. *Abh. Hamb. Sternw.*, **10**, *Heft 3*, 119.
de Vegt, C., Florkowski, D., Johnston, K.J., Wade, C.M., 1985. Stellar radio astrometry I, II, III. *Astr. J.*, **90**, 2381, 2387, 2390.
de Vegt, C., Schramm, J., Johnston, K.J., 1987. Precise optical and radio astrometry of the quasar 1928+738. *Astr. J.*, **93**, 261.

Part VI

Evolution on Large Scales

Edwin Loh

Vincent Icke

Dietmar Hömberg

Michael Hawkins

Voronoi Foam as a Model of the Medium-Scale Universe

Vincent Icke
Sterrewacht Leiden
Leiden, The Netherlands

Abstract

The large-scale morphology of the high-density baryonic material in the universe, consisting of "clusters" in the form of pancakes, filaments, and nodes is obtained when matter streams away from a distribution of low-density expansion centres ("nuclei") and collects in the interstices of a close packing of spheres.

This naturally leads to a partitioning of space generated by a process known as *Voronoi tessellation.* We have studied the statistical properties of specific instances of these tessellations, which we call Voronoi foams, for several model distributions of expansion centres. We derive the statistical properties of the regions where baryonic matter accumulates, and of the voids between the galaxies. These can be compared with observations, leading to indirect constraints on the initial spectrum of the density perturbations that produced the matter distribution we observe today.

The appearance of a Voronoi foam closely resembles the mass distribution found in numerical hydrodynamic experiments, and the bubble structure observed in the galaxy distribution.

1 Introduction

Once upon a time, in a galaxy far, far away, a famous astronomer wrote a paper called *Cosmology, a search for two numbers.* That, it seems to me, was quite naïve, but perhaps this volume will properly display the efforts of a generation of cosmologists who are rather more ambitious. My ambition here is to describe a model for the medium-scale Universe.

In choosing the middle ground, I am motivated by the following considerations. First, the large scales (on the order of the size of the particle horizon) are very well understood theoretically, in the framework of the Friedmann–Robertson–Walker solutions of the Einstein equations. However, the observations of that regime are very uncertain. Galaxies beyond a redshift of 2 are practically unheard of, and quasars do not, at present, extend beyond $z = 4.5$ or so. The only observable is the microwave background, and there are some deep puzzles associated with that; in particular, there is the baffling discrepancy between its isotropy and the graininess of the visible baryon distribution. Second, the small scales (on the order of small groups of galaxies) are very well charted, but the theory here is in a dismal state: nobody really knows how galaxies form, and what detailed processes produce their present properties.

Thus, it seems prudent to first consider the medium-scale mass distribution. For this presentation, the corresponding length scale is roughly the harmonic mean of the scales mentioned above, say between 50 and 500 Mpc. On larger scales, it is believed that the Universe is asymptotically of FRW type; on smaller scales, in particular below 5 Mpc, dissipative effects are expected to dominate (for example photon trapping during protogalaxy collapse, or tides or mergers later on). In the medium-scale regime, we have a good theoretical tool (Newtonian gravity), and we are beginning to get good observations, as shown elsewhere in this volume.

2 Some history

Said observations date back to the work of Shapley and Ames (1932), whose catalogue of bright stellar systems showed the dramatic excess of galaxies in some regions of the sky, notably the Virgo Cluster (see Oort 1983). Even in those early days, it was amply evident that the distribution of luminous matter on intermediate length scales (up to about 20 Mpc, in the Shapley–Ames work) is distinctly non-Poissonian.

But until about 1970, this remarkable fact was hardly a subject of study, even though the catalogue of galaxy positions and redshifts compiled by Humason *et al.* (1956) contained a wealth of evidence about the curious structure of the nearby Universe (it is interesting to note that appreciable improvements on these data were not published until the mid-1980's, thirty years later). The clustering first studied by Abell (1958), and most apparent in the galaxy counts by Shane and Wirtanen (1967), was thought to be analogous to that which is observed in open clusters and associations of galactic stars. With hindsight, we see that only the high-density Abell clusters must be regarded as dynamically distinct entities, as is evident in the "thermal broadening" of their distribution in velocity space (*e.g.* Giovanelli and Haynes 1982).

Interpretation of the non-Poissonian galaxy distribution took shape only slowly. Totsuji and Kihara (1969) and Peebles (1980, and references therein) studied the small-scale deviations by means of the two-point correlation; Oort (1970) emphasised the importance of larger structures because these, due to their long evolutionary timescales, can be expected to give information about the earliest times of structure formation. In keeping with this, Icke (1972, 1973) hypothesised that such structures form before galaxies, and tried to find evidence for coherent objects on a scale of 15–20 Mpc by studying the distribution of galaxies in position-velocity space, a technique that has since become known as "slicing".

Icke's efforts to corroborate his hypothesis were largely unsuccessful, due to a conspicuous lack of good observations (the data by Humason *et al.* (1958) were the only show in town, in 1970!) It was not until the work by Kirshner *et al.* (1981) that it became clear just how real, and in particular how large, these intermediate structures can be. Finally, Einasto *et al.* (1980) proposed that the voids glimpsed by Kirshner *et al.* are a real and typical element of the medium-scale structure of our Universe. Since then, most authors have regarded this structure as a void-and-filament or sponge-like arrangement.

3 Theory

On large scales, the formation of structure must consider perturbations of a FRW-type Universe (Lifschitz 1946), which is difficult and not securely linked to observations. On small scales, galaxy formation is dominated by dissipative processes (Jones 1976, Efstathiou and Silk 1983) and is theoretically quite intractable. But on intermediate scales, Newtonian gravitational instability should suffice for describing the formation of structure (Jeans 1902). In fact, pressure effects are unlikely to be important in the progenitors of structures in the 50–500 Mpc regime, so we can restrict ourselves to "dust" collapse.

The potential Φ near any point (x, y, z) of a selfgravitating medium can be written as

$$\Phi = \sum_{ijk} a_{ijk}\, x^i y^j z^k \,. \tag{1}$$

Near a density maximum, the leading terms are the quadratic ones, which, by a suitable orientation of Cartesian coordinates, can be written as

$$\Phi = Ax^2 + By^2 + Cz^2 + \cdots \,. \tag{2}$$

Neglecting terms of higher than second order, this is the potential of a homogeneous ellipsoid. That should be no surprise: the smallest closed contours in any topographical map are ellipses!

The collapse of high-density regions can thus be approximated by considering the motion of homogeneous ellipsoids. Suppose that a particle of such a body were initially located at (a, b, c), and that at some later time t it were at $(aX(t), bY(t), cZ(t))$, then the density ρ would evolve according to

$$\rho(t) = \rho_0 / XYZ \,. \tag{3}$$

The equations of motion for the scaling functions X, Y, and Z are found as follows (Lin *et al.* 1965, Icke 1972). The potential Φ obeys

$$\begin{aligned}\Phi &= k(\alpha x^2 + \beta y^2 + \gamma z^2) = \\ &= k(\alpha a^2 X^2 + \beta b^2 Y^2 + \gamma c^2 Z^2) \,,\end{aligned} \tag{4}$$

and Poisson's Equation prescribes that

$$k(\alpha + \beta + \gamma) = 2\pi G \rho \,. \tag{5}$$

The components of the gravitational force are $-\partial\Phi/\partial x = -\frac{1}{X}\partial\Phi/\partial a$ *et cycl.*, so that the equations of motion become

$$-\frac{1}{X}\frac{d^2X}{dt^2} = 2\pi G \rho \alpha \,, \tag{6}$$

$$-\frac{1}{Y}\frac{d^2Y}{dt^2} = 2\pi G \rho \beta \,, \tag{7}$$

$$-\frac{1}{Z}\frac{d^2Z}{dt^2} = 2\pi G \rho \gamma \,, \tag{8}$$

where the functions α, β, and γ are defined as

$$\alpha = abc \int_0^\infty \frac{ds}{(a^2+s)\Delta} \quad et\ cycl.\ , \tag{9}$$

$$\Delta^2 \equiv (a^2+s)(b^2+s)(c^2+s)\ , \tag{10}$$

(*cf.* Chandrasekhar 1969, Ch. 3); here a, b, and c are identified with the axes of the ellipsoid.

Now comes a crucial observation, first made by Lynden-Bell (1964): without loss of generality, we can order the axes according to $a > b > c$, in which case $\alpha < \beta < \gamma$, so that Eqns. (6–8) give

$$-\frac{1}{X}\frac{d^2X}{dt^2} < -\frac{1}{Y}\frac{d^2Y}{dt^2} < -\frac{1}{Z}\frac{d^2Z}{dt^2}\ . \tag{11}$$

Consequently, the axial ratios $a : b : c$ always increase with time, and *slight initial asphericities are amplified during the collapse.* Note also that, for a quadratic potential and a homologous contraction as described, the velocities inside the ellipsoid are linear functions of position: *the collapse produces a Hubble-type velocity field.*

The secular increase of aspherical perturbations provides an explanation for the filamentary appearance of the medium-scale structure, but the above approximation must break down as soon as the filaments grow to nonlinear proportions. However, there is a better way to view the development of structure in a selfgravitating pressure-free medium, namely by "turning the Universe inside out" and to consider the evolution of the *low*-density regions. These are the progenitors of the observed voids. The arguments presented so far can still be applied, except that the sense of the final effect is reversed: because a void is effectively a region of negative density in a uniform background, the voids expand as the overdense regions collapse, while *slight asphericities decrease as the voids become larger* ("Bubble Theorem", Icke 1984). Moreover, the density in the voids becomes smaller in the course of time, so that the linear approximation will remain good for a longer period, except, of course, near the outer parts of the voids, where the matter gets swept up.

Just as in the case of growing filaments, the velocity field in the voids is proportional to the distance inside them: voids are thus expected to be "superhubble bubbles". This has non-trivial observational consequences, because the linearity makes it impossible to separate local homologous void expansion from true cosmic Hubble motion. Hence, considerable deviations from the mean Hubble flow may go undetected until the observations encompass a scale larger than that of the voids. When such scales are reached, the void expansion will appear in the form of large scale streaming motions. The corresponding velocities are evidently on the order of

$$v \approx \epsilon D H_0\ , \tag{12}$$

where D is the size of the void and ϵ is the local excess of the expansion over the mean Hubble motion. The latter is not known in detail, but might be about 0.1–0.2

(Centrella and Melott 1983). If D is about 50 Mpc (De Lapparent *et al.* 1986), we find that $v \approx 500\,\mathrm{km\,s^{-1}}$, taking $H_0 = 75\ \mathrm{km\,s^{-1}Mpc^{-1}}$.

According to the above, we may think of the structure of the medium-scale Universe as a close packing of spheres of different sizes, out of which matter flows in a slightly super-Hubble expansion towards the interstices of the spheres. *The importance of the Bubble Theorem is, that it provides a specific physical mechanism for producing the non-Poissonian matter distribution in the medium-scale Universe.*

4 Voronoi foam

We have now found at least a partial answer to the question: if the distribution of galaxies is not Poissonian, then what is it? For, continuing the above argument, we can construct the "skeleton" of the mass distribution by considering the locus of points towards which the matter streams out of the voids. Suppose that some cosmic process produces a collection of regions where the density is slightly less than average (the origin of the requisite fluctuations is a very important unsolved problem). As we have seen, these regions are the seeds of the voids, because underdense patches become expansion centres, from which matter flows away until it encounters similar material flowing out of an adjacent void. If the excess Hubble parameter is the same in all voids, the matter must collect on planes that perpendicularly bisect the axes connecting the expansion centres (otherwise, the matter collects on hyperboloids which are cylindrical about these axes).

For any given set of expansion centres, or *nuclei*, the arrangement of these planes define a unique process for the partitioning of space, a *Voronoi tessellation* (Voronoi 1908). The planes generate walls that enclose polyhedral cells, which are observed as voids. A particular realisation of this process (*i.e.* a specific subdivision of N-space according to the Voronoi tessellation) may be called a *Voronoi foam* (Icke and Van de Weygaert 1987). In 3-space, such a foam is built out of three geometrically distinct elements: polyhedron walls (*pancakes*; *cf.* Zel'dovich 1970), *filaments* where three walls intersect, and *nodes* where four filaments come together. In this picture, the filaments are identified with the elongated "super"clusters (Icke 1972, Oort 1983), and the nodes correspond to the virialised Abell clusters (Giovanelli and Haynes 1982; Giovanelli *et al.* 1986).

5 Observations

The Bubble Theorem shows that a specific physical mechanism (pressure-free self-gravitational collapse) generates a specific statistical process (Voronoi tessellation). Is there observational evidence for or against this view? Theoretical study of the Voronoi process (Icke and Van de Weygaert 1987) shows that the *variances* of several observable geometrical quantities, notably the angles between the filaments, are good indicators of the statistical properties of the underlying distribution. But observations do not yet encompass enough structure to allow a definite assessment of the Voronoi model. However, slices of the Universe do strongly suggest that Voronoi foam is a sensible framework (Giovanelli *et al.* 1986, De Lapparent *et al.* 1986).

One possible way to sample the Universe over large distances is, to observe correlations in quasar absorption lines (*cf.* Weymann *et al.* 1981). Assuming that each intersection with a Voronoi wall produces an absorption line, we can calculate the mean number

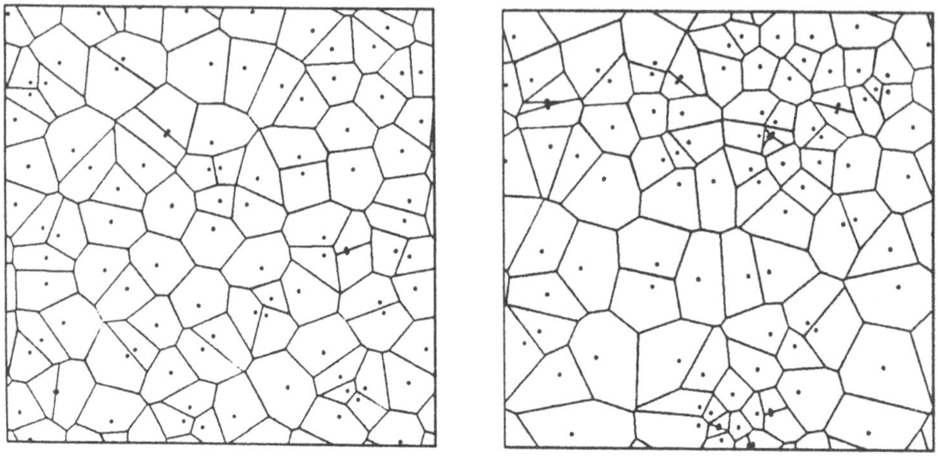

Fig. 1. Two examples of two-dimensional Voronoi foams, for different amounts of correlation between the nuclei, which are indicated by dots.

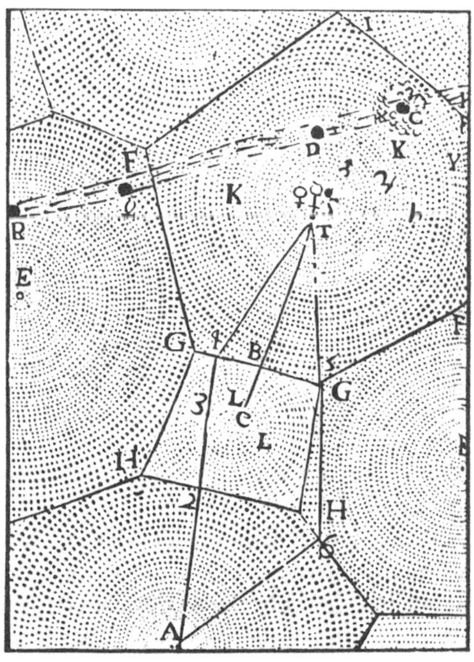

Fig. 2. Illustrating the long history of the idea of cosmic fragmentation: the drawing indicates the disposition of matter in the Solar System and its environs (Descartes 1664).

N of absorption lines per unit interval of redshift z as follows. Let a Voronoi foam be attached to a comoving coordinate system; let the radial comoving coordinate be r. Then

$$dN \propto r^2\, dr\ , \tag{13}$$

$$r = R_h \left(1 - \frac{1}{\sqrt{1+z}}\right) \quad \text{(Einstein – De Sitter)}\ , \tag{14}$$

where R_h is the horizon radius. Differentiation of r with respect to z then gives

$$dN/dz \propto (1+z)^{-3/2}\ . \tag{15}$$

This relationship is different from that which is expected for absorption lines caused by galaxies, because these have a *fixed* interception cross section σ, so that we must use rR instead of r, where R is the cosmic scale factor:

$$dN \propto r^2\, dr \propto (rR)^2 R^{-2}\, dr = \sigma R^{-2}\, dr\ . \tag{16}$$

Because $R \propto 1/(1+z)$, we obtain

$$dN/dz \propto \sqrt{1+z}\ . \tag{17}$$

The region where $d\log N/d\log(1+z)$ changes slope from $1/2$ to $-3/2$ corresponds to the epoch of galaxy formation; however, this point is probably out of our present range (beyond $z \approx 5$ or so). Moreover, the observed slope seems to be closer to 2, at least for absorption lines due to Ly α (Weymann *et al.* 1981), probably due to the evolution of the absorbing hydrogen clouds. Thus, dN/dz is unlikely to provide direct information about the disposition of Voronoi cells.

An alternative approach is, to calculate the two-point correlation function of absorption lines. But extensive Monte-Carlo simulations show that this method is quite insensitive to the distribution of the expansion centres (Van der Valk *et al.* 1988).

Fortunately, the same work reveals that correlations between the absorption spectra of quasar pairs with comparable redshifts are a promising probe of void-and-filament structure.

6 Conclusions

The structure of our Universe on a scale of about 50–500 Mpc can be plausibly modeled by means of pressure-free Newtonian gravitational collapse. This mechanism produces a mass distribution which, asymptotically, is described by a statistical process known as Voronoi tessellation. Visual comparison between Voronoi foams and the observed distribution of galaxies shows a promising similarity, but a detailed statistical assessment of this appearance has not yet been made.

Acknowledgements

It is a pleasure to thank Rien van de Weygaert and Marc van der Valk for their enthousiastic collaboration on this subject. I am especially indebted to J.H. Oort for his criticism and encouragement.

References

Abell, G.O., 1958. *Astrophys. J. Suppl.*,**3**, 211.
Centrella, J.M., Melott, A.L., 1983. *Nature*, **305**, 196.
Chandrasekhar, S., 1969. *Ellipsoidal Figures of Equilibrium*, Yale Univ. Press, New Haven.
De Lapparent, V., Geller, M.J., Huchra, J.P., 1986. *Astrophys. J.*, **302**, L1.
Descartes, R., 1664. *Le Monde, ou Traité de la Lumière*, Paris (posthumous publication of work dating to about 1620).
Efstathiou, G., Silk, J., 1983. *Fundam. Cosm. Phys.*, **9**, 1.
Einasto, J., Jôeveer, M., Saar, E., 1980. *Mon. Not. R. astr. Soc.*, bf 193, 353.
Giovanelli, R., Haynes, M.P., 1982. *Astr. J.*, **87**, 1355.
Giovanelli, R., Haynes, M.P., Chincarini, G.L., 1986. *Astrophys. J.*, **300**, 77.
Humason, M.L., Mayall, N.U., Sandage, A.R., 1956. *Astr. J.*, **61**, 97.
Icke, V., 1972. *Formation of Galaxies Inside Clusters*, Ph. D. Thesis, Leiden.
Icke, V., 1973. *Astr. Astrophys.*, **27**, 1.
Icke, V., 1984. *Mon. Not. R. astr. Soc.*, **206**, 1p.
Icke, V., Van de Weygaert, R., 1987. *Astr. Astrophys.*, **184**, 16.
Jeans, J., 1902. *Phil. Trans. Roy. Soc.*, **199A**, 49.
Jones, B.J.T., 1976. *Revs. Mod. Phys.*, **48**, 107.
Kirshner, R.P., Oemler, A., Schechter, P.L., Shectman, S.A., 1981. *Astrophys. J.*, **248**, L57.
Lifschitz, E.M., 1946. *J. Phys. U.S.S.R.*, **10**, 116.
Lin, C.C., Mestel, L., Shu, F.H., 1965. *Astrophys. J.*, **142**, 1431.
Lynden-Bell, D., 1964. *Astrophys. J.*, **139**, 1195.
Oort, J.H., 1970. *Astr. Astrophys.*, **7**, 381.
Oort, J.H., 1983. *Ann. Rev. Astr. Astrophys.*, **21**, 373.
Peebles, P.J.E., 1980. *The Large-Scale Structure of the Universe*, Princeton Univ. Press, Princeton.
Shane, C.D., Wirtanen, C.A., 1967. *Publ. Lick Obs.*, **XXII**, Part 1.
Shapley, H., Ames, A., 1932. *Ann. Harvard Obs.*, **88**, 43.
Totsuji, H., Kihara, T., 1969. *Publ. Astr. Soc. Japan*, **21**, 221.
Van der Valk, M., van de Weygaert, R., Icke, V., 1988. *In preparation.*
Voronoi, G., 1908. *J. reine angew. Math.*, **134**, 198.
Weymann, R.J., Carswell, R.F., Smith, M.G., 1981. *Ann. Rev. Astr. Astrophys.*, **19**, 41.
Zel'dovich, Ya.B., 1970. *Astr. Astrophys*, **5**, 84.

Prospects for Measuring the Evolution of the Luminosity Function and the Angular Correlation Function

Edwin D. Loh
Department of Physics and Astronomy
Michigan State University
East Lansing, MI, USA

Abstract

The angular correlation function has been measured at redshift 0.5 to 30 % precision and the characteristic luminosity to 20 % precision with a catalog of photometric redshifts of 1000 field galaxies. In the near future, the number of galaxies can be increased a hundredfold. Then 7 % precision in the characteristic luminosity for each of ten color classes and 5 % precision in the correlation function are attainable. If a nearby sample of 100 000 galaxy redshifts becomes available and if the same selection criteria are used for both samples, then one can measure the evolution of these statistical properties of galaxies over a period of a third the Hubble time to high accuracy.

1 Introduction

In 1981 we began a project to obtain a complete, magnitude-limited sample of galaxies with redshifts and magnitudes. The aims were to measure the volume element as a function of redshift and so the cosmology and to study evolution in a statistical fashion. We built a camera with a charge-coupled detector for the $f/2$ prime focus of the Wyoming 2.3 m telescope, and collected a sample of 1000 galaxies in 1983. The median redshift of the galaxies is 0.5. The cosmological density parameter Ω measured to a precision of 30 % (Loh and Spillar 1986b, Loh 1988b), the characteristic luminosity of galaxies at $z = 0.5$ (Spillar and Loh 1988), and the galaxy correlation function at $z = 0.5$ (Loh 1988a, Loh and Spillar 1988) — all these measurements have come from this rather small sample.

The aim of this paper is to discuss a planned experiment to increase the number of galaxies by a hundredfold. Both for studying cosmology and evolution of galaxies, one must compare a distant sample with a nearby one. The projects at Cambridge, Edinburgh/Durham and Münster, to collect large numbers of galaxies at $z \leq 0.2$, which were discussed at this meeting, are crucial for this project. Therefore I discuss in particular what is required of the low-redshift surveys.

2 Photometric redshifts

The technical advance that enables this experiment is the photometric method (Loh and Spillar 1986a) for measuring redshifts, which Baum (1962) invented 30 years ago. In approximate terms one determines the redshift of every galaxy in a field by finding the wavelength of the 400 nm break with six broad-band filters. More precisely, every object with a greater than minimum flux is photometered. The data for each object are compared with the colors, which are computed from spectra, of 100 types of stars and 11 types of galaxies at various redshifts. The object is identified as the star or the galaxy represented by the fiducial object that matches best. The 400 nm break (and therefore the redshift) is apparent even in the calibrated picture, as one can see for the cluster 0024+1654 at $z = 0.4$ as shown in *Fig. 1*. In 3 hours on the Wyoming 2.3 m telescope, we obtain the redshifts of 200 galaxies, for which the median redshift is 0.5 and the median distance is 0.4 of the Hubble distance. Spectroscopy yields redshifts with higher accuracy, but photometry is faster.

This technique for measuring redshifts has been tested (Loh and Spillar 1986a) by comparing the photometric redshifts of the cluster 0024+1654 at $z = 0.4$ with the spectroscopic redshifts of Dressler *et al.* (1985). Since the goal of the spectroscopy was to study the Butcher-Oemler effect, half of the spectroscopic redshifts were of blue galaxies. For 30 of 34 galaxies, the photometric and spectroscopic redshifts agree. (For the others, two photometric redshifts are probably wrong, and two spectroscopic redshifts are probably wrong.) From this test, one draws these conclusions: (a) The mean of $x = (z_{\mathrm{photo}} - z_{\mathrm{spect}})/\sigma_{\mathrm{photo}}$ is consistent with zero for both the red and the blue galaxies; *i.e.*, the photometric redshifts are not biased. (b) The mean of x^2 is consistent with 1; *i.e.*, the errors σ_{photo} in the photometric redshifts are estimated accurately. (c) Stars and galaxies are separated correctly.

3 The new experiment

A new experiment is planned to increase the detector area by a factor of 20. This, combined with a longer time at the telescope, enables a new sample of 10^5 galaxies. The new sample and the old are alike in other respects, namely the redshift range and the depth. The key to the new experiment is a new Tektronix charge-coupled detector, which is a square, 5.5 cm on a side, and contains 4×10^6 elements. Other essential items are a Wynne corrector to remove the coma of the mirror over a large field and an array processor to process the data. The new camera on the Wyoming telescope will have a 35 arcmin field.

4 Luminosity evolution

Using the correlation between color and the ratio of current to ancient star formation, Tinsley (1980) and Bruzual and Kron (1980) computed the evolution of the luminosity of galaxies. In these models the galaxies are $-0\overset{\mathrm{m}}{.}1$ to $1\overset{\mathrm{m}}{.}0$ brighter at $z = 0.5$ than at the present, the redder galaxies showing larger evolution. These must be considered first-order models, since one cannot account for the light in the U, B, and V bands of even the simplest galaxies, the ellipicals (Gunn *et al.* 1981).

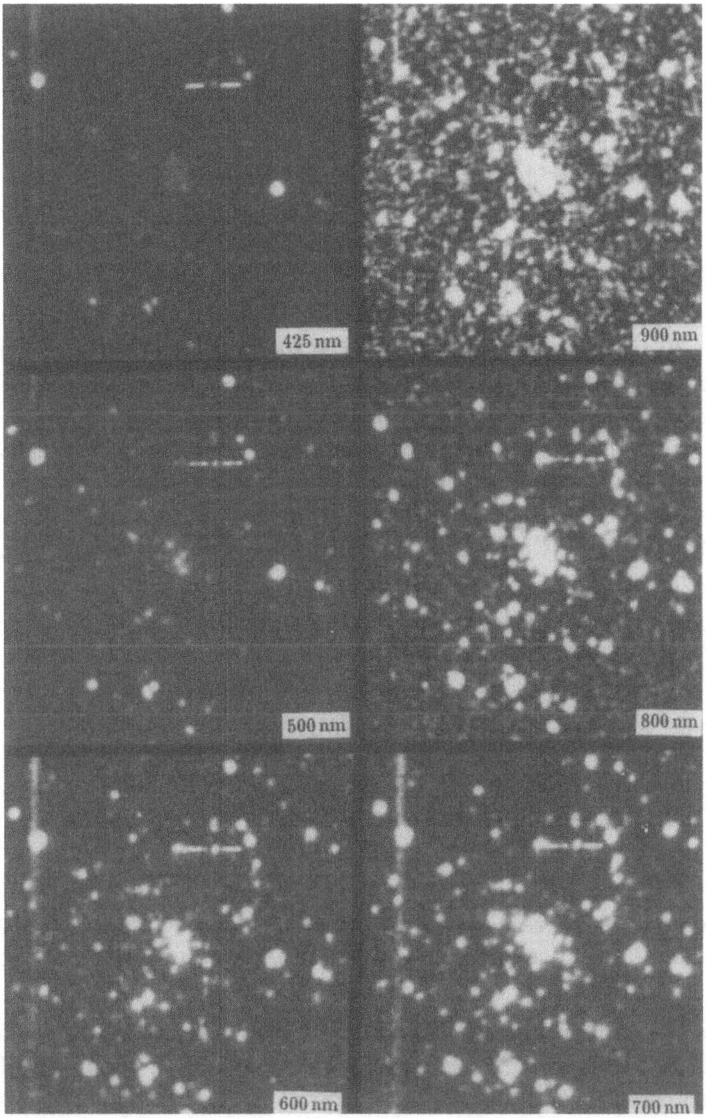

Fig. 1. The cluster 0024+1654 at $z = 0.4$ shown at 425 nm (upper left), 500 nm, 600 nm, 700 nm, 800 nm, and 900 nm (in counter clockwise sequence). The pictures are scaled so that an object with the same flux in ergs $\sec^{-1}$ cm^{-2} Hz^{-1} appears equally bright in all six pictures. The redshift of the cluster is readily apparent – the objects are much brighter in the 600 nm picture than in the 500 nm picture, and they brighten to a lesser extent at longer wavelengths, since the 400 nm break at $z = 0.4$ appears at 560 nm. A break is clear even for a blue galaxy (marked by a dash), which has colors of an Im galaxy. Each picture is a 3.3 arcmin square; north is to the right and east is down. The exposure time is 5 min for each filter on the Wyoming 2.3 m telescope.

Spillar and Loh (1988) find that the red galaxies, those with rest-frame $B - V > 0.7$ (the mean color of Sb galaxies), are brighter by $0^{\mathrm{m}}.7 \pm 0^{\mathrm{m}}.3$ and the blue galaxies are brighter by $0^{\mathrm{m}}.5 \pm 0^{\mathrm{m}}.4$ at $z = 0.5$, assuming an Einstein–deSitter universe. For an empty universe, the evolution is $0^{\mathrm{m}}.3$ greater. For this result, one measures the spatial density of galaxies with luminosity L, which is commonly assumed to have the Schechter form, $\phi^* e^{-x}, x^{\alpha} dx$, where $x = L/L^*$. One assumes α does not evolve and determines the characteristic luminosity L^*. The sample of Kirshner *et al.* (1983) is used to find L^* nearby, and half of the error is due to the nearby sample. This result is consistent with the models.

What can one learn with a hundredfold increase in the number of galaxies? The error in the luminosity scales as $N^{-1/2}$, where N is the number of galaxies. The ultimate error is probably limited by the systematic errors in matching photometric systems, which I take to be $0^{\mathrm{m}}.1$. One can learn more by using the color information. Assume the galaxies are split into 10 classes by color, with approximately the same number in each class. (The angular resolution is insufficient to classify the galaxies in this sample by morphology.) Then the error in measuring the characteristic luminosity of a single color class is $0^{\mathrm{m}}.07$. The fraction of the galaxies that belong to a particular class can be measured to about 3 %.

To discuss any question of evolution, one needs a sample of nearby galaxies with comparable characteristics, and these are: (1) magnitudes that are accurate to 10 %, (2) redshifts of the entire sample or at least of a sufficient subsample, and (3) classification of the galaxies by color. At the least, one must have enough redshifts to disentangle distant, intrinsically bright galaxies from nearby, intrinsically faint ones and to compute the mean shift in spectral band as a function of color class. To measure a 10 % evolutionary shift in the populations of 10 color classes requires enough redshifts to classify objects that make up only 1 % of the entire sample.

The J band of the nearby surveys, the median redshift of which is 0.15 (Schuecker 1987), measures the same rest-frame band as the sensitive 700 nm and 800 nm bands of the distant survey. Therefore, one need not invoke spectral models to compare the samples.

The shape of the luminosity function can be measured at $z = 0.5$ to 4^{m} fainter than L^*. Phillipps and Shanks (1987) measured the shape of the luminosity function of field galaxies at $z \ll 1$ with an ingenious scheme. The excess number of galaxies as a function of magnitude around a small sample of galaxies with known redshifts is determined from a larger and deeper sample without redshifts. The same method was developed and is used by Schuecker *et al.* (1988). The excess number is a direct and inexpensive (because it requires few redshifts) measurement of the luminosity function. This same idea can be used with the proposed experiment, if the redshift data are augmented with deeper pictures at 700 nm. The deeper pictures are easy to obtain compared with the redshift data because only one band, rather than six, is required and the 700 nm band is relatively efficient.

Having measured the joint evolution of luminosity and population frequency of the color classes, one can answer these questions in a statistically way. Are the first-order models confirmed? Are they deficient in certain color classes? If there is a rapid depletion of the gas in spiral galaxies (Larson *et al.* 1980), then spiral galaxies evolve from blue to red classes. Is this shift observed?

5 Correlation function

The form of the spatial correlation function is

$$\xi(r_p) = (r_0/r_p)^{\gamma}(1+z)^{-3}$$

for $r_p < r_0$, where r_p is the proper distance, r_0 is a parameter, and $\gamma = 1.8$ (Peebles 1980). At $r_p \geq r_0$, ξ appears to fall faster than the power law (Groth and Peebles 1977); this is known as the 'break'.

Models of the evolution of the clustering have been constructed. In the BBGKY solution (Davis *et al.* 1977), small virialized groups, which dominate ξ for small r_p, neither contract nor expand, so that r_0 is a constant. Clustering does grow in the regions where $\xi \approx 1$, and the radius of the break changes as $(1+z)^{-1.67}$. In models in which galaxies have extended halos, dynamical friction changes the shape of the correlation function (Tremaine 1987).

The angular correlation function at $z = 0.5$ has been measured to 30 % accuracy (Loh 1988a, Loh and Spillar 1988). Extrapolating with the present data, one expects these errors for the angular correlation function $w(\vartheta)$ at $z = 0.5$ with angular bins of width $d\vartheta = \frac{1}{4}\vartheta$: at a projected distance of $100\,h^{-1}$ kpc, the relative error is 0.05. At a projected distance of $2.5\,h^{-1}$ Mpc, where the power law equals one, the relative error is 0.25. Therefore the data are sufficient to measure the amplitude of the angular correlation function for each of 10 color classes to 15 %, the shape of the correlation function, and the projected distance of the break if it exists.

Measurement of the evolution of the correlation function, also as a function of the color class, adds yet another dimension to the study of evolution. The evolution of luminosity and color involve the stellar evolution time scale and the gas evolution time scale, but the evolution of the correlation function is an entirely different process with a different timescale. Whether the two time scales are indeed different is an interesting question.

6 Test of object classifications

The classification of objects by color is likely not to be perfect. K stars and galaxies of all types at $z = 0.3$ may be confused, and M stars may be confused with early galaxies at $z = 0.8$. Furthermore, types of galaxies that are not among the fiducial objects, *e.g.* starburst galaxies and hitherto unknown extragalactic objects, will be misclassified.

The same technique of measuring the excess number around galaxies with known redshifts can be used to probe for misclassified objects in suspect samples. For example, one can find the correlation between galaxies and objects that are classified as K stars to determine the fraction of 'K stars' that are actually galaxies. Another group of objects, those that are classified as intrinsically faint galaxies at low redshift, appear anomolous because they are too numerous. The nature of these objects can be discovered by this technique. Extrapolating the measurement of the correlation function with the Loh-Spillar data, one can look for misclassified galaxies as rare as $0.02\sqrt{f}$ of all galaxies (or $0.02/\sqrt{f}$ of the sample), in a sample that is f times as numerous as galaxies. For example, suppose we wish to find the fraction of objects that are classified as M stars but are actually galaxies. M stars are about 25 % as numerous as galaxies. Then the galaxy contamination among the M stars can be measured to 4 %, and the fraction of the galaxies misclassified as M stars can be measured to 1 %.

7 Cosmology

The new experiment will measure the sum of the cosmological density parameter Ω and the dimensionless form λ of the cosmological constant to $\sigma(\Omega + \lambda) = 0.4$ and the difference to $\sigma(\Omega - \lambda) = 0.08$. I am assuming that the systematic error due to the evolution of the characteristic luminosity and the redshift errors are measured from the new data, and furthermore that the data from Cambridge, Edinburgh/Durham, and Münster are used to find the local density ϕ^*.

At the conclusion of this experiment when the data are at hand, will we know the geometry of the universe? I think that potentially the greatest problem is that the luminosity function may evolve in a way that prevents a reliable estimate of the evolution of ϕ^*. If galaxies of luminosity L at $z = 0.5$ evolves to be L^γ, where $\gamma - 1 = -0.2$, then with the redshift-volume test, one finds $\Omega = 1$ for an universe with $\Omega = 0.1$ (Loh 1988b). A strong hint for this is the very short time for which the gas in a spiral galaxy, if not augmented, is converted into stars. The measurements of the luminosity function with the proposed experiment may settle this problem. Questions about the photometric redshifts are not fundamental. Photometric redshifts will have been tested against the spectroscopic redshifts of Koo and Kron, and the fraction of galaxies that are misclassified will be estimated accurately by the method of correlation. Therefore, if the measurements of the luminosity functions are easily interpreted, we will know the geometry of the universe.

Suppose these results show a substantial evolution that is difficult to model. Perhaps a substantial number of objects, not classified as galaxies, are found to be galaxies by their correlation with known galaxies. Perhaps the shape of the luminosity function has evolved substantially. Then this experiment will have failed to measure the geometry of the universe, but it will have contributed substantial data on the global properties of galaxies at $z \leq 1$ — the luminosity function of galaxies of different color classes, the correlation function, the density of all objects that cluster as galaxies regardless of correct identifications by color, and limits to the spectral evolution of field galaxies.

Acknowledgement

This research was supported by the National Science Foundation under grant PHY 81–18157 and by Michigan State University.

References

Baum, W.A., 1962. In *IAU Symp. No. 15, Problems of Extragalactic Research*, ed. McVittie, G.C., MacMillan, New York, p. 390.

Bruzual, A.G., Kron, R., 1980. *Astrophys. J.*, **241**, 25.

Davis, M., Groth, E.J., Peebles, P.J.E. 1977. *Astrophys. J.*, **212**, L107.

Dressler, A., Gunn, J.E., Schneider, D.P. 1985. *Astrophys. J.*, **294**, 70.

Groth, E.J., Peebles, P.J.E., 1977. *Astrophys. J.*, **217**, 385.

Gunn, J.E., Stryker, L.L., Tinsley, B.M., 1981. *Astrophys. J.*, **249**, 48.

Kirshner, R.P., Oemler, A., Jr., Schechter, P.L., Schectman, S.A., 1983. *Astr. J.*, **88**, 1285.

Larson, R.B., Tinsley, B.M., Caldwell, N., 1980. *Astrophys. J.*, **237**, 692.

Loh, E.D., 1988a. In *Towards Understanding Galaxies at High Redshift*, ed. Kron, R., Renzini, A., Kluwer, Dordrecht, p. 197.

Loh, E.D., 1988b. *Astrophys. J.*, **329**. In press.

Loh, E.D., Spillar, E.J., 1986a. *Astrophys. J.*, **303**, 154.

Loh, E.D., Spillar, E.J., 1986b. *Astrophys. J.*, **307**, L1.

Loh, E.D., Spillar, E.J., 1988. In preparation.

Peebles, P.J.E., 1980. *The Large Scale Structure of the Universe*, Princeton Univ. Press, Princeton.

Phillipps, S., Shanks, T., 1987. *Mon. Not. R. Astr. Soc.*, **227**, 115.

Schuecker, P., 1987. Preprint, Universität Münster.

Schuecker, P., Horstmann, H., Seitter, W., 1988. In *IAU Symp. 130, The Structure of the Universe*, eds. Audouze, J., Szalay, A., Kluwer, Dordrecht, in press.

Spillar, E.J., Loh, E.D., 1988. In preparation.

Tinsley, B.M., 1980. *Astrophys. J.*, **241**, 41.

Tremaine, S., 1987. Private communication.

Galaxy Redshift–Number Counts with MRSP Data: A Method of Estimating q_0

H.-A. Ott
Astronomisches Institut
Westfälische Wilhelms-Universität
Münster, F.R. Germany

Abstract

An $N(M,z)$ test, using 6300 galaxy redshifts from the MRSP, leads to a statistical determination of q_0. While the current range of $0 < q_0 < 0.25$ is only preliminary, the method seems promising when applied to a forthcoming larger sample of galaxy redshifts.

1 Introduction

The Muenster Redshift Project (MRSP) combines medium redshift range, large coverage of solid angle across the sky and sufficient number density of galaxies to permit us to apply the galaxy $N(M,z)$ test of observational cosmology. The redshift–number test, used here in bins of *absolute magnitude*, is more sensitive to the geometry of the universe than the more familiar flux–number test and less affected by galaxy evolution (Weinberg 1972, Loh and Spillar 1986, Sandage 1987). In *physical* terms, it is a redshift–volume test, sensitive to all kinds of gravitating matter (Loh 1988).

2 Theoretical framework

In standard Friedmann models **differential number counts** in given intervals of absolute magnitude M and redshift z are given by

$$N(M,z) = dM\, d\omega\, dz\, \mathrm{LF}(M,z)\, g(z,H_0,q_0)\,. \tag{1}$$

$d\omega$ is the solid angle covered; in the case of no evolution and of galaxy number conservation in a comoving volume

$$\mathrm{LF}(M,z) = \mathrm{LF}(M,0)\ (1+z)^3\,, \tag{2}$$

where $\mathrm{LF(M,0)}$ is the present galaxy luminosity function; the term g describes the geometry, here of the matter dominated universe:

$$g = \left(\frac{c}{H_0}\right)^3 \frac{1}{q_0^4}\, \frac{\left(zq_0 + (q_0-1)\left(\sqrt{2q_0 z+1}-1\right)\right)^2}{(1+z)^6\sqrt{2q_0 z+1}}\,. \tag{3}$$

All equations are taken from Weinberg (1972).

The number counts open the possibility of estimating q_0 when all other quantities are known or can be properly estimated. In particular, our method does not require an explicit expression for the LF; instead we only assume its *invariance* over a certain limited redshift range.

3 The galaxy data

A homogeneous (sub)sample of the MRSP data is used, obtained from the combined UKST objective prism and direct plates of the ESO/SRC Atlas field No. 411. It contains 6 300 galaxies with measured redshifts $z < 0.3\,(\pm 0.008\,m.e.)$ and apparent magnitudes $16^{\mathrm{m}}\!.5 < m_B < 20^{\mathrm{m}}\!.5\,(\pm 0.1\,m.e.)$, distributed over a solid angle of $30\,\mathrm{deg}^2$. *Fig. 1* is the M_B *vs.* z diagram, showing all measured galaxies. Two strips for counting are included, one of width 0.5 in M extending over all z values, one of width 0.01 in z, extending over all M values. The data show a sharp cutoff at the limiting magnitude of the survey and a stochastic dying out of numbers for bright galaxies. The latter is due to a real decrease in density but also to the upper brightness limit of the z measuring method. For technical details of the measurements and reductions of the raw data, see Horstmann (1988) and Schuecker (1988a).

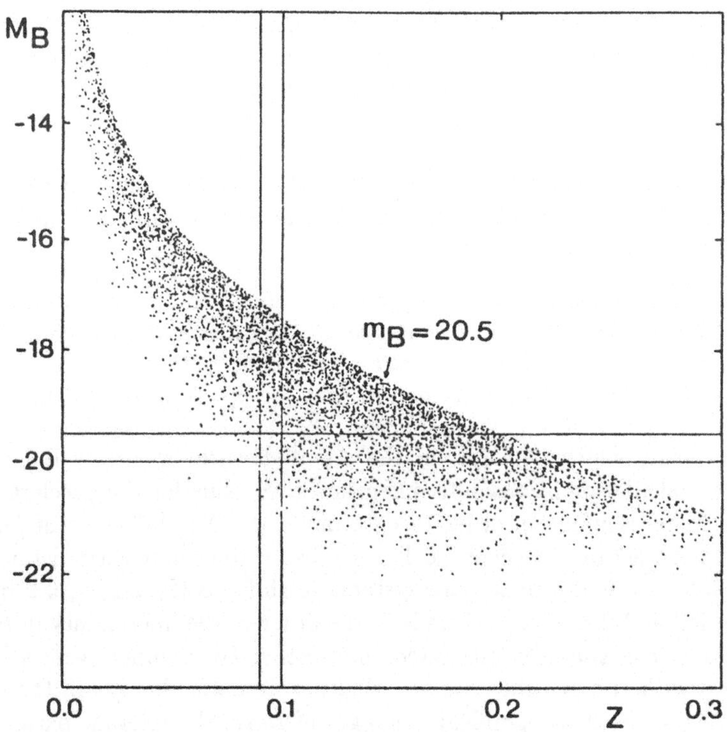

Fig. 1. Absolute magnitudes M_B *vs.* redshift z for all galaxies investigated. Indicated are two counting strips of widths $\Delta M = 0.5$ and $\Delta z = 0.01$. The data range is limited by faint and bright apparent magnitudes.

4 Data corrections

Several systematic effects bias the *empirical* galaxy counts $n(M,z)$. The most important ones depend on apparent brightness.

1) Due to the decreasing likelihood of including a galaxy into the counts near the bright as well as near the faint magnitude limit, *edge effects* are found at both low and high z values; $n_M(z) := n\,(M = const, z)$ and $n_z(M) := n\,(z = const, M)$ become too small – the curve thins out at both ends. As a simple but effective correction we use appropriate cuts.

2) The $n_M(z)$ distributions suffer from a slowly changing effectivity of z measurability with apparent brightness, as shown by Schuecker (1988b, his *Fig. 1a, b*). A linear correction function $s(m)$ is applied, obtained from the ratio of galaxy numbers on the direct plate, which are not affected by this bias, and the numbers on the objective prism plate. An updated version of Schuecker's redshift measuring method is noticeably less influenced by this effect so that for future data no corrections are needed.

3) Due to observational errors in redshift and magnitude there is a net *migration* of galaxies across the M,z-plane, affecting the counts in the bins. The observed galaxy distribution is a convolution of the real underlying distribution with the observational errors. For steep gradients in the counts the bins are systematically *filled up* in the direction of decline and *emptied* in the direction of increasing numbers. Corrections depend on the affected parameter

M : If the LF does not vary too much over the redshift range, there should be no pronounced effects in the counts $n_M(z)$; the *effective* LF remains the same and does not require correction.

z : Numerical simulations (convolution of assumed underlying distributions with redshift errors) show major effects at the beginning and the end of the $N_M(z)$ distributions. We apply a correction by cutting both ends appropriately. Also, the selection procedure described in Sect. 5 helps to avoid this bias.

4) Another bias, which depends on the adopted world model, will be introduced through the calculation of absolute magnitudes. A value for the deceleration parameter q_0 must be assumed in order to determine M. The influence of *false* adopted values has been estimated theoretically: By using the luminosity distance given by Mattig (1958), we found maximum systematic shifts $\Delta M(\max) < 0.2$ in the range $0.03 < z < 0.2$ and $0 < q_0 < 1$. This is of the order of the uncertainty in the absolute magnitudes. It has also been estimated empirically: By assuming values for q_0 of 0.1, 0.5, and 1.0 with the present data set, slight systematic changes in the corresponding $n_M(z)$ curves, but no significant changes of *slope*, the relevant parameter in this context, were found.

In addition to these more technical problems, a *physical* bias in the $n_M(z)$ counts may arise from the fact that evolution is not entirely negligible, even in a z-interval as small as 0.2, and from real density fluctuations (clusters of galaxies, voids) which change the count distributions. A possible "correction" for fluctuations is filtering to obtain a smoothed distribution. This, however, touches on the fundamental problem of how to deal with large inhomogeneities. Are we allowed to smooth, cut, neglect them as we like?

Here, we chose to ignore all possible perturbations of this kind. No corrections are needed for biases in *absolute* number counts, because only the slope of the $n_M(z)$ curves, *i.e. relative* numbers are important in the following analysis.

5 Analysis of the data

Because the MRSP is a magnitude-limited survey whose degree of completeness remains constant within certain limits of *apparent* magnitude, galaxies of different *absolute* magnitudes show different coverage of the redshift ranges and corresponding counts cannot simply be added without introducing an appreciable bias.

In order to be able to combine galaxies in all magnitude strips we adopted a **normalization** procedure consisting of three steps:

1. **Division** of empirical counts $n_M(z)$ obtained in regions overlapping in z and adjacent in M (*z-overlapping, M-adjacent*), without an explicit assumption about the LF.
2. **Selection** of the most *coherent* parts from the distribution of ratios obtained in 1), assuming invariant (or covariant) LFs with redshift.
3. **Normalization** of all counts to an *arbitrary constant* value LF_0, using the selected mean ratios obtained from 2).

Comments:

ad 1)

We expect

$$\frac{N(M_i,z)}{N(M_j,z)} = \frac{\mathrm{LF}(M_i,z)}{\mathrm{LF}(M_j,z)} =: K_{ij}(z) \tag{4}$$

with $K_{ij}(z) = \mathrm{const}$ for invariant LF(z) or covariance of $\mathrm{LF}(M_i)$ and $\mathrm{LF}(M_j)$ with redshift. We then expect the empirical values

$$k_{ij}(z) := \frac{n_{M_i}(z)}{n_{M_j}(z)} \tag{5}$$

to be constant within stochastic fluctuations, when the actual LFs do not change with redshift or when they change in the same sense, and when no systematic bias affects the counts.

Figure 2 shows a test of the expectation $k_{ij}(z) = \mathrm{const}$. Four ratios $k_{ij}(z)$ of *z-overlapping, M-adjacent* counts *vs.* redshift are shown. Adjacency is not a necessary condition, but it provides the largest overlapping regions. The coherence requirement helps to avoid an arbitrary selection of data points.

We find high noise (stochastic and/or systematic) at the beginning and at the end of the distributions and fairly flat central parts. Slight systematic shifts are present (differential density fluctuations between the curves or different shapes of LFs?), but the small number of data points does not permit conclusions. In order to avoid the noise we use cuts at both ends.

ad 2)

Selection of the coherent parts in the curves of *Fig. 2* is presently performed interactively and thus somewhat arbitrary. In the future statistical selection criteria will be used automatically.

ad 3)

Using the selected ratios $k_{ij}(z)$ we derive mean conversion factors for all pairs of *z-overlapping, M-adjacent* counts. They depend only on the assumption of invariant (or covariant) LFs with redshift, which is justified by the present distribution of the ratios.

The counts are normalized to an arbitrary value LF_0 after the mean conversion factors have been applied, leading to $n_0(z)$ and, averaged over all counts within a given red-

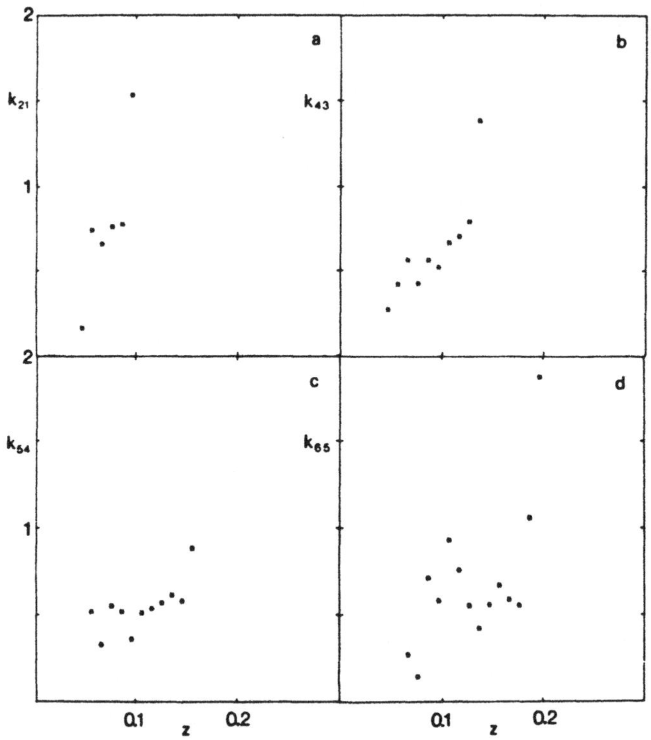

Fig. 2. Ratios $k_{ij}(z)$ from four *z-overlapping, M-adjacent* counts *vs.* redshift (definitions see text). The expectation $k_{ij}(z)$ = const, indicating an invariant (or covariant) LF(z) for pairs is a fair assumption for the central parts of the point distributions.

shift interval, to $\bar{n}_0(z)$. The normalized counts do not contain any information about the LF, but fully conserve their dependence upon the universal geometry $g(z, H_0, q_0)$, in which we are interested.

6 Results

The normalized counts are tracers of the universal geometry and can be compared with the theoretically expected values

$$N(M,z) = A \cdot N(z, q_0) = A \cdot g \cdot \left(\frac{H_0(1+z)}{c}\right)^3 , \tag{6}$$

where $A := d\omega\, dz\, c^3\, H_0^{-3}\, \mathrm{LF}_0$ includes all parameters assumed to be constant and g represents the global geometry.

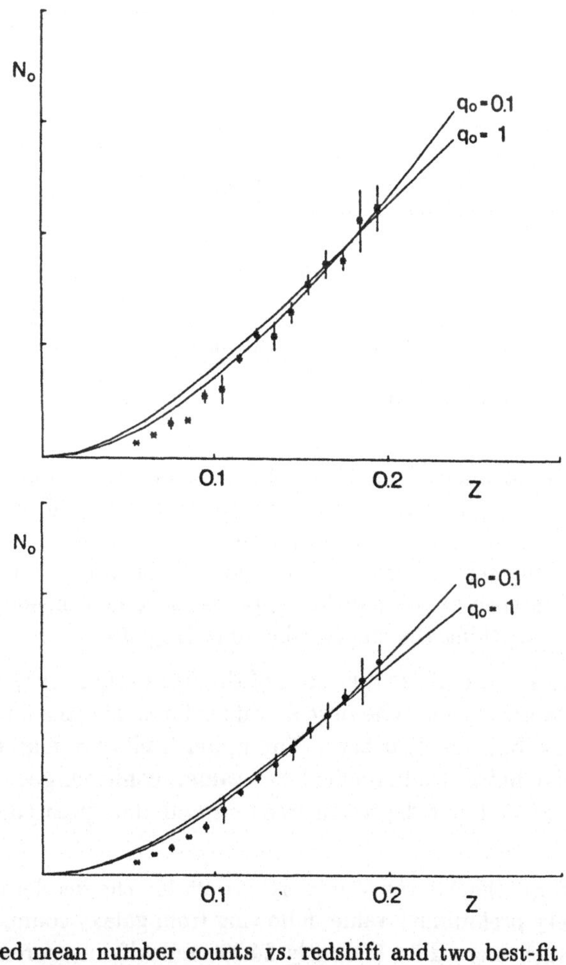

Fig. 3. Normalized mean number counts *vs.* redshift and two best-fit theoretical lines for $q_0 = 0.1$ and $q_0 = 1$ through the weighted points with $z > 0.1$.
3a: original distribution of data points; 3b: smoothed distribution of data points.

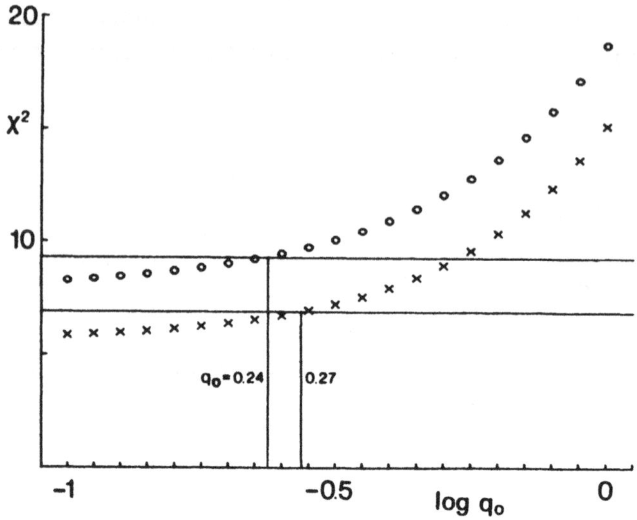

Fig. 4. χ^2 *vs.* $\log(q_0)$ for weighted (circles) and unweighted (crosses) fits through all points of *Fig. 3* with $z > 0.1$. Indicated are the upper 1σ limits. The minima of q_0 lie too close to zero to be represented. The fits to the weighted data define a smaller q_0 range and show χ^2 values consistent with the statistical expectation.

The comparison is made using least squares fits, weighted and unweighted, of the relation

$$\bar{n}_0(z) = A \cdot N(z, q_0) \; . \tag{7}$$

The assumption of Poisson statistics attributes to each count n the weight $p(n) = (\mathrm{var}\,(n))^{-1} = n^{-1}$.

Figure 3a shows the normalized mean number counts *vs.* redshift and two best-fit theoretical lines for $q_0 = 0.1$ and $q_0 = 1$, respectively, through all weighted points in the most reliable range $0.1 < z < 0.2$. The indicated error bars are obtained from Poisson statistics of the actual numbers involved. The fit for $q_0 = 0.1$ is better than for $q_0 = 1$. The difference of the two fits appears more conspicuous, when a block filter is applied for smoothing the empirical counts (*Fig. 3b*).

In *Fig. 4* the χ^2-values of the fits as functions of the free parameter q_0 for the weighted and unweighted data are shown. The best-fit values for q_0 and the lower $1\,\sigma$ limits are not apparent because both lie near zero. The upper limits are marked in the figure. The weighted fits give better (not smaller!) χ^2 values, confirming with the value 1.04 the expectation of $\chi^2 = 1$ per degree of freedom and thus justifying our weighting procedure.

As the result we adopt the $1\,\sigma$ range $0 < q_0 < 0.25$ for the deceleration parameter. This represents a very preliminary value, following from galaxy counts of *one* Schmidt plate only. With much more data (recently 24 000 redshift measurements have been completed) it seems not unrealistic to expect a better estimate of q_0 in the near future.

References

Horstmann, H., 1988. *These proceedings*, p. 111.
Loh, E.D., 1988. *Astrophys. J.*, **239**, in press; see also *these proceedings*, p. 267.
Loh, E.D., Spillar, E.J., 1986. *Astrophys. J.*, **307**, L1.
Mattig, W., 1958. *Astr. Nachr.*, **284**, 109.
Sandage, A., 1987. In *IAU Symp. 124, Observational Cosmology*, eds. Hewitt, A., Burbidge, G., Li, Z.F., Reidel, Dordrecht, p. 1.
Schuecker, P., 1988a. *These proceedings.*, p. 142.
Schuecker, P., 1988b. *These proceedings.*, p. 160.
Weinberg, S., 1972. *Gravitation and Cosmology*, Wiley, New York.

Part VII

Methods and Tools

Jörg Pfleiderer

Fionn Murtagh

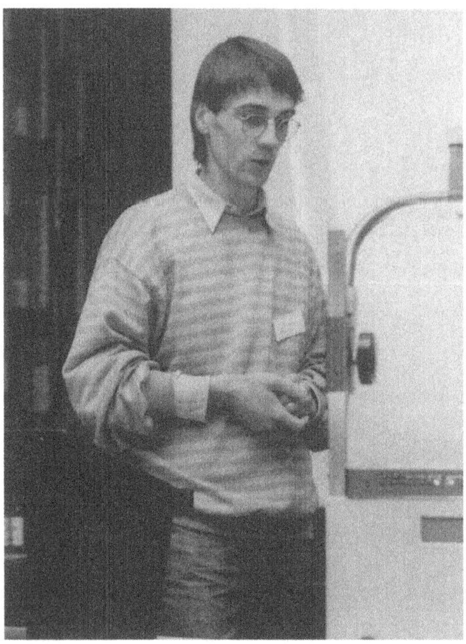

Dieter Teuber

Hans–Martin Adorf

Photometry from Schmidt Plates

M.R.S. Hawkins
Royal Observatory, Blackford Hill
Edinburgh, Scotland, UK

Abstract

With the advent of fast measuring machines such as COSMOS, there has been a revival of interest in photometry of wide field Schmidt plates. The very large datasets which have now become available are typically used for two types of programme, large scale statistical surveys of the distribution of various astronomical populations, and searches for very rare objects. In each case, good photometry is essential, and a number of techniques have been developed to attain acceptable accuracy over the large fields covered. A number of different approaches are used for detecting objects or defining samples, including magnitude, colour index and variability, in addition to non-photometric methods such as proper motions and objective prism spectra. Each detection method presents its own particular problems, and a number of calibration algorithms have been developed to optimise photometric accuracy.

1 Historical background

Before the advent of the photoelectric photometer, the only method of quantitative photometry was the measurement of photographic plates. Despite the inherent non-linearity of the photographic process, sophisticated algorithms and techniques were developed which, at least in a relative sense, produced stellar magnitudes with quite small formal errors. The introduction of photoelectric detectors enabled the identification and removal of a number of systematic effects, and enabled the production of well calibrated data for large numbers of stars. It was accepted at this time that a large plate scale and several plates in each colour were essential for obtaining accurate and reliable results. Schmidt plates were rarely used for photometry due to their small plate scale, and a number of other problems which will be discussed below.

In the early 1970s fast measuring machines specifically designed to measure Schmidt plates were developed at the Royal Observatory, Edinburgh, and Cambridge University. These machines were designed to detect images above a pre-computed threshold related to the sky background, and output a number of parameters for each image, including position, integrated density, image size, and measures describing the shape of the image.

Early work with COSMOS measures of Schmidt plates was hampered by the small number of plates available at that time. It was rare for there to be more than one plate in any one field which meant that establishing the magnitude and nature of photometric errors, especially as a function of position in the field, was not easy. As a result, much of the early published photometric work was based on only one set of measures of one plate. This break with traditional practice in photographic photometry made for some lack of credibility in early results from the Schmidt/COSMOS combination.

There are however other more fundamental difficulties associated with photometry from Schmidt plates. The small plate scale and hence image size significantly reduces the accuracy obtainable, especially for faint objects. This is clearly a fundamental limitation which can be offset by using a small ($\sim 8\mu$m) measurement aperture although this in itself produces problems associated with the satisfactory definition of the aperture on some measuring machines.

2 Photometric problems

There are a number of problems associated with change of photometric performances across the field of a Schmidt telescope. Perhaps the best known, and least serious, is *geometrical vignetting* caused by obstruction of the light beam to the outer parts of the field by the structure of the telescope. The attenuation of the beam can quite easily be calculated as a function of position of the plate, and appropriate corrections made to the photometry. In fact the corrections are insignificant over most of the field, and only important in the corners. A more serious field effect is caused by *differential desensitization* of the plate while it is in the plate holder. The plate is kept in a curved shape during the exposure to follow the focal plane of the telescope. The filter however is planar, and so traps a layer of damp air of varying thickness over the plate. The damp has the effect of desensitizing the plate, and since the air gap is thickest towards the edge of the plate holder, the effect is proportionally worse towards the edge of the field. This rather serious effect is present on all UK Schmidt plates until about 1981, when the problem was solved by flushing the plate holder with dry nitrogen during the exposure.

Field effects due to intrinsic changes in sensitivity of the emulsion across the plate are generally quite small, of the order of $1-2\,\%$, which for most purposes is negligible compared with other photometric errors. A more serious problem is associated with *changes of image structure* across the field after a long exposure, due to field rotation. Also, any *local defocussing* or *astigmatism* would have a similar effect. With a strictly linear detector, change in image structure would not present serious difficulties, as the total amount of incident flux is not changed, but for a non-linear detector such as a photographic plate, combined with the *thresholding procedure* of COSMOS, the effect on the measured magnitude can be large ($> 0.1\,$mag). There is no satisfactory cure for this problem, apart from ensuring that where possible plates are taken close to the meridian, and that the telescope alignment is carefully monitored to obtain optimum image shape.

3 Modern photographic photometry

The COSMOS measuring machine has revolutionised photographic photometry in a number of ways, but especially in making possible the measurement of large (35 ×

35 cm) Schmidt plates in a reasonable length of time (about 4 hours). The scanning system comprises a flying spot cathode ray tube with a photomultiplier detector. The machine measures plates in lanes 128 pixels across, with a range in pixel size from 8 to 32 μm. The sky background is measured in grid form in a pre-scan, and used to define the threshold above which images are to be detected. This is typically 7 – 10 % above the local night sky level. The images detected above the threshold are analysed and a number of parameters output to tape, including position, image area, maximum intensity, integrated density and several shape parameters defining ellipticity, and major and minor axes. The positional accuracy of the measures depends to some extent on the size of the image, but is about 2 μm.

The integrated density parameter (COSMAG) should in principle be proportional to incident flux. In practice, brighter images are saturated or nearly saturated at the centre, while for the fainter images, the thresholding procedure leads to a systematic reduction in proportion to the flux recorded. There is thus only a relatively small regime where COSMAG provides a strictly monotonic measure of star magnitudes, which is suitable for empirical calibration using a photoelectric or CCD sequence. The photometric accuracy of the calibrated measures depends on the quality of the plate (seeing, background uniformity *etc.*) and also on the size of the image. In favourable circumstances (well exposed images on a good plate) the accuracy is about $\pm 0\overset{\mathrm{m}}{.}06$ increasing to about $\pm 0\overset{\mathrm{m}}{.}10$ near the detection limit ($B = 21$ for a IIIa-J survey plate).

4 Diagnostics and cures

Each COSMOS measure is accompanied by a comprehensive set of diagnostics showing changes in the relation of the various image parameters across the plate. These may be used to check for field effects.

If significant field effects are found to be present, several remedies are available. If only local changes in magnitude are important, such as for variability studies, then the plate to plate transformations may be made as a function of position. If colour effects only are important, then a requirement to keep the main sequence in the same position across the plate will suffice (although in some cases this may not be an acceptable assumption). If it is necessary to maintain the magnitude zero point over the whole field then a grid of photoelectric standards is necessary, although in most cases there is a strong correlation between bright and faint stars, and so bright standards will suffice.

5 Conclusion

As the measurement accuracy and consistency of fast measuring machines improve, new reduction techniques are developed and large numbers of Schmidt plates become available, it is now possible to attain the accuracy of photographic photometry in its heyday, but with an increase in speed of several orders of magnitude.

Mathematical Aspects of Internal Magnitude Calibration

D. Hömberg
Astronomisches Institut
Westfälische Wilhelms-Universität
Münster, F.R. Germany

Abstract

The internal calibration method, introduced by Bunclark and Irwin, is presented as a non-linear optimization process. Two different approaches to solve the problem are presented, the conjugate gradient method, a deterministic optimization algorithm, and simulated annealing, belonging to the class of Monte Carlo methods. Tests with model data show that the conjugate gradient method is superior to simulated annealing in computational effort, if the latter is defined by the number of equivalent functional evaluations. Both methods lead to the same average residual of the objective function. Considering the simplicity of the algorithm, simulated annealing suggests itself for solving optimization problems of high complexity with small theoretical effort.

1 Introduction

The most attractive feature of *internal calibration* is that it uses the objects on the photographic plate itself as gauging objects. It overcomes the problems of varying plate sensitivity and the lack of standards on the plate area under consideration.

The method was first developed by Bunclark and Irwin (1983). Fundamental to it is the concept, that the light of all stars on a well-defined area of the photographic plate is scattered by the same point-spread function. This means, the intensity profiles of the stars all have the same shape in logarithmic measure, while additive constants take care of the difference in stellar magnitudes.

In continuation of an earlier publication (Hömberg 1988) this paper describes internal calibration in terms of a nonlinear optimization process. Two different minimizing algorithms are introduced and their results are compared using model data.

2 Data reduction

To extract the data needed for the internal calibration procedure, picture frames of the direct plate are used (Horstmann 1988). For all frames containing stars, the centre of density is computed and by discrete integration over concentric areas and subsequent normalization one-dimensional profiles are obtained, in the following referred to as *density profiles*.

3 Internal calibration as a nonlinear optimization process

Let k be the characteristic curve transforming density to logarithmic intensity, *i.e.* $\log I = k(d)$. As derived in Hömberg (1988),

$$k'(D_1)D_1'(r) - k'(D_2)D_2'(r) = 0 \tag{1}$$

is the defining criterion for an internal characteristic curve. This means, every function, solving Eqn. 1 for arbitrary density profiles D_1 and D_2 will be called an *internal characteristic curve.*

This definition implies a corollary:

Every linear transformation $\tilde{k}(d) = ak(d) + c$ $(a \neq 0)$ of an internal characteristic curve $k(d)$ again is an internal characteristic curve.

To define the optimization problem, the analytical representation of the characteristic curve by Honeycutt and Chaldu (1970) is used which is Eqn. 2:

$$\log I = a_1 d + a_2 \ln(\exp(bd^{c_1}) - 1) + a_3 \exp(bd^{c_2}) + a_4 \,. \tag{2}$$

Assuming $a_1 \neq 0$, Eqn. 2 may be written this way:

$$\log I = a_1 \left(d + \frac{a_2}{a_1} \ln(\exp(bd^{c_1}) - 1) + \frac{a_3}{a_1} \exp(bd^{c_2})\right) + a_4 \,. \tag{2'}$$

Eqn. 2′ makes it quite clear, that any choice of parameters a_1 and a_4 will represent an internal characteristic curve, provided that the remaining parameters have been estimated correctly.

Thus, in order to get an unique parameter representation, one chooses a_1 and a_4 as constants and defines Eqn. 3 to be the general representation of the internal characteristic curve:

$$k(x, d) = 0.1d + x_1 \ln(\exp(x_3 d^{x_4}) - 1) + x_2 \exp(x_3 d^{x_5}) \,, \tag{3}$$

with the parameter vector $x \in \mathbb{R}^5$ being the parameter vector.

Let the density profiles be marked $D_i(r), 1 \leq i \leq m$, with m equal to the total number of density profiles used for the procedure. As they are only known at certain discrete values r_j of the radius, one defines

$$d_{i,j} = D_i(r_j) \quad , \qquad 1 \leq i \leq m, 1 \leq j \leq n_i$$

with

$$n_i = max\{j | \, D_i(r_j) \neq 0\}$$

defined to be the length of the density profile D_i. The first derivative of the density profile D_i is approximated by the symmetric difference quotient

$$\Delta_{i,j} = \frac{1}{2}(d_{i,j+1} - d_{i,j-1}) \quad , \qquad 2 \leq j \leq n_i - 1 \,.$$

Now solving the internal calibration problem means searching for a vector $x \in \mathbb{R}^5$, so that Eqn. 1 is fulfilled for all pairs of density profiles D_i and D_j $(i \neq j)$. Thus one takes an arbitrary profile, names it D_1 and defines Eqn. 4 to be the objective function

$$f(x) = \frac{1}{m} \sum_{i=2}^{m} \sum_{j=2}^{N_i - 1} \left(k'(x, d_{1,j})\Delta_{1,j} - k'(x, d_{i,j})\Delta_{i,j}\right)^2 , \tag{4}$$

with $N_i = min(n_i, n_1)$ being the minimal length of the density profiles D_1 and D_i.

In Eqn. 4 the differences between D_1 and all other profiles are summed up according to Eqn. 1. Hence finding a parameter vector x for which the objective function $f(x)$ takes its minimal value, hopefully close to zero, becomes equivalent to solving the internal calibration problem.

4 Remarks about the objective function

Before discussing algorithms for solving this optimization problem, some remarks have to be made about the objective function.

i) The choice of a reference profile D_1 is not difficult. On the one hand its length should not be too short in order to use as much information about the data as possible. On the other hand it should not be overexposed, because this would falsify the results.

ii) In order to obtain useful characteristic curves, all parameters have to be larger or equal to zero. Therefore the domain of the objective function has to be restricted to the set

$$\mathcal{S} = \{x \in \mathbb{R}^5 | x_i \geq 0, 1 \leq i \leq 5\} .$$

Instead of solving the constrained optimization problem

$$\text{minimize } f(x), x \in \mathcal{S} ,$$

another approach is used here. A new objective function is defined to be the sum of the old one and a penalty function

$$P(x) = \sum_{i=1}^{5} (\min(0, x_i))^2 .$$

Whenever one component of x becomes less than zero, the function $F(x)$ is penalized by an increase of its function value. Thus one has to solve the unconstrained optimization problem

$$\text{minimize } F(x) = f(x) + \mu P(x) \quad , \qquad x \in \mathbb{R}^5 . \tag{5}$$

The greater μ, the better the constrained problem will be approximated by the unconstrained one.

iii) The minimum of the objective function is not defined sharply. Perhaps it may be better described as a broadenend valley with slowly increasing gradient (*Fig. 1*).

In the next two paragraphs optimization algorithms for solving the internal calibration problem (*i.e.* for minimizing the objective function as defined in Eqn. 5) will be presented.

5 Conjugate gradient methods

The simplest algorithm for minimizing nonlinear functions is the method of steepest descent. It is based on the fact that the gradient of a function always points at the direction of the steepest growth of a function, which implies that the negative gradient points at the direction of steepest descent of a function.

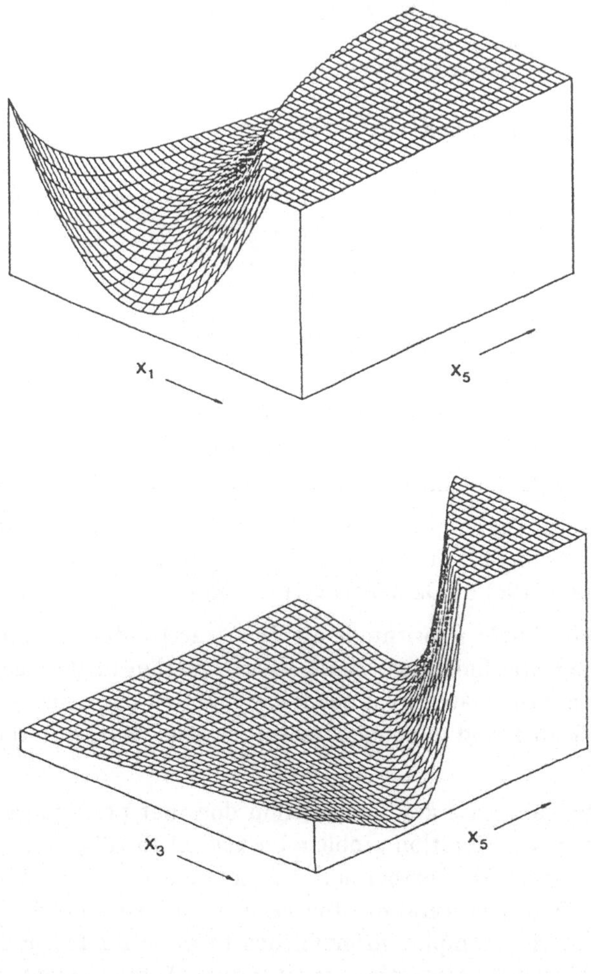

Fig. 1. Two-dimensional sections of the graph of the objective function in the neighbourhood of the minimum.

This means starting at a point x_k one searches along the direction of the negative gradient (grad $F)(x_k)$ to a minimum point of the line which is taken to be x_{k+1}. Thereby the five-dimensional minimization problem is reduced to a one-dimensional one which can easily be solved by standard line search techniques like quadratic and cubic curve fitting (*e.g.* Luenberger 1972).

The method of conjugate gradients differs from steepest descent only in the choice of the minimizing direction which is a linear combination of the old minimizing direction and the gradient. More precisely, the algorithm works as follows:

Conjugate Gradient Algorithm

Step I. Given x_0 compute $g_0 := (\text{grad } F)(x_0)$
set $d_0 := g_0$.

Step II. $x_{k+1} = x_k - \alpha_k d_k$, where α_k minimizes $F(x_k - \alpha d_k)$
$g_{k+1} = (\text{grad } F)(x_{k+1})$
$d_{k+1} = g_{k+1} + \beta_k dk$.

If $k+1$ is a multiple of 5, then

$$\beta_k = 0 \, ,$$

otherwise

$$\beta_k = \frac{\langle g_{k+1}, g_{k+1} \rangle}{\langle g_k, g_k \rangle} \tag{FR}$$

or

$$\beta_k = \frac{\langle g_{k+1} - g_k, g_{k+1} \rangle}{\langle g_k, g_k \rangle} \tag{PR}$$

or

$$\beta_k = \frac{\langle g_{k+1}, H_{k+1} d_k \rangle}{\langle d_k, H_{k+1} d_k \rangle} \, , \tag{DAN}$$

with $H_k := (\text{Hess } F)(x_k)$ defined to be the Hessian of the objective function.

Hence every five steps the conjugate gradient method is restarted with a pure steepest descent step. Depending on the choice of β_k the algorithm is named Fletcher-Reeves (FR), Polak-Ribière (PR) or Daniel (DAN) method.

This comparatively simple algorithm is highly attractive due to its good convergence in the case of quadratic functions. Since every function in the neighbourhood of a minimum can be approximated by a quadratic function one expects that these properties can also be transferred to nonquadratic functions, at least in the neighbourhood of the minimum.

Unfortunately, the conjugate gradient method does not perform satisfactorily when applied to the internal calibration problem because of the ill-defined minimum of the objective function (Sect. 4). Furthermore this method is comparatively expensive in computing time. For each iteration, the gradient of the objective function, which is given through a very complex formula, has to be evaluated and the line search algorithm has to be called. Altogether, one iteration of the conjugate gradient method in connection with an efficient line search technique costs as much computing time as 21 evaluations of the objective function.

6 Simulated Annealing

In contrast to the deterministic optimization algorithm discussed in the last section, simulated annealing belongs to the class of Monte Carlo methods. Based on the so-called Metropolis Algorithm in statistical mechanics, it was developed independently by Kirkpatrick *et al.* (1983) and Cèrny (1985). The algorithm is very simple and is described in the following.

Simulated Annealing Algorithm

Step I. Choose starting point $x_0 \in \mathbb{R}^5$,
starting temperature $T_0 \in \mathbb{R}_+$,
cooling rate $q \in]0,1[$,
number of iterations
per fixed temperature $N \in \mathbb{N}$;
initialize $k = 0$
and $T_k = T_0$

Step II. Choose $\Delta x \in D$ 'at random', $D \subset \mathbb{R}^5$, being a compact
cube centered at the origin

$\Delta F = F(x_k + \Delta x) - F(x_k)$

if $\Delta F \leq 0$ then

$x_{k+1} = x_k + \Delta x$

else

$P = \exp(-\frac{\Delta F}{T_k})$

choose $z \in [0,1]$ 'at random'

if $z \leq P$ then

$x_{k+1} = x_k + \Delta x$

else

$x_{k+1} = x_k$

end if

end if

$k = k + 1$

Step III. If k is a multiple of N then

$T_{k+1} = qT_k$

else

$T_{k+1} = T_k$

end if

go to step II.

After the initialization in step I has been carried out, for each iteration of step II a small Δx is chosen at random and the function value at the new point $x_k + \Delta x$ is computed. If this function value is smaller than the old one, $x_k + \Delta x$ is accepted to be x_{k+1}. Otherwise the worse value $x_k + \Delta x$ is accepted with the Boltzmann probability P. This is realized by choosing a random number z between zero and one. If z is less or equal to P, then $x_k + \Delta x$ is accepted to be x_{k+1}, otherwise not.

After step II has been carried out N times, the temperature is lowered in step III. Then the algorithm starts again with step II. In other words, the probability to accept an increased function value at x_{k+1} decreases exponentially until the system freezes when T approaches zero.

The name Simulated Annealing is derived from the analogy to the cooling process of a physical system. In that case free energy assumes the role of the function to be minimized. If the system is cooled too fast, it freezes before reaching its ground state.

For a detailed introduction concerning mathematical theory and applications of simulated annealing the author refers to the textbook of van Laarhoven and Aarts (1987). An astronomical application is given by Jeffrey and Rosner (1986).

7 Discussion of results

For an objective comparison of both methods model data were used. A set of logarithmic Gaussian profiles (*i.e.* parabolas) were transformed back to two-dimensional density profiles. For a better representation of reality, Poisson noise was added to the profiles. Using the routines described in Sect. 2, two sets of one-dimensional density profiles were thus obtained (*Fig. 2*).

Table 1. Results for the conjugate-gradient-methods

a) model data without noise

algorithm	PR	FR	DAN
min. res.	0.02	0.01	0.04
aver. res.	0.05	0.05	0.08
# It	11.5	12.6	5
# f	242	1264	195

b) model data with Poisson noise

algorithm	PR	FR	DAN
min. res.	0.37	0.37	0.38
aver. res.	0.46	0.49	0.51
# It	8.4	7.6	6
# f	176	160	234

It : average number of iterations

f : average number of function evaluations

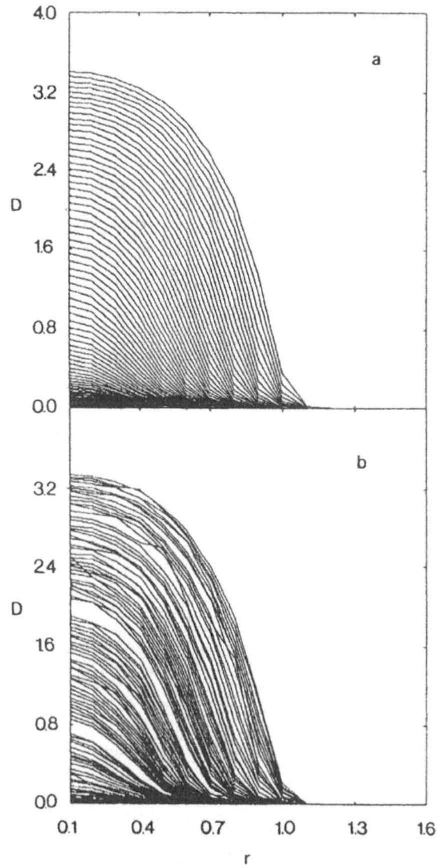

Fig. 2. Model density profiles $D(r)$: (a) without noise; (b) with Poisson noise.

Table 1 gives the results for the three different versions of the conjugate gradient method, when applied to the internal calibration problem. At first sight the results do not differ much. All methods arrive at nearly the same average residual of the objective function. The number of iterations for (DAN) is less then the one for both of the other methods. But in every iteration in (DAN) the Hessian of the objective function has to be evaluated. This additional effort nearly compensates, or in the case of noisy data overcompensates the advantage of a small number of iterations. *Figs. 3* and *4* illustrate the results of such a minimization procedure.

For simulated annealing tests have been carried out with varying choices for the three control parameters, starting temperature T_0, number of iterations per fixed temperature N and cooling rate q. The results are shown in Table 2.

Two general conclusions are: the number of function evaluations increases with increasing cooling rate q, and the average residual of the objective function decreases with increasing cooling rate.

The best results were obtained with small cooling rate q and large N (number of iterations per fixed temperature). This result is also supported by theory.

In this case, the results of simulated annealing and the conjugate gradient method arrive at nearly the same average residual of the objective function. Although the number of function evaluations for simulated annealing exceeds that of the deterministic conjugate gradient method, it is of the same order of magnitude. Thus it becomes difficult to decide which algorithm should be preferred. To draw a final conclusion, one might say that simulated annealing positively compares with deterministic optimization algorithms. It gains further attraction when the simplicity of the algorithm is also taken into account.

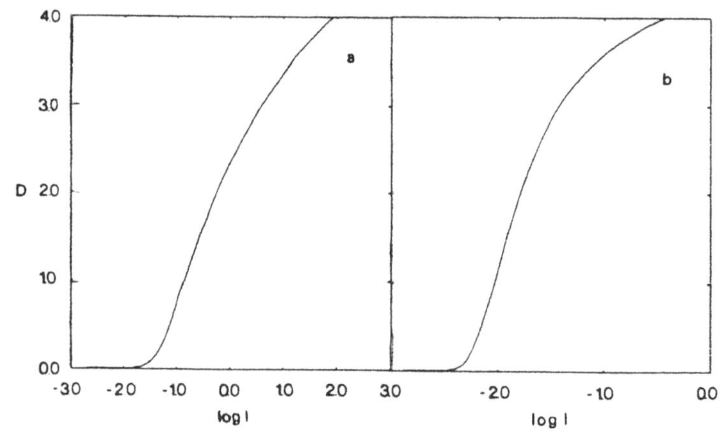

Fig. 3. Characteristic curves: (a) for starting parameters; (b) for solution parameters.

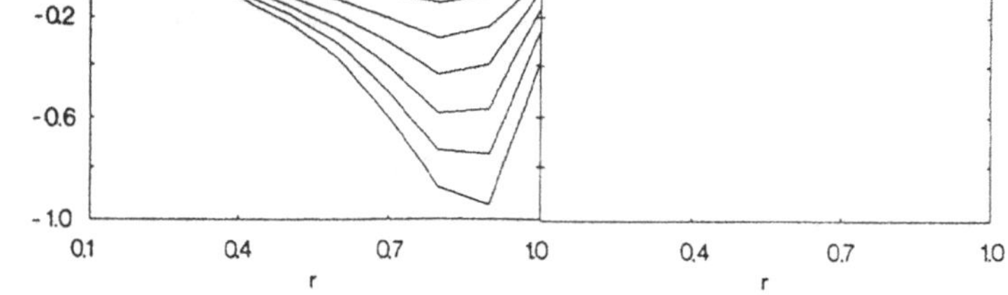

Fig. 4. Differences ΔS between all profiles and a reference profile taken from the same sample: (a) using the characteristic curve with starting parameters; (b) with solution parameters (see *Fig. 3*).

Table 2. Results for simulated annealing

a) model data without noise

T_0	N	q	# f	min. res.	max. res.	aver. res.
100	20	0.1	187	0.01	1.18	0.24
100	20	0.3	184	0.04	1.83	0.24
100	20	0.5	277	0.02	2.11	0.22
100	20	0.7	380	0.03	6.33	0.31(0.06*)
100	20	0.9	410	0.03	2.08	0.24
30	40	0.1	315	0.01	0.13	0.05
30	40	0.3	290	0.03	0.51	0.06

b) model data with Poisson noise

T_0	N	q	# f	min. res.	max. res.	aver. res.
100	20	0.1	141	0.40	10.08	0.24
100	20	0.3	158	0.40	20.86	2.24
100	20	0.5	188	0.40	4.45	0.91
100	20	0.7	244	0.39	3.15	0.63
100	20	0.9	411	0.48	7.20	0.76(0.49*)
30	40	0.1	221	0.40	4.52	0.67(0.47*)
30	40	0.3	240	0.38	4.49	0.69(0.51*)

* without maximal residual

References

Bunclark, P., Irwin, M.J., 1983. In *Statistical Methods in Astronomy*, ESA SP-201, p. 195.

Cèrny, V., 1985. *J. Optim. Theory and Appl.*, **45**, 41.

Hömberg, D., 1988. In *IAU Workshop Astrophotography*, ed. Marx, S., Springer-Verlag, Berlin, p. 77.

Honeycutt, R.K., Chaldu, R.S., 1970. *AAS Photo-Bull.*, **2**, 14.

Horstmann, H., 1988. *These proceedings*, p. 111.

Jeffrey, W., Rosner, R., 1986. *Astrophys. J.*, **310**, 473.

Kirkpatrick, S., Gelatt Jr., S.D., Vecchi, M.P., 1983. *Science*, **220**, 671.

Laarhoven, P.J.M. van, Aarts, E.H.L., 1987. *Simulated Annealing: Theory and Applications*, Reidel, Dordrecht.

Luenberger, D.G., 1972. *Introduction to Linear and Nonlinear Programming*, Addison-Wesley, Reading, MA.

Methods of Deconvolution

Jörg Pfleiderer
Institut für Astronomie der Universität
Innsbruck, Austria

Abstract

We describe the mathematical problem of deconvolution and discuss 3 classes of methods to solve it: (1) methods working in Fourier space. (2) methods working in image space, with smoothness constraints, and (3) methods for image improvement without true deconvolution. The highest resolution can be achieved by class-2 methods.

1 Introduction

Deconvolution as a means of improving the resolution of instruments is well established for more than a century, Lord Rayleigh being the first to treat the problem mathematically. Soon it was recognized that the most reliable method of resolution increase was to use better instruments. The interest in mathematical methods has been revived in recent years (see, *e.g.*, Pfleiderer and Reiter 1982). This is because the performance of observing instruments has reached a maximum in many fields, and cannot be improved much by building larger or better instruments (or only with excessively high costs). So one should extract as much information from the data as possible. Also, the increase in computer performance permits the use of methods with large numerical effort. One of the most favoured methods presently used is the maximum entropy method MEM. For radiointerferometric observations, the method most often applied is CLEAN, of which several improvements over the original version of the seventies (Högbom 1974) exist.

In Sect. 2, we state the mathematical problem. Sects. 3 - 5 describe classes of deconvolution methods which are discussed in Sect. 6.

2 The deconvolution problem

The convolution/deconvolution equation in image space is, in discrete form,

$$f_i = \sum_j g_j h_{ij} + e_i \ , \tag{1}$$

where i and j are pixel numbers (which, in a two-dimensional problem, would each have two components), f is the observable signal (data), h the point spread function PSF or beam (depending, in a true convolution, only on the distance between pixels i and j), e the error, and g the true signal, approximated by a set of point sources at pixels j. The data pixels $i = 1, \dots, I$ and the signal pixels $j = 1, \dots, J$ need not necessarily have the same size. It is by no means necessary at this stage to assume a special form of the error distribution, e.g., that the noise data e_i are independent from each other or from the source distribution or the data. In the convolution, the signal g is convolved with the PSF h and noise e is superimposed to give the data f. In the deconvolution, the solution of the problem consists of finding a model set of signals $\{m_j\}$ such that the residuals

$$r_i := f_i - \sum_j m_j h_{ij} \tag{2}$$

are left equal to the errors. The true solution $r_i = e_i$ cannot be found because e is not known. A possible or acceptable solution is one for which the statistics of the set $\{r_i\}$ is sufficiently similar to what is expected for the error distribution $\{e_i\}$.

The Fourier transform of Eqn. 1 is

$$F_k = G_k \cdot H_k + E_k \ , \tag{3}$$

where the index k stands for a spatial frequency u_k or frequency range $[u_k, u_k + du_k]$. F, G, H, and E are the coefficients of the transform of f, g, h, and e, respectively.

Consider a point source of unit strength at pixel j, with Fourier transform $G_k(j)$. Then

$$H_{kj} := G_k(j) \cdot H_k \tag{4}$$

is the Fourier response to that source. From the linearity theorem, it follows that the response to a source of strength g_j is $g_j H_{kj}$, and the response to a set of point sources is $\sum_j g_j H_{kj}$. That is, another and entirely equivalent form of Eqn. 3 is

$$F_k = \sum_j g_j H_{kj} + E_k \ , \tag{5}$$

which comprises, if seen as a deconvolution problem, a linear system of K equations

$$R_k = F_k - \sum_j m_j H_{kj} \ , \quad k = k_1, \ldots, k_K \tag{6}$$

for the J unknowns m. Here, the R's are the Fourier residuals. Again, the true solution $R_k = E_k$ cannot be found but only possible ones in which the statistical properties of the residuals are accceptable. Accordingly, there are infinitely many possible solutions, both for Eqn. 2 and Eqn. 6.

3 Fourier methods

They work completely in Fourier space and use Eqn. 3. Many PSFs, in particular Gaussian ones, have Fourier coefficients which are noticeably different from zero only within a finite frequency interval $[-u_c, u_c]$, thus comprising a low pass filter. A good example is a single-dish radio observation. Higher frequencies are not transmitted by the instrument. Low frequencies are well transmitted while intermediate frequencies are partially filtered out.

Bracewell and Roberts (1954) introduced, for that case, the so-called principal solution

$$G_k = F_k/H_k \quad \text{for} \quad |u_k| < u_c\,; \tag{7a}$$

$$G_k = 0 \qquad \text{otherwise}\,. \tag{7b}$$

It tends to show oscillations of frequency $\sim u_c$ in the image-space solution (Fourier transform of G) remindful of interference patterns, and negative values around pointed positive ones - a kind of Gibbs phenomenon. This is why a constraint of non-negativity can much improve the solution.

For a more complicated PSF, the principal solution can be generalized to be

$$G_k = F_k/H_k \quad \text{for} \quad |H_k| > C\,, \tag{8a}$$

$$G_k = 0 \qquad \text{otherwise}\,, \tag{8b}$$

where C is a suitably small constant. Eqns. 7a and 8a reverse the partial filtering of coefficients caused by low values of H and are, therefore, called "inverse filtering".

The sharp cut-off can be somewhat smoothed. In terms of a least-squares optimization, the best approach is the Wiener filter (Wiener 1942, Helstrom 1967) which was introduced into astronomy by Brault and White (1971):

$$G_k = \frac{F_k H_k^\star}{H_k H_k^\star + \Phi_k}\,, \tag{9}$$

where Φ is the ratio of the spectral densities of the signal and the noise. Φ is not known but can be reasonably estimated.

Most Fourier methods have the disadvantage that they put $G = 0$ for those frequencies which were lost in the process of observation, zero being the most non-committal default value. They are, therefore, unable to recover very steep features and to avoid Gibbs phenomena.

Some iterative methods work in image space but nevertheless are Fourier methods insofar as they converge to the inverse-filter solution (Frieden 1975). In order to avoid explosion of high-frequency components, the iteration must be stopped before convergence at some optimum point, or some other measure must be introduced (Jansson *et al.* 1970). We mention two methods:

Van Cittert's (1931) algorithm can (with use of the same pixels i for data and signal) be written as

$$m_i(0) = 0\,, \quad m_i(n+1) = m_i(n) + r_i(n)\,, \tag{10}$$

where n is the iteration number. The method has been used, *e.g.*, in solar physics (Wittmann 1971), in molecular beam scattering (Siska 1973), in geophysics (Ioup and Ioup 1983).

The algorithm of Lucy (1974) is, actually, not a pure Fourier method. It iteratively estimates the inverse beam $k := h^{-1}$ which reproduces the signal when the data is convolved with it:

$$m_j(n+1) = \sum_i f_i k_{ji}(n)\,, \quad k_{ji}(n) = \frac{m_j(n) h_{ij}}{\sum_l m_l(n) h_{il}}\,. \tag{11}$$

With the improvement of other methods and computer availability, it seems that pure Fourier methods are somewhat outdated and will be used only in some special applications (Subrahmanya 1980).

4 Image-space methods with smoothing constraint

As long as $J \geq I$ or $J \geq K$, Eqns. 2 or 6 can be solved for exactly vanishing residuals. This noise-fitting procedure would give an excessively oscillating solution, coinciding with the inverse-filter solution without cut-off: The high frequencies of the solution, after being strongly damped by the convolution with the beam, still have to reproduce the finite high-frequency amplitudes of the noise. Obviously, such solutions are not acceptable.

In order to find an acceptable solution, one has to use Eqn. 2 or 6 in a slightly modified form: The lefthand sides are first artificially neglected in order to have a well-defined linear system of equations but then the system is not solved exactly but only approximately. For selecting one of the infinite number of acceptable solutions, one has to introduce a constraint. Since small-scale structure in the true signal is damped out by the convolution, it cannot be recovered from the data with any certainty. Therefore, the constraint should suppress such features, selecting essentially the smoothest solution compatible with the data. A non-negativity constraint is also quite helpful (Biraud 1969) but is, depending on the problem, not always possible.

There are no standard methods for solving a linear system of equations approximately. Also, a nonlinear constraint destroys the linearity. Therefore, each method uses a different kind of iterative algorithm, adapted to the constraint. In most methods, the underlying philosophy is a least-squares fit. Then the residuals should resemble a normal distribution with average zero and a given variance which is known or can be estimated from the measuring error. For example, MEM deconvolves to a certain value of χ^2 or to a more refined error statistics (E^2 distribution: Bryan and Skilling 1981; position-independent distribution: Reiter and Pfleiderer 1986).

While the algorithm actually used is generally only of marginal influence on the result, the choice of a good smoothness constraint is essential. Some authors stress the necessity for a convincing constraint philosophy, as non-committance or simplicity, while others consider every constraint which provides a sufficiently smooth result as a good one (Nityananda and Narayan 1982).

Oscillatory solutions will always result in large values of the second derivative of the source distribution. Minimizing the second derivative in a least-squares sense is, therefore, a good constraint and has been used by a number of authors (Phillips 1962, Tikhonov 1963, Twomey 1963, Turchin and Turovtseva 1974, Tikhonov and Arsenin 1977, Subrahmanya 1980, Basistov *et al.* 1979, Jonas 1985). That this constraint seems to be not much used nowadays is probably not a result of the constraint being inferior but rather a result of the numerical algorithms used being inferior.

Another feature of oscillatory solutions is that some pixels have unnecessarily large content, which is recognizable, for example, by a large square. Minimization of the sum of squared pixel contents will again avoid such cases and thus produce a smooth solution. This is the constraint of the so-called Smoothness-Stabilized CLEAN or SSC (Cornwell 1983). The method can also be described as an (unconstrained) deconvolution with a modified beam which is the PSF with a central peak added. This is why it is also called Prussian-Helmet CLEAN or Prussian-Hat CLEAN. Other powers of the pixel contents than the second have also been discussed. For example, a maximization of the sum of square- rooted pixel contents is about as good.

Similarly, one can consider differences in the contents of adjacent or nearby pixels as unsmooth and try to minimize those differences as a function of the distance of pixels. This constraint can be formulated as giving a minimum of information on small-scale structure which was partly or wholly lost in the data-collecting process by the smoothing effect of the convolution with the PSF, hence the name Minimum Information Method MIM (Pfleiderer 1985, 1988). The derivation of a corresponding expression for structural information from general premises, such as invariances, will be given in a forthcoming paper (Pfleiderer, in preparation). The method is related to SSC. In particular, it also uses a kind of Prussian Helmet PSF.

Maximum entropy is characterized by a different approach to the question what "structure" is and what kind of structure should be suppressed. The philosophy of MEM has been described in a large number of papers (see, *e.g.*, Jaynes 1957, Frieden 1972, Ables 1974). The original main disadvantage of MEM, *viz.* the large size of the computer program which made MEM inaccessible to the average user, is now much eased by the availability of more compact programs and larger computers.

MEM has been the most successful deconvolution routine so far, with applications to a wide field of problems, as main beam deconvolution, photography (Bryan and Skilling 1981), interferometry (Wernecke 1977, Gull and Daniell 1978, Nityanda and Narayan 1982, Sanromà and Estalella 1984), incomplete data (Gull and Daniell 1978), spectral analysis and time series (Jensen and Ulrych 1973, Komesaroff *et al.* 1981), computer tomography, seismology.

5 CLEAN

This method, dating back to Högbom (1974), was specially devised for handling incomplete interferometric radio data. In image space, the incompleteness of Fourier data can be described by a beam with marked and extended side lobes ("dirty beam"), giving rise to a distorted image ("dirty map"). CLEAN removes, by deconvolution, the side lobes without, however, being a true deconvolution method. The result is not a model of the source distribution (to be observable with a perfect very large instrument) but rather a model of what would have been observed with a single dish of the same size as the interferometer ("clean map"). That is, it is a data model and not a source model which would need convolution with a beam to reproduce data. One could also say that the missing Fourier coefficients are interpolated but not extrapolated.

The dirty beam h_{ij} is divided into two parts: The "clean beam" $h_{ij}^{(1)}$ which is essentially the main lobe, or the response of a correspondingly large single dish to a point source, and the sidelobes and main-beam distortions, or "dirt" $h_{ij}^{(2)}$:

$$h_{ij} = h_{ij}^{(1)} + h_{ij}^{(2)} . \tag{12}$$

The original image-space data $f_1^{(0)}$ ("dirty map") is deconvolved to a source map $\{m_j\}$ but (in the original version) without additional constraint. The deconvolution result cannot be used directly because it is not smooth enough. Owing to the fact that sidelobes tend to be more extended or at least not less extended than the main lobe, a smooth "image", more or less free of sidelobe effects, can be recovered by convolving the source map with the clean (or "restoring") beam. One actually ends up with an improved (or "restored") data map $f_1^{(1)}$ ("clean map")

$$f_i^{(1)} = \sum_j m_j h_{ij}^{(1)} + r_i = f_i^{(0)} - \sum_j m_j h_{ij}^{(2)} . \tag{13}$$

The method is quite ingenious as it avoids such difficult questions as whether or not a smooth image is also a true image. It was also the first method not to neglect missing Fourier coefficients but to choose them according to a reasonably smooth image. Nevertheless, it definitely does not increase the resolution. Therefore, several improvements have been proposed of which we mention only two. First, one can restore with a clean beam that is decreased in size. This is equivalent to including the outer parts of the main lobe into the dirt. Second, one can introduce a constraint such that the deconvolution result is smooth enough to be directly used. This is done in the SSC.

CLEAN has as yet mostly been used in interferometry but at least some of the modern versions are suitable for other problems as well (Becker and Duerbeck 1980).

6 Comparison of methods

There are many deconvolution methods, of which we have mentioned only some, and all have different difficulties. It would certainly be wrong to try to make a linear order of successfulness for the available methods. Even if a method gives a result that looks "good" (meaning that it does not contain obviously improbable or impossible features), it may not be the most reliable one. In general, the best advice as to which methods should be used is to try several ones, and compare the results. Such procedure will quite often provide more information on the probable source distribution than the selection of just one method.

However, some general statements are nevertheless possible. First, no method is hitherto sufficiently understood to know exactly all the advantages and disadvantages, and to know how would be the best interpretation of the results in terms of reliability (as the question whether a slightly extended feature should be interpreted as an unresolved (nearly pointlike) source or as a resolved one). Or to give another example: Inspite of a wealth of theoretical papers on MEM, there is still not even agreement on which form of the entropy should be used. More dangerously, it is not known how much one part of the map may influence the results on other parts of the map.

This is because entropy is a "universal" constraint, not dependent on any details of the measurement. One only knows, from many practical examples, that the mutual influence seems, in most cases, small enough to be neglected.

There is always a competition between smoothness and resolution. The grand design of a map is most easily recognized if the map is very smooth but some essential details may be lost. High resolution tends to overresolve noisy data. The best compromize is probably MEM, with a very smooth image and some superresolution (= resolution beyond that of the data). The claim of some MEM theorists that MEM yields a maximum in possible superresolution is not true. The best resolution so far has been obtained by MIM which, on the other hand, tends to yield a noisier result than other smoothing-constraint methods. The resolution of optimized versions of CLEAN is comparable to that of MEM.

The opinion is widely held (see, *e.g.*, Koch and Anderssen 1987) that the result of a deconvolution should be unique (concave problem). It has been shown that the one-constraint MEM (but not two-constrained versions as that of Reiter and Pfleiderer 1986) as well as the basic CLEAN (Schwarz 1978, Marsh and Richardson 1986) do indeed converge to a unique result, independent of the actual realization of the method in the form of a specific numerical procedure. However, uniqueness is probably quite unimportant. Different methods do give different results, and still we are often unable to choose one as being better than another. The only criterion for the goodness of a solution is whether or not it looks "good" enough in the sense stated above - unless one can compare with better data. However, the most interesting use of deconvolution is, of course, that for the best available data where such comparison is not possible. Non-unique methods do, however, have the disadvantage that the result may depend on the numerical procedure. If different procedures produce different results within one method, one could consider them as varieties of a method and try to find out which variety, if any, works best. The iterational Fourier methods are not unique, the cut-off point of the iterations being empirically determined.

One mandatory feature of uniqueness in constraint methods is that the optimum Lagrange parameter connecting the data fit and the smoothing constraint must be determined by the method itself. It seems to the present writer that this is not necessarily a good approach. Depending on the questions asked, one and the same set of data may be used to emphasize the grand design (large smoothing) or fine details (little smoothing). Some methods therefore allow the choice of the degree of smoothing. A consistent theory of structure (Pfleiderer, in preparation) seems to make the free choice even mandatory.

Unfortunately, all methods which are such simple that they would easily be programmed are also inferior to others. In practice, one therefore has, generally, to rely on what methods an available program library has to offer.

Acknowledgments

Thanks are due to many colleagues for very helpful and clarifying discussions, especially to E.E. Baart, T.J. Cornwell, L. Ericson, S. Gottesman, J.A. Högbom, G.E. Ioup, J. Reiter, U. Schwarz, S. Tan.

References

Ables, J.G., 1974. *Astr. Astrophys. Suppl.*, **15**, 383.
Basistov, Y.A., Goncharski, A.V., Lekht, E.E., Cherepashchuk, A.M., Yagola, A.G., 1979. *Sov. Astr.*, **23**, 248.
Becker, H.J., Duerbeck, H.W., 1980. *Publ. Astr. Soc. Pacific*, **92**, 792.
Biraud, Y., 1969. *Astr. Astrophys.*, **1**, 124.
Bracewell, R.N., Roberts, J.A., 1954. *Austr. J. Phys.*, **7**, 615.
Brault, J.W., White, O.R., 1971. *Astr. Astrophys.*, **13**, 169.
Bryan, R.K., Skilling J., 1980. *Mon. Not. R. astr. Soc.*, **191**, 69.
Cornwell T.J., 1983. *Astr. Astrophys.*, **121**, 281.
Frieden, B.R., 1972. *J. Opt. Soc. Am.*, **62**, 511.
Frieden, B.R., 1975. In *Picture processing and digital filtering (Topics in Appl. Physics No. 6)*, ed. Huang, T.S., Springer, Berlin, p. 177.
Gull, S.F., Daniell, G.J., 1978. *Nature*, **272**, 686.
Helstrom, C.W., 1967. *J. Opt. Soc. Am.*, **57**, 297.
Högbom, J.A., 1974. *Astr. Astrophys. Suppl.*, **15**, 417.
Ioup, G.E., Ioup, J.W., 1983. *Geophys.*, **48**, 1287.
Jansson, P.A., Hunt, R.H., Plyler, E.K., 1970. *J. Opt. Soc. Am.*, **60**, 596.
Jaynes, E.T., 1957. *Phys. Rev.*, **106**, 620.
Jensen, O.G., Ulrych, T., 1973. *Astr.J.*, **78**, 1104.
Jonas, J., 1985. *South Afr. J. Phys.*, **8**, 67.
Koch, I., Anderssen, R.S., 1987. *Astr. Astrophys.*, **183**, 170.
Komesarov, M.M., Narayan, R., Nityanda, R., 1981. *Astr. Astrophys.*, **93**, 269.
Lucy, L.B., 1974. *Astr. J.*, **79**, 745.
Nityananda, R., Narayan, R., 1982. *J. Astrophys. Astr.*, **3**, 419.
Pfleiderer, J., 1985. *South Afr. J. Phys.*, **8**, 73.
Pfleiderer, J., 1988. *Astr. Astrophys.*, **194**, 344.
Pfleiderer, J., Reiter, J., 1982. *Sitz.-Ber. Oest. Akad. Wiss. Math.-Nat. Kl.*, **191**, 591.
Phillips, D.L., 1962. *J. Assoc. Comp. Mach.*, **9**, 84.
Reiter, J., Pfleiderer, J., 1986. *Astr. Astrophys.*, **166**, 381.
Sanromà, M., Estalella, R., 1984. *Astr. Astrophys.*, **133**, 299.
Schwarz, U., 1978. *Astr. Astrophys.*, **65**, 345.
Siska, P.E., 1973. *J. Chem. Phys.*, **59**, 6052.
Subrahmanya, C.R., 1980. *Astr. Astrophys.*, **89**, 132.
Tikhonov, A.N., 1963. *Sov. Math. Dokl.*, **4**, 1035 and 1624.
Tikhonov, A.N., Arsenin, V.Y., 1977. *Solutions of Ill-posed Problems*, Winston & Sons.
Turchin, V.F., Turovtseva, L.S., 1974. *Opt. Spectrosc.*, **36**, 162.
Twomey, S., 1963. *J. Assoc. Comp. Mach.*, **10**, 97.
Twomey, S., 1965. *J. Frank. Inst.*, **279**, 95.
van Cittert, P.H., 1931. *Z. Phys.*, **69**, 298.
Wernecke, S.J., 1977. *Radio Sci.*, **12**, 831.
Wiener, N., 1942. *Extrapolation, Interpolation, and Smoothing of Stationary Time Series*, MIT Press, Cambridge, MA.
Wittmann, A., 1971. *Solar Phys.*, **21**, 237.

Comparison of Different Mathematical Methods for the Investigation of Object Distributions

Konrad Rudnicki
Jagiellonian University Observatory
Kraków, Poland

Abstract

The suitability of different mathematical methods with respect to particular problems is shown.

Mathematical methods

Looking over papers which deal with the distribution of objects on the celestial sphere or in space, one has the impression that each author clings to a certain mathematical method, his favorite one, and does not pay regard to any of the other well-known procedures, even if those might be more appropriate for his particular problem.

Table 1 lists some of the methods available, indicating their respective suitability for a particular purpose. Only the most obvious properties of the methods are taken into consideration here, since this contribution is not meant to be a deep comparative study of the methods considered. It also refrains from comparing the numerical complexity of calculations, as well as their response to small effects. These are certainly important factors for users, but require a more comprehensive analysis, appropriate for an extensive monograph, and thus beyond the scope of this paper.

The following methods are included in Table 1:

CF - the Correlation Function Method as presented by Peebles (1980)

TC - the Three-Circle Method as presented by Garncarek (1986)

P - the Percolation Method as presented by Klypin (1988)

SR - the Method of Statistical Reduction as presented by Garncarek et al. (1988) and Zięba (1988)

LG - the Local Grouping Method as presented by Bereś (1986)

I hope that this comparison may be helpful for new studies in the field.

Table 1. The suitability of different methods for particular purposes.

Purpose	Methods CF	TC	P	SR	LG
Determination of:					
characteristic sizes and strength of clustering	2	1	1	2	1
numbers of individual clusters within the assumed working definition of a cluster	0	0	2	1	2
regions of enhanced clustering	0	1	1	2	1
anisotropies, e.g. general gradients of clustering or background	0	0	0	2	0
shapes of individual clusters	0	2	0	0	2
General properties:					
Possibility of:					
independent studies of effects in different directions	0	0	1	2	1
direct comparison of samples including substantially different numbers of objects	2	2	0	2	0

0 – not suitable; 1 – suitable; 2 – very suitable.

References

Bereś, K., 1986. *Acta Cosmologica*, **14**, 109.
Garncarek, Z., 1986. *Acta Cosmologica*, **14**, 101.
Garncarek, Z., Kuklewski, J., Rudnicki, K., 1988. *Acta Cosmologica*, **15**, 63.
Klypin, A.A., 1988. *Acta Cosmologica*, **15**, 102.
Peebles, P.J.E., 1980. *The Large-Scale Structure of the Universe*, Princeton University Press, Princeton.
Zięba, S., 1988. *Acta Cosmologica*, **15**, 85.

Multivariate Analysis Methods: Background and Example

Fionn Murtagh[1]
Space Telescope — European Coordinating Facility
European Southern Observatory
Garching bei München, F.R. Germany

Abstract

Multivariate statistical methods deal with the inherently very difficult problem of detecting patterns in data. These patterns can take many forms — natural groups, inherent dimensionality, correlations, dependencies, and so on. Often, therefore, different methods bring different features of the data to light.

Following a brief overview of some prominent multivariate methods, we illustrate their use on IRAS data. We indicate how different multivariate methods can be "chained together" to yield powerful tools for uncovering structure in data.

1 Multivariate data analysis methods

When faced with large quantities of multiple-parameter data, multivariate data analysis algorithms can offer considerable time-savings, together with ensuring consistency and "objectivity" of treatment. Being multivariate (multidimensional), they allow the simultaneous treatment of many variables.

There are many types of multivariate algorithms, but among the most commonly used are algorithms for *cluster analysis*, *discriminant analysis* and *principal component analysis.*

Given a set of objects, each characterised by the same set of variables, clustering methods will produce groups of the objects. The objects in the resulting groups will either be more similar in feature space to one another than to non-group members, or satisfy some other homogeneity or compactness criterion. "Similarity" is most often defined by the Euclidean distance, but other metrics may well merit consideration. The question of "standardization" or "normalization" (centring the objects in the multidimensional space and rescaling them to have unit variance) may also have to be addressed before carrying out the clustering. Of course the groups of objects found by a clustering algorithm are in the parameter/variable space, and this will not necessarily have a direct relationship with positional, 2- or 3-dimensional space.

A method of clustering (closely related to widely used hierarchical clustering methods and percolation methods; for the latter, see, *e.g.*, Schulman and Seiden 1986), is the

[1] Affiliated to Astrophysics Div., Space Science Dept., European Space Agency.

minimal spanning tree. It is a graph theoretic representation of the set of points and has been used in studies of galaxy clustering (*e.g.* Barrow *et al.* 1985).

One of the aims of discriminant analysis methods is to assess a known assignment of objects to groups. Thus, such methods can be used to study the results of a cluster analysis. Discriminant methods also can be used for assignment of objects to already existing groups. When used for this second objective (*i.e.* assignment), discriminant analysis has been referred to as "supervised classification" (because of the need to define the training set, — perhaps by a visual study of a relatively small number of objects), while cluster analysis has been termed "unsupervised classification".

Principal components analysis is used for dimensionality reduction The best linear combinations of the axes in the initial parameter space are sought. Thereby new and often fewer coordinate axes (the underlying "principal components" of the data) are determined. These may be used for interpreting the data or for providing the best possible planar projection(s) of the data.

Comprehensive background material on multivariate methods is available in Murtagh and Heck (1987b). Other general references on this area include the MIDAS Users' Guide (1985) and Murtagh and Heck (1987a). MIDAS software includes all methods discussed here. In particular a storage-economic hierarchic clustering method is available: in-core storage of an $n \times m$ input matrix (where n and m are number of rows and columns, respectively) is required rather than the usual $O(n^2)$ storage. Also a

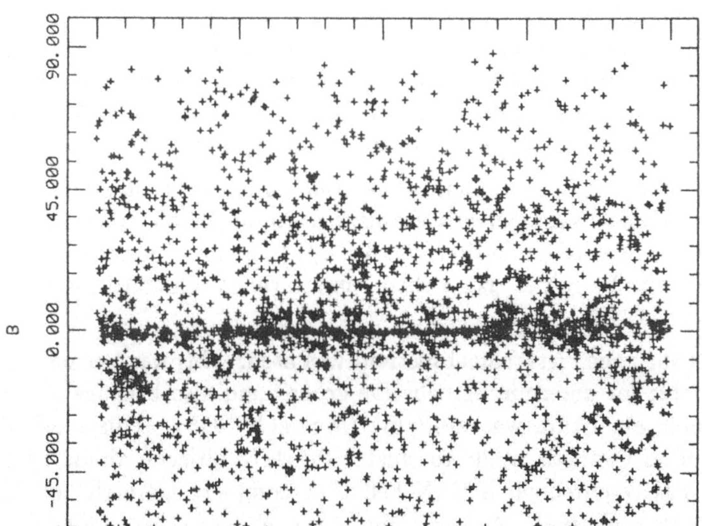

Fig. 1. Galactic longitude and latitude locations of all 3178 objects.

very efficient routine for the minimal spanning tree is available, based on an initial preprocessing of parameter space (Rohlf 1978).

2 Application to IRAS Point Source Catalog data

The data used consisted of a sample of 3181 IRAS PSC objects. The article by Adorf and Meurs (1988) should, in particular, be referred to for more background on the data used (further references may be obtained in Meurs *et al.* 1988). The 3181 objects were taken from approximately 30 000 non-stellar objects for computational/storage convenience; the selection was random (every tenth object from the collection of 30 000-odd objects was taken). The aim was to investigate the main classes of these objects with a view towards finding relatively well defined groups of objects for further study.

Three colours and a log *flux* value (see Meurs *et al.* 1988) were used to characterize these objects. In the notation of the last-mentioned reference, these are: c_{12}, c_{23}, c_{34} and $\log f_{100}$. Three objects having missing values were deleted, leaving a set of 3178 objects characterisable as points in a four-dimensional parameter space. Again, in the last-mentioned reference, it is described how clusters corresponding to "thin (galactic) plane", "cirrus" and "galaxies" are of interest: these can be represented in the plot of galactic longitude and latitude positions associated with the 3178 objects (*Fig. 1*).

A principal components analysis (PCA) was carried on the 3178×4 matrix. The PCA was carried out on a correlation matrix, *i.e.* the 3178 objects were *centred* and *reduced* in the parameter space. A plot of the objects in the principal plane (*i.e.* the plane defined by principal components 1 and 2) did not seem particularly interesting. For instance, no grouping of the objects was visible in this optimal planar representation.

However, the three principal components accounted for more than 96.5 % of the variance. These three new parameters were used as input for a cluster analysis (*i.e.* the input data matrix was of dimensions 3178×3). If, as an alternative, clustering had been employed on the initial data, then some form of standardization would have been necessary.

The minimum variance hierarchical method was used. This was because this method is recommendable for determining cohesive groups and also because an efficient and storage-economic algorithm was available (Murtagh 1985). This method took the longest time of all methods used: to cluster the 3178 objects, about 10–15 minutes elapsed time was required on a VAX 8600. A complete hierarchy of partitions is provided by such a method. Knowing that three classes were primarily of interest, the three-cluster partition alone was examined.

The galactic longitude and latitude coordinates associated with the 3178 objects were plotted for the three different groups found. These are shown in *Fig. 2*. As can be seen, from a visual point of view they are quite satisfactory.

To assess these results, a multiple discriminant analysis (MDA; or canonical discriminant analysis) was carried out. MDA may be informally described as a PCA on the

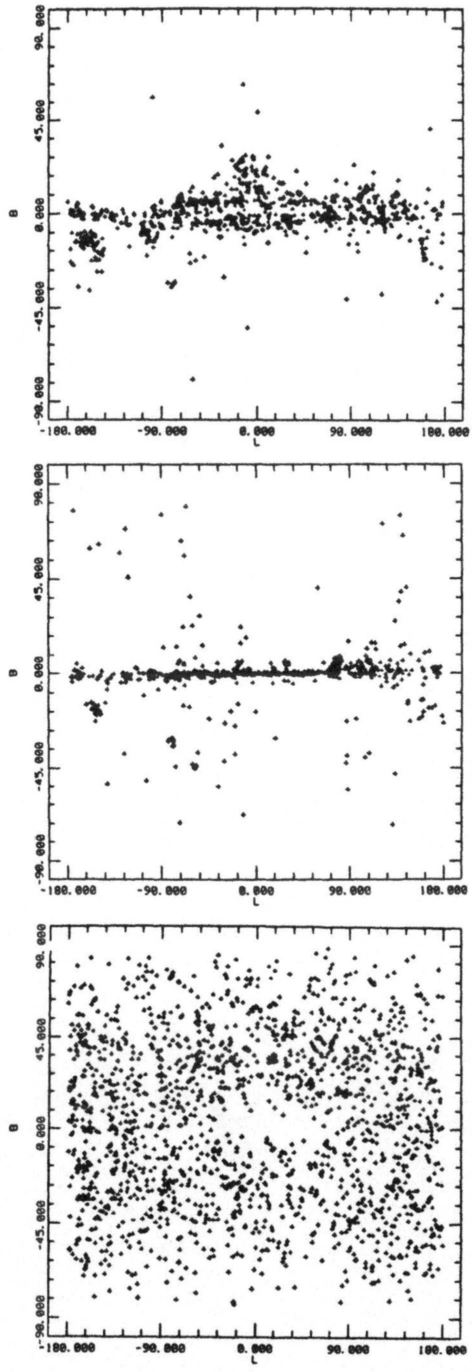

Fig. 2. Galactic longitude and latitude locations of (from top down) "galaxy","thin plane" and "cirrus" groups.

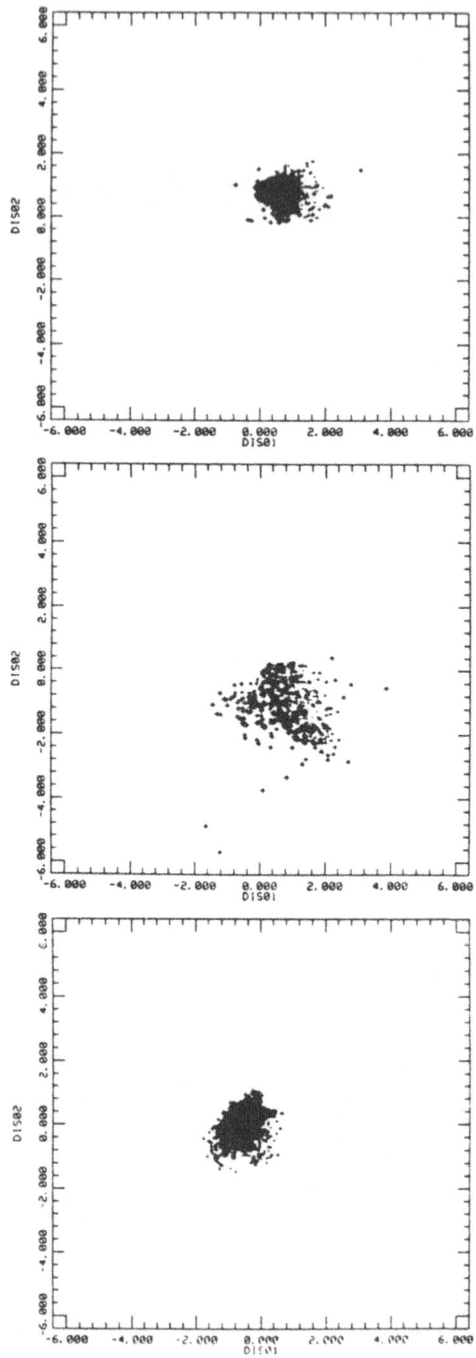

Fig. 3. Discriminant factor plane showing (from top down) "galaxy","thin plane" and "cirrus" groups.

groups. It attempts to separate the groups optimally using hyperplanes. The MDA implemented took the known assignments of the 3178 objects to the three groups (found by the cluster analysis above) in the original four-dimensional space (c_{12}, c_{23}, c_{34} and $\log f_{100}$). This was done because conclusions derived would be more meaningful for these parameters rather than the derived principal components.

Two discriminant factors (*i.e.* axes) of nearly equal "discriminating power" (measured by the eigenvalues) were obtained. Discriminant factors 1 and 2 were found to be defined as

$$f_1 = -0.5c_{12} + 0.1c_{23} - 1.8c_{34} + 0.1 \log f_{100}$$
$$f_2 = 0.9c_{12} - 2.0c_{23} - 0.5c_{34} - 0.0 \log f_{100}$$

The above equations indicate the relative importance of these parameters for these discriminating factors. The projections of the three groups in the plane defined by discriminant factors 1 and 2 are shown in *Fig. 3* (f_1 and f_2, above, are denoted by $DIS01$ and $DIS02$). Some overlap is evident, but also some clear "regions" of unequivocal group membership. Such unequivocally classed objects could be used as "pure" samples of "galaxy", "thin plane" or "cirrus".

It may be noted in the above equations that the parameter $\log f_{100}$ is almost irrelevant from the point of view of the discriminating factors. This points to the redundancy of this parameter if one uses the above equations for the assignment of new objects to one or other of the three classes.

The most time consuming algorithm used was the cluster analysis one. Its computational requirements are $O(n^2)$ when n objects are being classified. Although a sample of 3000-odd objects was used in the foregoing, a clustering was also carried out on the 30 000–odd set of objects. Clustering (hierarchic clustering using Ward's minimum variance criterion) of 31 760 objects took, on a VAX 8600 machine, 11.11 hours of CPU time. Note that (i) most widely available unsupervised clustering methods require in-core storage of dissimilarities and therefore would not work for such a large number of objects; and (ii) no special speed-up techniques were availed of (Murtagh 1985).

Acknowledgements

H.-M. Adorf and E.J.A. Meurs made their data available for the application described here, and helpfully discussed the issues involved.

References

Adorf, H.-M., Meurs, E.J.A., 1988. *These proceedings*, p. 315.

Barrow, J.D., Bhavsar, S.P., Sonoda, D.H., 1985. *Mon. Not. R. astr. Soc.*, **216**, 17.

Meurs, E.J.A., Adorf, H.-M., Harmon, R.T., 1988. In *Astronomy from Large Databases*, eds. Murtagh, F., Heck, A., European Southern Observatory, Garching, p. 49.

The MIDAS Users Guide, 1985 (and subsequent versions). Chapter 13, *Multivariate Statistical Methods*, European Southern Observatory, Garching.

Murtagh, F., 1985. *Multidimensional Clustering Algorithms*, Physica–Verlag, Heidelberg and New York.

Murtagh, F., Heck, A., 1987a. *Astr. Astrophys. Suppl.*, **68**, 113. (*copy of bibliography available from author*).

Murtagh, F., Heck, A. 1987b. *Multivariate Data Analysis*, Kluwer Academic, Dordrecht.

Rohlf, F.J., 1978. *Information Processing Letters*, **7**, 44.

Schulman, L.S., Seiden, P.E., 1986. *Science*, **233**, 425.

Supervised and Unsupervised Classification – The Case of IRAS Point Sources

H.-M. Adorf
Space Telescope – European Coordinating Facility
European Southern Observatory
Garching b. München, F.R. Germany

E.J.A. Meurs
European Southern Observatory
Garching b. München, F.R. Germany

Abstract

Progress is reported on a project which aims at mapping the extragalactic sky in order to derive the large scale distribution of luminous matter. Our approach consists in selecting from the IRAS Point Source Catalog a set of galaxies which is as clean and as complete as possible. The decision and discrimination problems involved lend themselves to a treatment using methods from multivariate statistics, in particular statistical pattern recognition. Two different approaches – one based on supervised Bayesian classification, the other on unsupervised data-driven classification – are presented and some preliminary results are reported.

1 Introduction

The Infrared Astronomical Satellite (IRAS) was launched in January 1983 and successfully operated for a period of about 300 days, during which more than 96 % of the sky was surveyed at an angular resolution between $\sim 0.5'$ and $\sim 2'$ depending on wavelength (Beichman *et al.* 1985). The Point Source Catalog (PSC) resulting from the IRAS mission constitutes an attractive database for classification pursuits. With a high level of homogeneity and almost complete sky coverage, the PSC provides positions and infrared fluxes at four wavelengths for a total of $\sim$ 250 000 sources. IRAS looked relatively unhampered through much of the Galaxy, but nearer the galactic centre the high source density causes noticeable source confusion along the galactic plane.

On the basis of their infrared colours (flux ratios), sources contained in the PSC can to a large degree be separated into four main categories, as has been demonstrated in a number of studies (Chester 1986, Lawrence *et al.* 1986, Wolstencroft *et al.* 1986, Habing 1987, Soifer *et al.* 1987). Exploiting this property, Meurs and Harmon (1988) have produced sky maps for these source categories, while aiming specifically at a homogeneous map of almost the entire extragalactic sky. The other source categories that could be distinguished are stars, a very thin galactic component (which may largely consist of H II regions) and a broader and more diffusely distributed galactic component that is related to star forming regions (where also the Gould Belt can be recognised).

A successful and reliable selection of extragalactic objects from the IRAS PSC will be highly interesting for various research projects. The importance of a homogeneous and complete all-sky sample of galaxies for a dipole anisotropy determination has already been demonstrated (Harmon *et al.* 1987). Although cosmological inhomogeneities may be less prominent in the sky distribution of spiral galaxies (mainly recorded in the PSC) when compared to ellipticals, the all-sky homogeneity of the IRAS data render them an important sample for studying sky distribution features. Besides this, the forthcoming Co-Added Catalog may be expected to contain ellipticals as well (*cf.* Knapp 1987). Other areas where such an extragalactic IRAS sample may be beneficial include studies of luminosity functions and of statistical relations involving IRAS data. Furthermore, regions much nearer the galactic plane become accessible for all kinds of research (see *e.g.* Dow *et al.* 1988).

2 A sidestep into data analysis methodology

Data analysis may be subdivided into two categories, *exploratory* and *confirmatory.* With confirmatory data analysis one tries to corroborate – or falsify – a specific preconceived hypothesis. Exploratory data analysis, on the other hand, aims at discovering regularity or structure inherent to a given data set "with no preconceived notions or precise questions in mind" (Friedman 1986). In exploratory mode, the data analyst must be open to several equally legitimate structures in the data. The exploration phase logically precedes the confirmation phase and is a prerequisite for forming any hypothesis to be tested.

Exploratory data analysis is commonly performed by constructing a classification scheme over the set of data points (objects), where the abstract clustering task can be defined as follows (Fisher and Langley 1986): "Given: A set of objects, O. Goal: Distinguish clusters (*i.e.* subsets of O) $s_1, \ldots, s_n$, such that the intra-cluster object similarity of each s_i tends to be maximized, and the inter-cluster object similarity over all s_j's tends to be minimized." If successful, such a classification procedure results in a data description, which is more condensed and therefore easier to communicate than the original set itself. Creating a classification is also a typical first step in developing a theory about a collection of observations (Stepp and Michalski 1986b).

When no *a priori* information is given for the association of the objects with categories, the classification process is said to be *unsupervised.* For quite some time unsupervised classification or "learning without a teacher" was widely felt to be impossible, and, indeed it is not uniquely possible in general (Cooper 1969a, b) because quite often the same data can be organised in different ways (Fisher and Langley 1986). In particular, difficulties may arise from overlapping or interleaved categories, or when the data stem from a continuous distribution displaying no natural decomposition into classes.

Nevertheless, unsupervised classification is not only possible in a wide range of situations, as has been shown for example by Cooper and Cooper (1964), but also of importance, because *supervised* classification may be inconvenient, too costly or even impossible, for one of the following reasons:

(i) classes may be unknown, *e.g.* because data are coming from a new instrument *(problem novelty)*;
(ii) the dimensionality of the feature space may be too high for easy visualisation *(problem complexity)*;
(iii) it may be difficult to separate the various populations from each other *(problem difficulty)*;
(iv) the number of objects to be considered may be very large *(problem size)*.

Given the need for *unsupervised* classification and the potential difficulties a human analyst may run into, the intriguing question is, to what extent the class formation task could be carried out by a computer, with its capability of analysing large data sets *automatically* and *objectively*.

Past work on automated generation of classes was performed under the headings of numerical taxonomy and cluster analysis. Numerical taxonomy offers a number of algorithmic clustering techniques: *e.g. optimisation*, which attempts to construct an optimal partition of the data set into mutually exclusive classes; *hierarchical clustering*, which forms a classification tree over the object set; and *clumping*, which allows for overlapping classes. Kurtz (1983), Murtagh (1986, 1987), and Murtagh and Heck (1987a, b) review the application of some of these techniques to astronomical problems.

Recently the automated generation of classification schemes has attracted researchers from the area of artificial intelligence (see the bibliography by Kedar-Cabelli and Mahadevan 1986). From an artificial intelligence perspective, numerical taxonomy can be viewed as a first step towards *conceptual clustering* (Fisher and Langley 1986, Michalski and Stepp 1983, Stepp and Michalski 1986a, b), an artificial intelligence technique which aims at identifying higher level (conceptual) descriptions of object groups. However, the claim that conceptual clustering is superior to numerical taxonomy has been criticised by Dale (1985).

Once a partition of feature space into classes has been established, *supervised* techniques from the well-founded theory of statistical inference, in particular decision theory, can be used to classify additional objects into classes derived from "training sets". (For various aspects of pattern recognition and classification see Watanabe 1969, Grasselli 1969, Duda and Hart 1973, Bock 1974, Batchelor 1978, Melsa and Cohn 1978, Fu 1980, Hand 1981, Sklansky and Wassel 1981, Kulkarni 1986, Jain 1987, Mantas 1987.)

3 Supervised classification of IRAS point sources

A first step towards a proper classification of the IRAS PSC, applying multivariate statistical methods and concepts from decision theory, was made by Meurs *et al.* (1988). The distributions of data points in four-dimensional feature space – three infrared colours and one flux, as in Meurs and Harmon (1988) – were represented by multivariate Gaussian distributions. These were fitted to training sets for each source

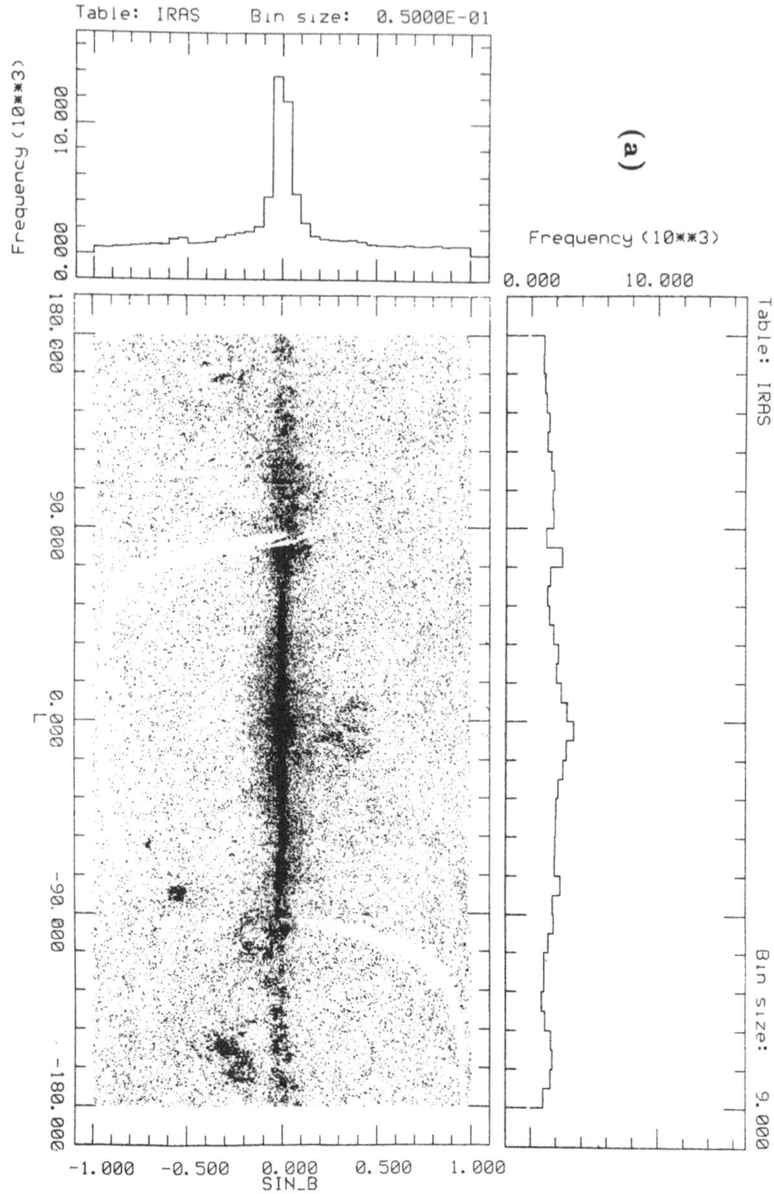

Fig. 1a–b. Sky distribution (a) of all IRAS Point Sources except stars (a sample similar to that in Meurs *et al.* 1988), and (b) of the set of "galaxies" as found by the maximum likelihood classifier. Histograms at the top and left show how source density varies with galactic longitude and (sine of) galactic latitude, respectively. (Note the different scales of the histograms!)

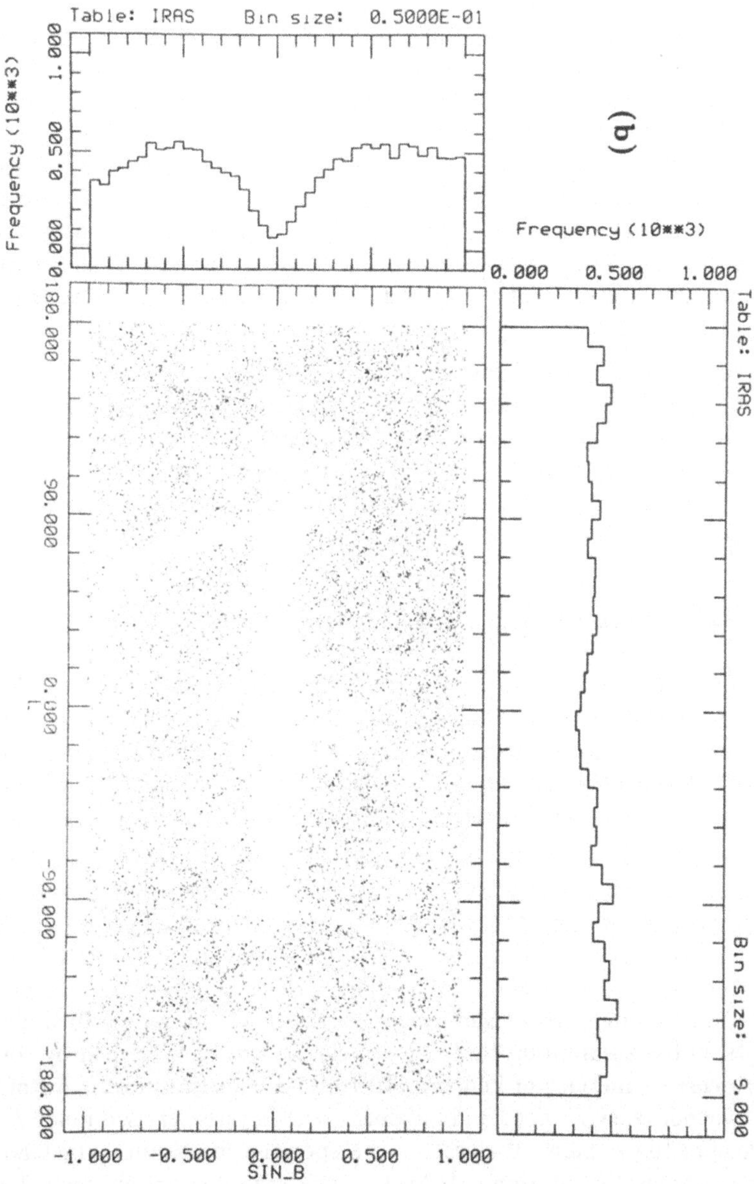

The latitude distribution in (b) clearly shows the effect of missing sources along the galactic plane in the galactic centre direction. There is a decrease in source density towards the southern galactic pole consistent with the north-south anisotropy found for IRAS galaxies (see Clowes *et al.* 1987); an additional slight decrease at high galactic latitudes, north and south, may be attributed to those parts of the empty strips (representing the 4 % of the sky not covered in the PSC) which run parallel to the longitude axis.

category obtained from sky regions where one category at a time could be expected to dominate. (From the beginning stars were essentially excluded from consideration, by applying an infrared colour cut and requiring good fluxes at longer wavelengths). The Gaussian approximation appears appropriate for the galaxy category; the distributions of the other two categories show some non-Gaussian structure, suggesting a different distribution type or a division into two subcomponents. A maximum-likelihood decision strategy established a category separation very comparable to, though probably better than, the more intuitive "handcrafted" approach of Meurs and Harmon (1988). The separation was only slightly modified when a maximum-a-posteriori decision strategy was used which takes into account the relative population of the categories considered.

From the sky map (*Fig. 1b*) displaying the resulting set of "galaxies" – the source distribution of the IRAS PSC (stars excluded) is shown for comparison in *Fig. 1a* – it is obvious that the apparent all-sky galaxy distribution is far from being homogeneous, reflecting the confusion problem in the area surrounding the galactic centre. Indeed it is remarkable that, near the galactic anti-centre, galaxies can be found practically in the middle of the galactic plane.

4 Unsupervised classification of IRAS point sources

Our steps into the area of unsupervised classification are mainly motivated by our interest to see whether the non-Gaussian distributions mentioned above would naturally split into two or more subcomponents. Also we are curious to see whether an unsupervised classification procedure, having access only to information intrinsic to the data, would find the same three categories used in supervised classification.

Various approaches are currently being pursued: 'Conventional' cluster analysis has been performed on a randomly selected subset of the IRAS PSC (stars excluded) with encouraging results (see Murtagh 1988). In another approach we are using the AutoClass program developed at the NASA/Ames Research Center (Cheeseman *et al.* 1987). AutoClass was designed for studying problems of machine learning and is implemented in Common Lisp. Similar to the Bayes classifier described above, AutoClass rests on the assumption that pattern classes can be described by conditional probability density functions of known (multivariate Gaussian) form. AutoClass can be used in two modes, supervised and unsupervised. In unsupervised mode AutoClass does not require the number of classes to be specified in advance. Instead, it uses an information-theoretic criterion that for a given data set simultaneously determines an optimal partition of feature space and an optimal number of classes. Thus AutoClass appears well-suited to the task of finding additional object categories beyond those already known, or to optimally splitting categories that show non-Gaussian distributions in feature space.

Preliminary investigations with AutoClass of a sample from the IRAS PSC are suggestive but inconclusive. Areas requiring further investigation include:
(i) *convergence properties:* it appears that the search for a globally optimal set of classes sometimes ends trapped in a local optimum;

(ii) *performance:* AutoClass in its present form appears to be significantly slower than, *e.g.* more conventional cluster analysis methods, and so may be inappropriate for very large datasets. On the other hand, the attractiveness of the method makes it important to assess whether this is intrinsic to the method or can be circumvented by a more efficient implementation.

5 Summary

Using supervised and unsupervised classification methods we have attempted an 'intrinsic classification' of a relevant subset of the IRAS Point Source Catalog (essentially excluding stars) with the immediate goal of selecting a maximally *clean*, *complete* and *unbiased* set of galaxy candidates. It appears that the extragalactic sources to a large degree can be separated from other, galactic sources contained in the IRAS PSC. This was achieved by constructing a Bayesian classifier from suitably chosen training sets for each source category, with sources being selected on the basis of three infrared colours and one infrared flux. A representative subset of the PSC sources considered was also subjected to unsupervised classification by feeding it to the AutoClass program. The results initially obtained with this program suggest its potential as a powerful tool for exploring unknown data sets, but reaching this state will without doubt require further development efforts.

Acknowledgements

Part of this work relied on the availability of the AutoClass program, which was kindly provided by Peter Cheeseman and his group at NASA/Ames and installed at ESO by Mark D. Johnston, STScI.

References

Batchelor, B.G., 1978. In *Pattern Recognition, Ideas in Practice*, ed. Batchelor, B.G., Plenum Press, New York and London, p. 65.

Beichman, C.A., Neugebauer, G., Habing, H.J., Clegg, P.E., Chester, T.J. (eds.), 1985. *Explanatory Supplement to the IRAS Catalogs and Atlases*, US Govmt. Print. Off.

Bock, H.H., 1974, *Automatische Klassifikation*, Vandenhoeck & Ruprecht, Göttingen.

Cheeseman, P., Stutz, J., Freeman, D., Self, M., 1987. In *Proc. 1987 GSFC Conf. on Space Applic. of AI* (forthcoming).

Chester, T.J., 1986. In: *Light on Dark Matter*, ed. Israel, F.P., Reidel, Dordrecht, p. 3.

Clowes, R.G., Savage, A., Wang, D., Leggett, S.K., MacGillivray, H.T., Wolstencroft, R.D., 1987. *Mon. Not. R. astr. Soc.*, **229**, 27p.

Cooper, D.B., Cooper, P.W., 1964. *Information and Control*, **7**, 416.

Cooper, P.W., 1969a. In *Automatic Interpretation and Classification of Images*, ed. Grasselli, A., p. 97.

Cooper, P.W., 1969b. In *Methodologies of Pattern Recognition*, ed. Watanabe, S., p. 97.

Dale, M.B., 1985. *IEEE Trans. Pattern Analysis Machine Intelligence*, **PAMI-7**, 241.

Dow, M.W., Lu, N.Y., Houck, J.R., Salpeter, E.E., Lewis, B.M., 1988. *Astrophys. J.*, **324**, L51.

Duda, R.O., Hart, P.E., 1973. *Pattern Classification and Scene Analysis*, John Wiley & Sons, New York.

Fisher, D., Langley, P., 1986. In *Artificial Intelligence & Statistics*, ed. Gale, W.A., Addison-Wesley, Reading, Mass., p. 77.

Friedman, J.H., 1986. In: *Data Analysis in Astronomy II*, eds. Di Gesù, V., Scarsi, L., Crane, P., Friedman, J.H., Levialdi, S. Plenum Press, New York, p. 3.

Fu, K.-S., 1980. *IEEE Trans. Comput.*, **C-29**, 845.

Grasselli, A., (ed.), 1969. *Automatic Interpretation and Classification of Images*, Academic Press, New York.

Habing, H.J., 1987. In *IAU Symp. 122, Circumstellar Matter*, ed. Appenzeller, I., Reidel, Dordrecht, p. 197.

Hand, D.J., 1981. *Discrimination and Classification*, John Wiley & Sons, Chichester.

Harmon, R.T., Lahav, O., Meurs, E.J.A., 1987. *Mon. Not. R. astr. Soc.*, **228**, 5p.

Jain, A.K., 1987. In *Pattern Recognition – Theory and Applications*, eds. Devijver, P.A., Kittler, J., Springer-Verlag, Berlin - Heidelberg, p. 1.

Kedar-Cabelli, S.T., Mahadevan, S., 1986. In *Machine Learning, An Artificial Intelligence Approach* **II**, eds. Michalski, R.S., Carbonell, J.G., Mitchell, T.M., Morgan Kaufmann, Los Altos, Cal., p. 671.

Knapp, G.R., 1987. In *Proc. Conf. Cooling Flows in Clusters and Galaxies* (in preparation).

Kulkarni, A.D., 1986. In *Electronic and Electron Physics*, **66**, ed. Harkes, P.W., Academic Press, Orlando, Fla., p. 309.

Kurtz, M.J., 1983. In *Statistical Methods in Astronomy*, **ESA SP-201**, p. 47.

Lawrence, A., Walker, D., Rowan-Robinson, M., Leech, K.J., Penston, M.V., 1986. *Mon. Not. R. astr. Soc.*, **219**, 687.

Mantas, J., 1987. *Pattern Recognition*, **20**, 1.

Melsa, J.L., Cohn, D.L., 1978. *Decision and Estimation Theory*, McGraw-Hill, New York.

Meurs, E.J.A., Adorf, H.-M., Harmon, R.T., 1988. In *Astronomy from Large Databases — Scientific Objectives and Methodological Approaches*, eds. Heck, A., Murtagh, F., ESO Conf. Proc., p. 49.

Meurs, E.J.A., Harmon, R.T., 1988. *Astr. Astrophys.* (submitted).

Michalski, R.S., Stepp, R., 1983. *IEEE Trans. Pattern Analysis Machine Intelligence*, **5**, 396.

Murtagh, F., 1986. In *Data Analysis in Astronomy II*, eds. Di Gesù, V., Scarsi, L., Crane, P., Friedman, J.H., Levialdi, S., Plenum Press, New York, p. 31.

Murtagh, F., 1987. In *Conf. Image Analysis and Processing*, Cefalù, (forthcoming).

Murtagh, F., 1988. *These proceedings*, p. 308.

Murtagh, F., Heck, A., 1987a. *Astr. Astrophys. Suppl.*, **68**, 113.

Murtagh, F., Heck, A., 1987b. *Multivariate Data Analysis*, Reidel, Dordrecht.

Sklansky, J., Wassel, G.N., 1981. *Pattern Classifiers and Trainable Machines*, Springer, New York.

Soifer, B.T., Houck, J.R., Neugebauer, G., 1987. *Ann. Rev. Astr. Astrophys.*, **25**, 187.

Stepp, R.E., Michalski, R.S., 1986a. *Artif. Intell.* **28**, 43.

Stepp, R.E., Michalski, R.S., 1986b. In *Machine Learning, An Artificial Intelligence Approach*, **II**, eds. Michalski, R.S., Carbonell, J.G., Mitchell, T.M., Morgan Kaufmann, Los Altos, Cal., p. 471.

Watanabe, S. (ed.), 1969. *Methodologies of Pattern Recognition*, Academic Press, New York.

Wolstencroft, R.D., Savage, A., Clowes, R.G., MacGillivray, H.T., Leggett, S.K., Kalafi, M., 1986, *Mon. Not. R. Astr. Soc.*, **223**, 279.

The Hardware and Software Support for the MRSP

Dieter Teuber
Astronomisches Institut
Westfälische Wilhelms-Universität
Münster, F.R. Germany

Abstract

The Muenster Redshift Project (MRSP) described by Horstmann (1988) and Schuecker (1988) relies on an arrangement of hardware and software which is referred to as the Astronomical Data Analysis System (ADAS). In this paper the hardware will be briefly introduced and the support software GAME will be discussed.

1 The hardware

Direct Schmidt plates from the ESO/SRC-Survey and corresponding objective prism plates are digitized with the improved microdensitometer PDS 2020 GM plus, using a sampling width of 15μm. Each plate yields approximately one gigabyte of data. These data are processed with the 32 Bit minicomputer PE 3220 under the operating system OS32. The respective application programmes either reduce the data online or store them on tape or disk. *Fig. 1* shows the hardware configuration currently in operation.

2 Designing software for the ADAS

The software of most ADASes has *grown* through the years driven by the (momentary) needs of interactive users causing incompatibility among application software. In order to avoid inconsistencies the MRSP software has been accompanied by a **designed** support software system from the beginning. For a successful design a thorough analysis of the data reduction process has to be carried out.

Processing of digitized data from wide angle plates results in hundreds of thousands to millions of object images. These images have to be detected, calibrated and classified by an expert system. The exchange of expert knowledge is the major purpose of a conference like this one. But how can we take expert knowledge to the computer and have it applied to the data?

A normal way to cope with complex problems is to break the task into steps which are less abstract. *Fig. 2* illustrates from left to right how a task is transferred to less abstract algorithmic levels until the computer hardware is reached. The bottom to top course of the enquiry leads from the sampled image to more abstract object levels.

In the astronomer's mind an (astronomical) *world model* is formed. From his current world model he derives questions which he tries to answer through astronomical ob-

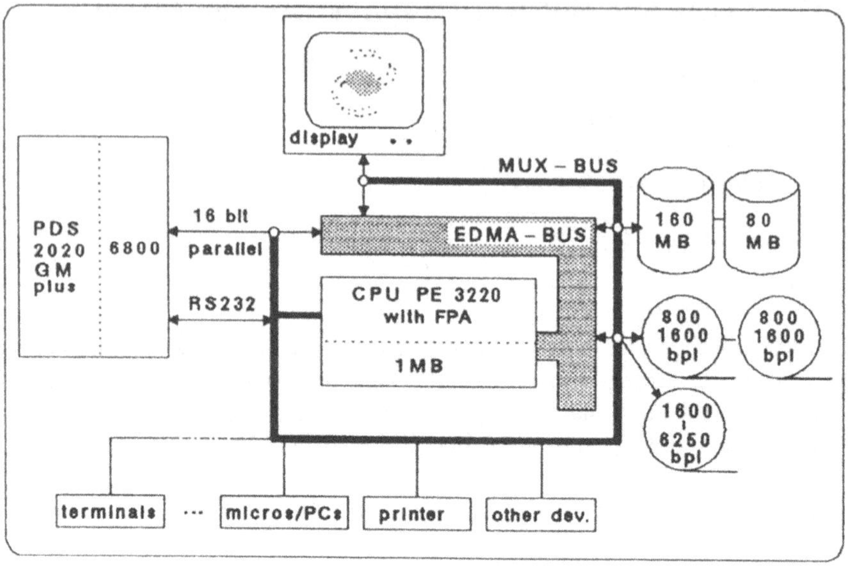

Fig. 1. The hardware configuration of the AIM.

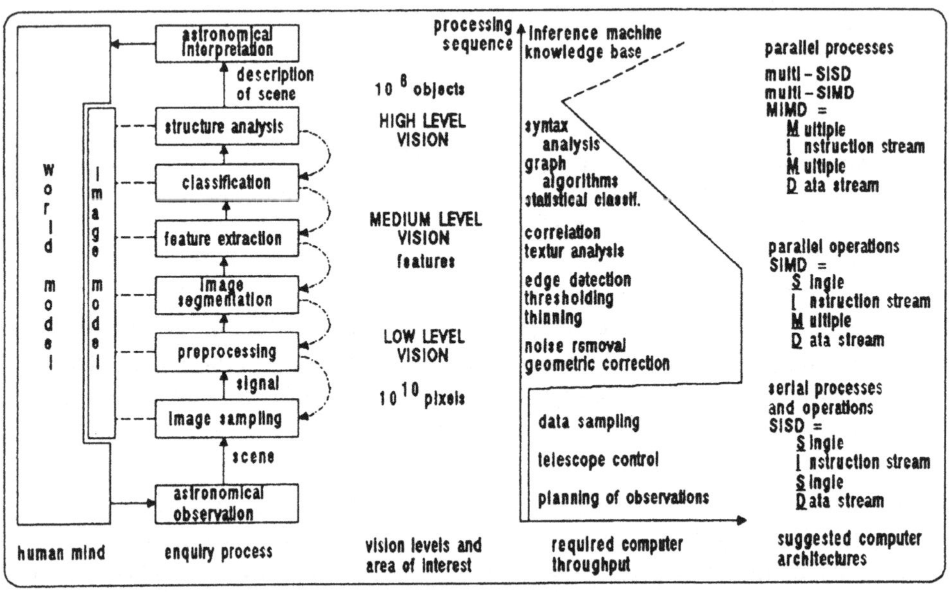

Fig. 2. How to take MRSP expert knowledge to the computer hardware (for explanation see text).

servations. He then processes and interprets the observed data following procedures which depend on his observational techniques. The answer to 'What is an image or object?' is given by the astronomer himself – often intuitively. As soon as he hands the decision to some device, he has to supply also a model which decribes the image of an object. This *image model* influences the whole analysis process. If an expert system requests a new observation, it will also influence the observation and the sampling process.

An enquiry may be carried out in various ways. In *Fig. 2* a potential sequence of analytical steps is shown. It should be noticed that in the course of enquiry these steps are often repeated in iterative loops. The steps can be grouped into *low*, *medium* and *high level vision* which are characterized by their area of interest: pixels at low level, image structures at medium level and astronomical objects at high level.

For each step algorithms must be found and coded in a programming language. From the programmes the resulting demands on support software and hardware can be determined. The two right hand columns of *Fig. 2* concern the coding and execution of data reduction procedures. Typical procedures and the required computer throughput are shown as they are used during each step of enquiry. For observations and subsequent steps not much computing power is needed until the observed scene is transferred to the computer. In the pixel regime array processors are the most suitable devices. Coming closer to the object regime, procedures can be divided into parallel processes, *i.e.* several procedures applied to one object or the same procedure applied to several objects. These processes run on multi–computers or transputer arrays.

Having looked at the process of enquiry, the model of a suitable ADAS can be outlined according to the demands of the user. Within the ADAS the user needs a facility to handle his procedures, data and devices in a way which reflects the current abstraction level of his work. But a high abstraction level requires a high degree of automisation. This statement shall be made evident by some examples.

A request from the user may expand into complex operations and automated sequencing of programmes becomes necessary. The facility shall determine the next action evaluating the status of the data and of the ADAS. But the facility should not only give support in local operations. New data often become more valuable when combined with existing data. This requires access to other computer systems, so that remote login and data retrieval from foreign data bases must be provided. Thus a request may lead to activities ranging from reading a parameter from a local storage location up to remotely running a programme on a foreign computer and transferring the result back into the application programme via communication networks. The details of the various routes of access to the data are of no concern (except for cost) to the user and should be hidden.

The amount of scientific data to be handled will also influence the structure of the facility. Large amounts of data which will be accessed by various selection criteria demand for organising the data by a data base system. For each data aggregate a log of its processing history must be recorded.

To avoid unwanted activities of the automated system the user must be informed, not only of the results, but of all actions taken during processing his request. The user must be able to interrupt and abort processing. This, of course, demands for a mechanism to return to the previous state. Problems like these are known from transaction systems.

The requirements of automised data analysis and those of interactive data analysis have both to be met for the MRSP. The approach introduced in this paper is to administrate the flow of information by maintaining and evaluating descriptions of all subjects and objects participating in the data analysis process:

Data analysis on high abstraction levels is achieved by embedding the participants in an administrative software environment.

3 GAME: The internal structure

A Generic Applications and Monitors Environment (GAME) has been developed and implemented at Muenster. According to the recommendation of the WGCAS (1983) the implementation language is FORTRAN 77. GAME forms the basis of the work in interactive and batch mode (= natural intelligence, NI modes) and expert system mode (= artifical intelligence, AI mode). GAME is the link between all parts of the ADAS, hardware and software. GAME screens the users and applications from the specific properties of the actual computer system and supports developments and data handling on a more abstract level. Some principles of the GAME concept were outlined by Teuber (1985). GAME matches the features attributed to virtual operating systems as defined by Tody (1987). *Fig. 3* illustrates the interfaces of GAME to the participants and the modular structure of the interior of GAME.

The functional units of GAME are called administrators (ADM). ADMs are sets of routines which offer support in organising data and programmes, but do themselves no data reduction. Each ADM possesses its proper data structure which is not visible outside the ADM. These data structures are optimized for the specific needs of the ADM operations. ADMs can be compared to *modules* as they are defined in MODULA-2.

GAME offers services to four sides. To the interactive (astronomical) user a session monitor offers individual communication methods. Application interfaces provide data access, environmental control and check points for application programmes. A report facility supports the system manager in maintaining a functioning ADAS. The operating system is supplied with the appropriate system calls.

To achieve a high degree of portability all dependencies on operating system and hardware are resolved at the bottom layer of GAME. The interfaces to the operating system contain all system calls to perform intertask control and communication, to steer the file system, to translate symbols of the operating system into GAME symbols and vice versa, and to perform the data transfer. Other system dependencies like language extensions are isolated in a separate library.

All references to the outside world issued by an application programme are made

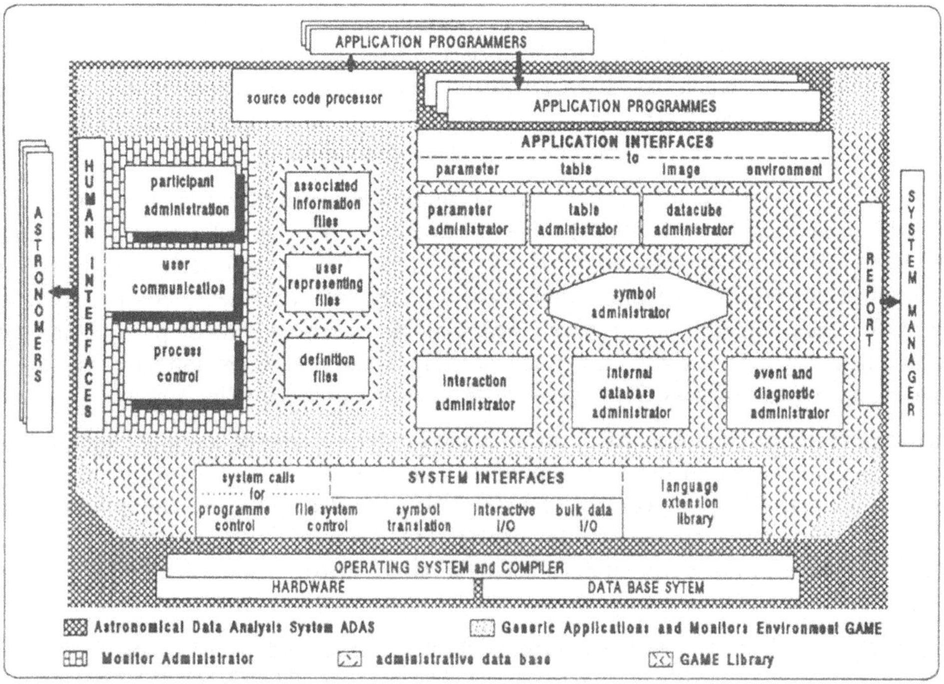

Fig. 3. GAME (interiors and interfaces).

using keywords. In GAME keywords may be connected to data aggregates or may translate into operations. The latter case leads into the domain of object oriented processing. Keywords may be regarded as messages sent to (data) objects which respond by executing methods.

4 GAME: Organisation of data

Data storage is organised in a kind of hierarchical data base which consists of so–called **structure trees** (not to be confused with simple tree structures). A structure tree is made recursively of nodes which themselves are structure trees, *i.e.*, each node contains in itself a structure tree, which is independent of the predecessor. A branch of the tree terminates with a leaf containing user (scientific) data together with a description.

Figure 4a shows an example of a structure tree. The root element is used to link the tree into the current GAME environment. In the example shown here the root opens a directory of contiguous type. When the user supplied path specification selects the middle element, the path continues into the next directory. When the element to the right is selected, the path leads into a list of data elements from which one element is chosen. The user has to specify only the *path* to find this value, he does not need to know the storage structures which have to be passed. The data could be stored in some different data structure, but addressed with the same path specification.

The path may be either a *key* to a certain location, as in the example, or a *directive* to search for a number of instances. It is apparent from the sketch of the administrator in *Fig. 4b* that the menu–like character of the administrator programme makes it easy to add new structures to the system. A dynamic path linker enables the redirection of a user specified path into another (part of a) structure tree.

Using the appropriate interface calls, the application programmes may store and retrieve data formatted as *image*, *table* or *parameter*. Requests for parameters may be redirected to the list and command channel of a process. This is the only way for an application programme to communicate directly with the user (expert system or human). *Image display* and *graphics interfaces* are available, but at present not yet fully integrated into GAME. A revised version of the respective software will meet existing standards such as IDI (Terret 1986) or GKS or AGL (Fini 1986).

Parameters are the simplest data units. A parameter consists of a name which is the key (when referenced within a structure) to a storage area which contains a control and a data segment. The control segment holds the description of the structure in which the values are arranged and the format of the coding. The structural description permits to handle the parameter as a single value, array, stack, ring buffer and/or menu.

Tables and *catalogues* of various formats may be processed using a corresponding definition file. Once the file is created all GAME users are able to access the respective catalogue.

An *image* is a compound format which uses a header made of parameters and tables to describe the data in the *datacube*. Several images may be packed into one file and

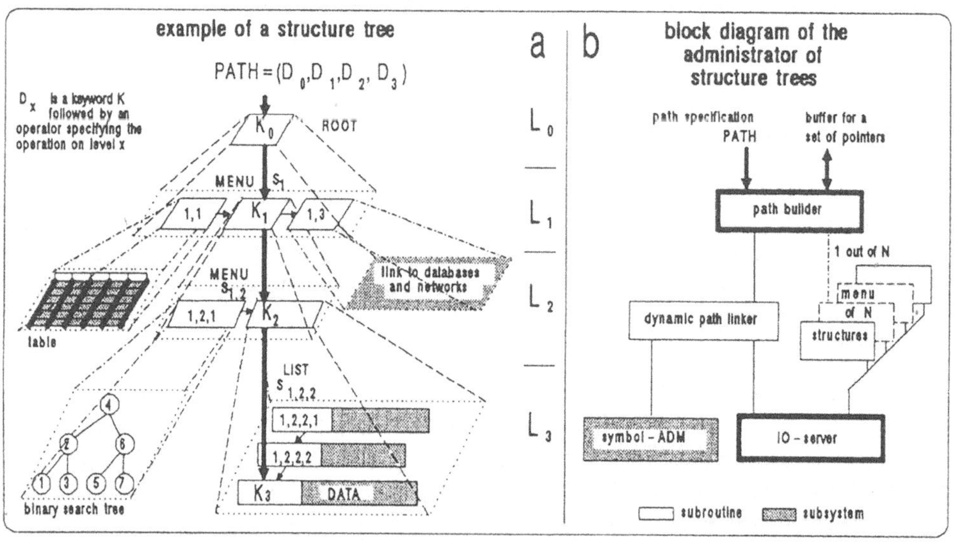

Fig. 4. a) a structure tree
b) scheme of the administrator of structure trees.

interrelated by a structure tree. In its simplest form a structure tree may serve as a list–structured or a hierarchical directory. It may be extended to form a kind of hierarchically organized database. Images written in FITS format according to Wells *et al.* (1979) may be processed on–line, *i.e.*, they do not need to be preprocessed by a FITS–reader.

As has been shown by Teuber (1988a) the application data aggregates are built from internal data structures of GAME. Those internal structures are transposed to the operating system and the hardware structures by a small number of system interfaces.

5 GAME: Organization of activities

When more than one process is active in an environment, the access to resources such as microdensitometers, colour displays or transmission lines has to be organised to avoid scrambling of data or deadlocks. In terms of computer science: a monitor is needed. For GAME the **Symbol Administrator** serves as the synchronizing mechanism. The data structure inherent to the Symbol ADM is the Symbol Queue. Every participant (=subject and object) in a data analysis process is assigned a slot in the Symbol Queue. Each slot has a descriptor for easy access by the human user, but is uniquely identified by a numerical index on the Symbol Queue to GAME. Multiple links may be established between the slots of the Symbol Queue by a set of pointers. One of them is used to form processes as a linked list.

The GAME organisation becomes visible to the application programme only through the handle returned after the respective participant has entered the reduction process. The environmental interfaces enter and remove a programme or a data aggregate onto/from the Symbol Queue.

The user may modify the response of GAME to requests over a wide range using the **event administrator**. The event mechanism is activated as soon as the event slot is included into the Symbol Queue. While obeying a command, GAME often encounters states, where continuation is possible in more than one way. By setting up the event list properly, the user can induce GAME to act in a specific way, instead of defaulting. The event administrator can also be used to check the processing history of a data aggregate or the user environment. More details are given by Teuber (1988b). The ADM of events also supports the detection of errors inherent to GAME.

The most complex unit is the **administrator of session monitors**. Based on definition files stored in the administrative data base, the monitor ADM generates the actual session monitors. A session monitor interfaces to the user via a communication method (language, menu, icons), supervises the processes initiated by the user and communicates results and status information to the user. The functional details of each monitor can be varied over a wide range by setting up the corresponding definition file. Thus the monitor ADM possesses properties of a user interface management system (UIMS, Green 1985). The monitor ADM is supported by the **interaction administrator**. Depending on the environment of the process, the interaction ADM either generates short prompts to the user or links the application programme to the monitor.

The aspects of registering application programmes with GAME are briefly described by Teuber (1988a).

6 Concluding remarks

When GAME supports an expert system, instructions to the system and messages from the system can also be considered as data (procedural data). An expert system may generate instruction to the ADAS and interpret messages from the ADAS. The session monitor can function either as an interactive user interface or as user port and ADAS interface to the expert system. Running an application programme in NI mode without session monitor is a matter of *convenience*. The session monitor and the related environment becomes a *necessity* when one proceeds towards an expert system. The definition (file) of such a monitor may well be elaborated using an AI language such as LISP.

GAME was originally developed on a Perkin Elmer 3220 computer under OS32. The above text refers to GAME version 3.0 which is scheduled to be ported to a UNIX-based computer system in 1988.

Acknowledgements

I thank Markus Tacke and Stefan Grefen who joined the Institute to support the MRSP from the side of computer science. They have begun to investigate the problems related to software portability when working with parallel processes and parallel processors and when using UNIX in a real time environment. Acknowledgements are also given to Randolf Budell who was instrumental in implementing the ADATA storage format and gave support in hardware maintenance.

References

Fini, L., 1986. *ASTRONET Graphic Library Reference Manual*, Version 2.1.

Green, M., 1985. In *Workshop on User Interface Management Systems, Proceedings*, ed. Pfaff, G.E., Springer, Berlin, p. 9.

Horstmann, H., 1988. *These proceedings*, p. 111.

Schuecker, P., 1988. *These proceedings*, p. 142.

Terret, D., Shames, M., 1986. *An Image Display Interface for Astronomy (draft)*, Version 0.4, STScI, Baltimore.

Teuber, D., 1985. In *Proceedings of the International Workshop on Data Analysis in Astronomy*, eds. DiGesu, V., Scarsi, L., Crane, P., Friedman, J.H., Levialdi, S., Plenum Press, New York, p. 235.

Teuber, D., 1988a. In *Astronomy from Large Databases*, eds. Murtagh, F., Heck, A., ESO Conference Proceedings, p. 261.

Teuber, D., 1988b. In preparation.

Tody, D., 1987. *The IRAF Data reduction and Analysis System*. In *IRAF System Handbook*, **3A**, Natl. Opt. Astr. Obs.

Wells, D.C., Greisen, E.W., Harten, R.H., 1981. *Astr. Astrophys. Suppl.*, **44**, 195.

WGCAS, 1983. *Working Group for Coordination of Astronomical Software*, ESO, Garching, unpublished.

List of Participants and Contributors

Adorf, Hans-Martin (P,C)[1]
Space Telescope/
European Coordinating Facility
Karl-Schwarzschild-Str. 2
D-8046 Garching b. München, F.R. Germany

Beuermann, Klaus, (P,C)
Institut für Astronomie und Astrophysik
Technische Universität Berlin
Hardenbergstr. 36
D-1000 Berlin 12, F.R. Germany

Budell, Randolf (P)
Astronomisches Institut
Westfälische Wilhelms-Universität Münster
Wilhelm-Klemm-Str. 10
D-4400 Münster, F.R. Germany

Chincarini, Guido (P,C)
Osservatorio Astronomico di Brera
Via Brera 28
Milano, Italy

Clowes, Roger G. (C)
Royal Observatory
Blackford Hill
Edinburgh EH9 3GHJ, Scotland UK

Collins, C.A. (C)
Department of Astronomy
University of Edinburgh
Blackford Hill
Edinburgh EH9 3HJ, Scotland UK

de Bernardis, Paolo (C)
Istituto di Fisica dell' Università
Roma, Italy

de Boer, Klaas S. (P)
Sternwarte der Universität Bonn
Auf dem Hügel 71
D-5300 Bonn 1, F.R. Germany

de Vegt, Christian (P,C)
Hamburger Sternwarte
Gojenbergsweg 112
D-2050 Hamburg 80, F.R. Germany

Duemmler, Rudolf (P)
Astronomisches Institut
Westfälische Wilhelms-Universität Münster
Wilhelm-Klemm-Str. 10
D-4400 Münster, F.R. Germany

Duerbeck, Hilmar W. (P)
Astronomisches Institut
Westfälische Wilhelms-Universität Münster
Wilhelm-Klemm-Str. 10
D-4400 Münster, F.R. Germany

Efstathiou, G. (C)
Institute of Astronomy
Cambridge University
Madingley Road
Cambridge CB3 0HA, England UK

Engels, Dieter (P,C)
Hamburger Sternwarte
Gojenbergsweg 112
D-2050 Hamburg 80, F.R. Germany

Flin, Piotr (P,C)
Obserwatorium Astronomiczne
Uniwersytet Jagiellonski
ul. Orla 171
PL-30-244 Kraków, Poland

Gericke, Volker (P,C)
Astronomisches Institut
Westfälische Wilhelms-Universität Münster
Wilhelm-Klemm-Str. 10
D-4400 Münster, F.R.Germany

Grewing, Michael (P)
Astronomisches Institut
Universität Tübingen
Waldhäuser Str. 64
D-7400 Tübingen, F.R. Germany

Groote, Detlef (P,C)
Hamburger Sternwarte
Gojenbergsweg 112
D-2050 Hamburg 80, F.R. Germany

Guzzo, Luigi (P)
European Southern Observatory, Garching
now at
Osservatorio Astronomico di Brera
Via Brera 28
Milano, Italy

Hagen H.J. (C)
Hamburger Sternwarte
Gojenbergsweg 112
D-2050 Hamburg 80, F.R. Germany

[1]Participant (P), Contributor (C)

Hawkins, Michael R.S. (P,C)
Royal Observatory
Blackford Hill
Edinburgh EH9 3HJ, Scotland UK

Heydon-Dumbleton, Neal (P,C)
Department of Astronomy
University of Edinburgh
Blackford Hill
Edinburgh EH9 3HJ, Scotland UK

Hickson, Paul (C)
Department of Geophysics and Astronomy
University of British Columbia
Vancouver, B.C., Canada

Hömberg, Dietmar (P,C)
Astronomisches Institut
Westfälische Wilhelms-Universität Münster
now at
Institut für praktische Mathematik
Universität/GH Essen, F.R. Germany

Hollósi, Joseph (P,C)
Space Telescope Science Institute
Baltimore, MD, USA
now at
Department of Atomic Physics
L. Eötvös University
Budapest, Hungary

Horstmann, Heinrich (P,C)
Astronomisches Institut
Westfälische Wilhelms-Universität Münster
Wilhelm-Klemm-Str. 10
D-4400 Münster, F.R. Germany

Icke, Vincent (P,C)
Sterrewacht Leiden
Postbus 9513
NL-2300 RA Leiden, Netherlands

Loh, Edwin D. (P,C)
Department of Physics and Astronomy
Michigan State University
East Lansing, Michigan 48824-1116, USA

Loveday, J. (C)
Institute of Astronomy
Cambridge University
Madingley Road
Cambridge CB3 0HA, England UK

MacGillivray, Harvey T. (C)
Royal Observatory
Blackford Hill
Edinburgh EH9 3HJ, Scotland UK

Maddox, Steve J. (P,C)
Institute of Astronomy
Cambridge University
Madingley Road
Cambridge CB3 0HA, England UK

Masi, Silvia (C)
Istituto di Fisica dell' Università
Roma, Italy

Meurs, E.J.A. (C)
European Southern Observatory
Karl-Schwarzschild-Str. 2
D-8046 Garching b. München, F.R. Germany

Murtagh, Fionn (P,C)
Space Telescope/
European Coordinating Facility
Karl-Schwarzschild-Str. 2
D-8046 Garching b. München, F.R. Germany

Nicklas, Harald (P)
Landessternwarte Heidelberg-Königstuhl
now at
AG-DGT, Universitäts-Sternwarte Göttingen
Geismarlandstr. 11
D-3400 Göttingen, F.R. Germany

Ott, Heinz-Albert (P,C)
Astronomisches Institut
Westfälische Wilhelms-Universität Münster
Wilhelm-Klemm-Str. 10
D-4400 Münster, F.R. Germany

Pfleiderer, Jörg (P,C)
Institut für Astronomie
Universität Innsbruck
Technikerstr. 25
A-6020 Innsbruck, Austria

Pittella, Giancarlo (C)
Osservatorio Astronomico di Roma,
sede Monte Porzio
Roma, Italy

Plionis, Manolis (P)
Astronomy Center, PG/PH MAPS
University of Sussex
Falmer, Brighton BN1 9HQ, England UK

Reimers, Dieter (P,C)
Hamburger Sternwarte
Gojenbergsweg 112
D-2050 Hamburg 80, F.R. Germany

Rohlfs, Kristen (P)
Astronomisches Institut
Ruhr-Universität Bochum
Universitätsstr. 150
D-4630 Bochum 1, F.R. Germany

Rudnicki, Konrad (P,C)
Obserwatorium Astronomiczne
Uniwersytet Jagiellonski
ul. Orla 171
PL-30-244 Kraków, Poland

Scheuer, Hans-Günter (P)
Astronomisches Institut
Westfälische Wilhelms-Universität Münster
Wilhelm-Klemm-Str. 10
D-4400 Münster, F.R. Germany

Schuecker, Peter (P,C)
Astronomisches Institut
Westfälische Wilhelms-Universität Münster
Wilhelm-Klemm-Str. 10
D-4400 Münster, F.R. Germany

Seitter, Waltraut C. (P,C)
Astronomisches Institut
Westfälische Wilhelms-Universität Münster
Wilhelm-Klemm-Str. 10
D-4400 Münster, F.R. Germany

Southerland, W.J. (C)
Institute of Astronomy
Cambridge University
Madingley Road
Cambridge CB3 0HA, England UK

Szalay, Alex S. (C)
Department of Atomic Physics
L. Eötvös University
Budapest, Hungary
and Johns Hopkins University
Baltimore, MD, USA

Tammann, Gustav A. (P)
Astronomisches Institut
Universität Basel
Venusstr. 7
CH-4102 Binningen, Switzerland

Teuber, Dieter (P)
Astronomisches Institut
Westfälische Wilhelms-Universität Münster
Wilhelm-Klemm-Str. 10
D-4400 Münster, F.R. Germany

Tritton, Susan B. (P)
Royal Observatory
Blackford Hill
Edinburgh EH9 3HJ, Scotland UK

Tóth, Gabor (P,C)
Space Telescope Science Institute
Baltimore, MD, USA
now at
Department of Atomic Physics
L. Eötvös University
Budapest, Hungary

Tucholke, Hans-Joachim (P,C)
Astronomisches Institut
Westfälische Wilhelms-Universität Münster
Wilhelm-Klemm-Str. 10
D-4400 Münster, F.R. Germany

Vettolani, P. (C)
Istituto di Radioastronomia CNR
Bologna, Italy

Wildenmann, Ms. (P)
Max-Planck-Institut für
Extraterrestrische Physik
Karl-Schwarzschild-Str. 1
D-8046 Garching b. München, F.R. Germany

Wolf, Bernhard (P)
Landessternwarte Heidelberg-Königstuhl
D-6400 Heidelberg, F.R. Germany

Wszołek, Bogdan (C)
Obserwatorium Astronomiczne
Universytet Jagielloński
ul. Orla 171
PL-30-244 Kraków, Poland

Lecture Notes in Mathematics

Vol. 1236: Stochastic Partial Differential Equations and Applications. Proceedings, 1985. Edited by G. Da Prato and L. Tubaro. V, 257 pages. 1987.

Vol. 1237: Rational Approximation and its Applications in Mathematics and Physics. Proceedings, 1985. Edited by J. Gilewicz, M. Pindor and W. Siemaszko. XII, 350 pages. 1987.

Vol. 1250: Stochastic Processes – Mathematics and Physics II. Proceedings 1985. Edited by S. Albeverio, Ph. Blanchard and L. Streit. VI, 359 pages. 1987.

Vol. 1251: Differential Geometric Methods in Mathematical Physics. Proceedings, 1985. Edited by P.L. García and A. Pérez-Rendón. VII, 300 pages. 1987.

Vol. 1255: Differential Geometry and Differential Equations. Proceedings, 1985. Edited by C. Gu, M. Berger and R.L. Bryant. XII, 243 pages. 1987.

Vol. 1256: Pseudo-Differential Operators. Proceedings, 1986. Edited by H.O. Cordes, B. Gramsch and H. Widom. X, 479 pages. 1987.

Vol. 1258: J. Weidmann, Spectral Theory of Ordinary Differential Operators. VI, 303 pages. 1987.

Vol. 1260: N.H. Pavel, Nonlinear Evolution Operators and Semigroups. VI, 285 pages. 1987.

Vol. 1263: V.L. Hansen (Ed.), Differential Geometry. Proceedings, 1985. XI, 288 pages. 1987.

Vol. 1265: W. Van Assche, Asymptotics for Orthogonal Polynomials. VI, 201 pages. 1987.

Vol. 1267: J. Lindenstrauss, V.D. Milman (Eds.), Geometrical Aspects of Functional Analysis. Seminar. VII, 212 pages. 1987.

Vol. 1269: M. Shiota, Nash Manifolds. VI, 223 pages. 1987.

Vol. 1270: C. Carasso, P.-A. Raviart, D. Serre (Eds.), Nonlinear Hyperbolic Problems. Proceedings, 1986. XV, 341 pages. 1987.

Vol. 1272: M.S. Livšic, L.L. Waksman, Commuting Nonselfadjoint Operators in Hilbert Space. III, 115 pages. 1987.

Vol. 1273: G.-M. Greuel, G. Trautmann (Eds.), Singularities, Representation of Algebras, and Vector Bundles. Proceedings, 1985. XIV, 383 pages. 1987.

Vol. 1275: C.A. Berenstein (Ed.), Complex Analysis I. Proceedings, 1985–86. XV, 331 pages. 1987.

Vol. 1276: C.A. Berenstein (Ed.), Complex Analysis II. Proceedings, 1985–86. IX, 320 pages. 1987.

Vol. 1277: C.A. Berenstein (Ed.), Complex Analysis III. Proceedings, 1985–86. X, 350 pages. 1987.

Vol. 1283: S. Mardešić, J. Segal (Eds.), Geometric Topology and Shape Theory. Proceedings, 1986. V, 261 pages. 1987.

Vol. 1285: I.W. Knowles, Y. Saitō (Eds.), Differential Equations and Mathematical Physics. Proceedings, 1986. XVI, 499 pages. 1987.

Vol. 1287: E.B. Saff (Ed.), Approximation Theory, Tampa. Proceedings, 1985–1986. V, 228 pages. 1987.

Vol. 1288: Yu. L. Rodin, Generalized Analytic Functions on Riemann Surfaces. V, 128 pages, 1987.

Vol. 1294: M. Queffélec, Substitution Dynamical Systems – Spectral Analysis. XIII, 240 pages. 1987.

Vol. 1299: S. Watanabe, Yu.V. Prokhorov (Eds.), Probability Theory and Mathematical Statistics. Proceedings, 1986. VIII, 589 pages. 1988.

Vol. 1300: G.B. Seligman, Constructions of Lie Algebras and their Modules. VI, 190 pages. 1988.

Vol. 1302: M. Cwikel, J. Peetre, Y. Sagher, H. Wallin (Eds.), Function Spaces and Applications. Proceedings, 1986. VI, 445 pages. 1988.

Vol. 1303: L. Accardi, W. von Waldenfels (Eds.), Quantum Probability and Applications III. Proceedings, 1987. VI, 373 pages. 1988.

Lecture Notes in Physics

Vol. 289: N. Straumann, Klassische Mechanik. XV, 403 Seiten. 1987.

Vol. 290: K.T. Hecht, The Vector Coherent State Method and Its Application to Problems of Higher Symmetries. V, 154 pages. 1987.

Vol. 291: J.L. Linsky, R.E. Stencel (Eds.), Cool Stars, Stellar Systems, and the Sun. Proceedings, 1987. XIII, 537 pages. 1987.

Vol. 292: E.-H. Schröter, M. Schüssler (Eds.), Solar and Stellar Physics. Proceedings, 1987. V, 231 pages. 1987.

Vol. 293: Th. Dorfmüller, R. Pecora (Eds.), Rotational Dynamics of Small and Macromolecules. Proceedings, 1986. V, 249 pages. 1987.

Vol. 294: D. Berényi, G. Hock (Eds.), High-Energy Ion-Atom Collisions. Proceedings, 1987. VIII, 540 pages. 1988.

Vol. 295: P. Schmüser, Feynman-Graphen und Eichtheorien für Experimentalphysiker. VI, 217 Seiten. 1988.

Vol. 296: M. Month, S. Turner (Eds.), Frontiers of Particle Beams. XII, 700 pages. 1988.

Vol. 297: A. Lawrence (Ed.), Comets to Cosmology. X, 415 pages. 1988.

Vol. 298: M. Georgiev, F' Centers in Alkali Halides. XI, 282 pages. 1988.

Vol. 299: J.D. Buckmaster, T. Takeno (Eds.), Mathematical Modeling in Combustion Science. Proceedings, 1987. VI, 168 pages. 1988.

Vol. 300: B.-G. Englert, Semiclassical Theory of Atoms. VII, 401 pages. 1988.

Vol. 301: G. Ferenczi, F. Beleznay (Eds.), New Developments in Semiconductor Physics. Proceedings, 1987. VI, 302 pages. 1988.

Vol. 302: F. Gieres, Geometry of Supersymmetric Gauge Theories. VIII, 189 pages. 1988.

Vol. 303: P. Breitenlohner, D. Maison, K. Sibold (Eds.), Renormalization of Quantum Field Theories with Non-linear Field Transformations. Proceedings, 1987. VI, 239 pages. 1988.

Vol. 304: R. Prud'homme, Fluides hétérogènes et réactifs: écoulements et transferts. VIII, 239 pages. 1988.

Vol. 305: K. Nomoto (Ed.), Atmospheric Diagnostics of Stellar Evolution: Chemical Peculiarity, Mass Loss, and Explosion. Proceedings, 1987. XIV, 468 pages. 1988.

Vol. 306: L. Blitz, F.J. Lockman (Eds.), The Outer Galaxy. Proceedings, 1987. IX, 291 pages. 1988.

Vol. 307: H.R. Miller, P.J. Wiita (Eds.), Active Galactic Nuclei. Proceedings, 1987, XI, 438 pages. 1988.

Vol. 308: H. Bacry, Localizability and Space in Quantum Physics. VII, 81 pages. 1988.

Vol. 309: P.E. Wagner, G. Vali (Eds.), Atmospheric Aerosols and Nucleation. Proceedings, 1988. XVIII, 729 pages. 1988.

Vol. 310: W.C. Seitter, H.W. Duerbeck, M. Tacke (Eds.), Large-Scale Structures in the Universe – Observational and Analytical Methods. Proceedings, 1987. II, 335 pages. 1988.

GPSR Compliance

The European Union's (EU) General Product Safety Regulation (GPSR) is a set of rules that requires consumer products to be safe and our obligations to ensure this.

If you have any concerns about our products, you can contact us on ProductSafety@springernature.com

In case Publisher is established outside the EU, the EU authorized representative is:

Springer Nature Customer Service Center GmbH
Europaplatz 3
69115 Heidelberg, Germany

Batch number: 10640288

Printed by Printforce, the Netherlands